GENETICS, GENOMICS AND BREEDING OF FORAGE CROPS

Genetics, Genomics and Breeding of Crop Plants

Series Editor
Chittaranjan Kole
Vice-Chancellor
Bidhan Chandra Agricultural University
Mohanpur, West Bengal
India

Books in this Series:

Published or in Press:

- Jinguo Hu, Gerald Seiler & Chittaranjan Kole: *Sunflower*
- Kristin D. Bilyeu, Milind B. Ratnaparkhe & Chittaranjan Kole: *Soybean*
- Robert Henry & Chittaranjan Kole: *Sugarcane*
- Kevin Folta & Chittaranjan Kole: *Berries*
- Jan Sadowsky & Chittaranjan Kole: *Vegetable Brassicas*
- James M. Bradeen & Chittaranjan Kole: *Potato*
- C.P. Joshi, Stephen DiFazio & Chittaranjan Kole: *Poplar*
- Anne-Françoise Adam-Blondon, José M. Martínez-Zapater & Chittaranjan Kole: *Grapes*
- Christophe Plomion, Jean Bousquet & Chittaranjan Kole: *Conifers*
- Dave Edwards, Jacqueline Batley, Isobel Parkin & Chittaranjan Kole: *Oilseed Brassicas*
- Marcelino Pérez de la Vega, Ana María Torres, José Ignacio Cubero & Chittaranjan Kole: *Cool Season Grain Legumes*
- Yi-Hong Wang, Tusar Kanti Behera & Chittaranjan Kole: *Cucurbits*
- Albert G. Abbott & Chittaranjan Kole: *Stone Fruits*
- Barbara E. Liedl, Joanne A. Labate, John R. Stommel, Ann Slade & Chittaranjan Kole: *Tomato*
- Byoung-Cheorl Kang & Chittaranjan Kole: *Peppers and Eggplants*

GENETICS, GENOMICS AND BREEDING OF FORAGE CROPS

Editors

Hongwei Cai
Department of Plant Genetics and Breeding
College of Agronomy and
Biotechnology, China Agricultural University
Beijing, China
and
Forage Crop Research Institute
Japan Grassland Agricultural and Forage Seed Association
Tochigi
Japan

Toshihiko Yamada
Field Science Center for Northern Biosphere
Hokkaido University
Sapporo
Japan

Chittaranjan Kole
Bidhan Chandra Agricultural University
Mohanpur, West Bengal
India

CRC Press
Taylor & Francis Group
Boca Raton London New York

CRC Press is an imprint of the
Taylor & Francis Group, an **informa** business
A SCIENCE PUBLISHERS BOOK

First published 2014 by Science Publisher

Published 2019 by CRC Press
Taylor & Francis Group
6000 Broken Sound Parkway NW, Suite 300
Boca Raton, FL 33487-2742

© 2014 Copyright reserved
CRC Press is an imprint of Taylor & Francis Group, an Informa business

No claim to original U.S. Government works

ISBN-13: 978-1-4822-0810-8 (hbk)

```
           Library of Congress Cataloging-in-Publication Data
Genetics, genomics and breeding of forage crops / editors: Hongwei
Cai, Toshihiko Yamada, Chittaranjan Kole.
      p. cm. -- (Genetics, genomics and breeding of crop plants)
   Includes bibliographical references and index.
   ISBN 978-1-4822-0810-8 (hardcover : alk. paper)  1.  Forage
plants--Breeding. 2.  Forage plants--Genetics.  I. Cai, Hongwei.
II. Yamada, Toshihiko. III. Kole, Chittaranjan. IV. Series:
Genetics, genomics and breeding of crop plants.
   SB193.5.G46 2014
   633.2--dc23
                                                    2013023191
```

Visit the Taylor & Francis Web site at
http://www.taylorandfrancis.com

CRC Press Web site at
http://www.crcpress.com

Preface to the Series

Genetics, genomics and breeding has emerged as three overlapping and complimentary disciplines for comprehensive and fine-scale analysis of plant genomes and their precise and rapid improvement. While genetics and plant breeding have contributed enormously towards several new concepts and strategies for elucidation of plant genes and genomes as well as development of a huge number of crop varieties with desirable traits, genomics has depicted the chemical nature of genes, gene products and genomes and also provided additional resources for crop improvement.

In today's world, teaching, research, funding, regulation and utilization of plant genetics, genomics and breeding essentially require thorough understanding of their components including classical, biochemical, cytological and molecular genetics; and traditional, molecular, transgenic and genomics-assisted breeding. There are several book volumes and reviews available that cover individually or in combination of a few of these components for the major plants or plant groups; and also on the concepts and strategies for these individual components with examples drawn mainly from the major plants. Therefore, we planned to fill an existing gap with individual book volumes dedicated to the leading crop and model plants with comprehensive deliberations on all the classical, advanced and modern concepts of depiction and improvement of genomes. The success stories and limitations in the different plant species, crop or model, must vary; however, we have tried to include a more or less general outline of the contents of the chapters of the volumes to maintain uniformity as far as possible.

Often genetics, genomics and plant breeding and particularly their complimentary and supplementary disciplines are studied and practiced by people who do not have, and reasonably so, the basic understanding of biology of the plants for which they are contributing. A general description of the plants and their botany would surely instill more interest among them on the plant species they are working for and therefore we presented lucid details on the economic and/or academic importance of the plant(s); historical information on geographical origin and distribution; botanical origin and evolution; available germplasms and gene pools, and genetic and cytogenetic stocks as genetic, genomic and breeding resources; and basic information on taxonomy, habit, habitat, morphology, karyotype, ploidy level and genome size, etc.

Classical genetics and traditional breeding have contributed enormously even by employing the phenotype-to-genotype approach. We included detailed descriptions on these classical efforts such as genetic mapping using morphological, cytological and isozyme markers; and achievements of conventional breeding for desirable and against undesirable traits. Employment of the *in vitro* culture techniques such as micro- and megaspore culture, and somatic mutation and hybridization, has also been enumerated. In addition, an assessment of the achievements and limitations of the basic genetics and conventional breeding efforts has been presented.

It is a hard truth that in many instances we depend too much on a few advanced technologies, we are trained in, for creating and using novel or alien genes but forget the infinite wealth of desirable genes in the indigenous cultivars and wild allied species besides the available germplasms in national and international institutes or centers. Exploring as broad as possible natural genetic diversity not only provides information on availability of target donor genes but also on genetically divergent genotypes, botanical varieties, subspecies, species and even genera to be used as potential parents in crosses to realize optimum genetic polymorphism required for mapping and breeding. Genetic divergence has been evaluated using the available tools at a particular point of time. We included discussions on phenotype-based strategies employing morphological markers, genotype-based strategies employing molecular markers; the statistical procedures utilized; their utilities for evaluation of genetic divergence among genotypes, local landraces, species and genera; and also on the effects of breeding pedigrees and geographical locations on the degree of genetic diversity.

Association mapping using molecular markers is a recent strategy to utilize the natural genetic variability to detect marker-trait association and to validate the genomic locations of genes, particularly those controlling the quantitative traits. Association mapping has been employed effectively in genetic studies in human and other animal models and those have inspired the plant scientists to take advantage of this tool. We included examples of its use and implication in some of the volumes that devote to the plants for which this technique has been successfully employed for assessment of the degree of linkage disequilibrium related to a particular gene or genome, and for germplasm enhancement.

Genetic linkage mapping using molecular markers have been discussed in many books, reviews and book series. However, in this series, genetic mapping has been discussed at length with more elaborations and examples on diverse markers including the anonymous type 2 markers such as RFLPs, RAPDs, AFLPs, etc. and the gene-specific type 1 markers such as EST-SSRs, SNPs, etc.; various mapping populations including F_2, backcross, recombinant inbred, doubled haploid, near-isogenic and pseudotestcross; computer software including MapMaker, JoinMap, etc. used; and different types of genetic maps including preliminary, high-resolution, high-density, saturated, reference, consensus and integrated developed so far.

Mapping of simply inherited traits and quantitative traits controlled by oligogenes and polygenes, respectively has been deliberated in the earlier literature crop-wise or crop group-wise. However, more detailed information on mapping or tagging oligogenes by linkage mapping or bulked segregant analysis, mapping polygenes by QTL analysis, and different computer software employed such as MapMaker, JoinMap, QTL Cartographer, Map Manager, etc. for these purposes have been discussed at more depth in the present volumes.

The strategies and achievements of marker-assisted or molecular breeding have been discussed in a few books and reviews earlier. However, those mostly deliberated on the general aspects with examples drawn mainly from major plants. In this series, we included comprehensive descriptions on the use of molecular markers for germplasm characterization, detection and maintenance of distinctiveness, uniformity and stability of genotypes, introgression and pyramiding of genes. We have also included elucidations on the strategies and achievements of transgenic breeding for developing genotypes particularly with resistance to herbicide, biotic and abiotic stresses; for biofuel production, biopharming, phytoremediation; and also for producing resources for functional genomics.

A number of desirable genes and QTLs have been cloned in plants since 1992 and 2000, respectively using different strategies, mainly positional cloning and transposon tagging. We included enumeration of these and other strategies for isolation of genes and QTLs, testing of their expression and their effective utilization in the relevant volumes.

Physical maps and integrated physical-genetic maps are now available in most of the leading crop and model plants owing mainly to the BAC, YAC, EST and cDNA libraries. Similar libraries and other required genomic resources have also been developed for the remaining crops. We have devoted a section on the library development and sequencing of these resources; detection, validation and utilization of gene-based molecular markers; and impact of new generation sequencing technologies on structural genomics.

As mentioned earlier, whole genome sequencing has been completed in one model plant (Arabidopsis) and seven economic plants (rice, poplar, peach, papaya, grapes, soybean and sorghum) and is progressing in an array of model and economic plants. Advent of massively parallel DNA sequencing using 454-pyrosequencing, Solexa Genome Analyzer, SOLiD system, Heliscope and SMRT have facilitated whole genome sequencing in many other plants more rapidly, cheaply and precisely. We have included extensive coverage on the level (national or international) of collaboration and the strategies and status of whole genome sequencing in plants for which sequencing efforts have been completed or are progressing currently. We have also included critical assessment of the impact of these genome initiatives in the respective volumes.

Comparative genome mapping based on molecular markers and map positions of genes and QTLs practiced during the last two decades of the last century provided answers to many basic questions related to evolution, origin and phylogenetic relationship of close plant taxa. Enrichment of genomic resources has reinforced the study of genome homology and synteny of genes among plants not only in the same family but also of taxonomically distant families. Comparative genomics is not only delivering answers to the questions of academic interest but also providing many candidate genes for plant genetic improvement.

The 'central dogma' enunciated in 1958 provided a simple picture of gene function—gene to mRNA to transcripts to proteins (enzymes) to metabolites. The enormous amount of information generated on characterization of transcripts, proteins and metabolites now have led to the emergence of individual disciplines including functional genomics, transcriptomics, proteomics and metabolomics. Although all of them ultimately strengthen the analysis and improvement of a genome, they deserve individual deliberations for each plant species. For example, microarrays, SAGE, MPSS for transcriptome analysis; and 2D gel electrophoresis, MALDI, NMR, MS for proteomics and metabolomics studies require elaboration. Besides transcriptome, proteome or metabolome QTL mapping and application of transcriptomics, proteomics and metabolomics in genomics-assisted breeding are frontier fields now. We included discussions on them in the relevant volumes.

The databases for storage, search and utilization on the genomes, genes, gene products and their sequences are growing enormously in each second and they require robust bioinformatics tools plant-wise and purpose-wise. We included a section on databases on the gene and genomes, gene expression, comparative genomes, molecular marker and genetic maps, protein and metabolomes, and their integration.

Notwithstanding the progress made so far, each crop or model plant species requires more pragmatic retrospect. For the model plants we need to answer how much they have been utilized to answer the basic questions of genetics and genomics as compared to other wild and domesticated species. For the economic plants we need to answer as to whether they have been genetically tailored perfectly for expanded geographical regions and current requirements for green fuel, plant-based bioproducts and for improvements of ecology and environment. These futuristic explanations have been addressed finally in the volumes.

We are aware of exclusions of some plants for which we have comprehensive compilations on genetics, genomics and breeding in hard copy or digital format and also some other plants which will have enough achievements to claim for individual book volume only in distant future. However, we feel satisfied that we could present comprehensive deliberations on genetics, genomics and breeding of 30 model and economic plants, and their groups in a few cases, in this series. I personally feel also happy that I could work with many internationally celebrated scientists who edited the book volumes on the leading plants and plant groups and included chapters authored by many scientists reputed globally for their contributions on the concerned plant or plant group.

We paid serious attention to reviewing, revising and updating of the manuscripts of all the chapters of this book series, but some technical and formatting mistakes will remain for sure. As the series editor, I take complete responsibility for all these mistakes and will look forward to the readers for corrections of these mistakes and also for their suggestions for further improvement of the volumes and the series so that future editions can serve better the purposes of the students, scientists, industries, and the society of this and future generations.

Science publishers, Inc. has been serving the requirements of science and society for a long time with publications of books devoted to advanced concepts, strategies, tools, methodologies and achievements of various science disciplines. Myself as the editor and also on behalf of the volume editors, chapter authors and the ultimate beneficiaries of the volumes take this opportunity to acknowledge the publisher for presenting these books that could be useful for teaching, research and extension of genetics, genomics and breeding.

Chittaranjan Kole

Preface

The production of forage for animals provides many complex challenges for plant improvement that are absent from single harvest grain crops.

Forages may be sown in complex pastures with mixtures of grass, legume and herb, or as pure grass swards and they may be continuously or intermittently grazed, or cut for hay or silage. They must fit into farming systems that supply an appropriate level of nutrition to maintain animal production. In addition some grasses may have association with beneficial endophytic fungi that further complicates the breeding process. Grasses are also widely used for turf and erosion control, and in recent year some grasses have received attention as feedstock for bioenergy. Often grasses are grown on poor soils unsuitable for cultivation of annual food crops. The large areas of natural and developed grasslands around the world also mean that they can influence carbon sequestration and potential climate change.

The genomic study of forage crops has lagged behind that of the major cereal crops such as rice and maize because of their outcrossing nature, relatively large size of genomes, and comparatively low economic value compared to cereal crops. However, forage crop improvement programs have entered the biotechnology era by the use of molecular biology tools. In recent years many researchers have reported very good progress in the field of molecular genetics and breeding in the forage crop sector.

In the volume of "Technical Crops" of the "Genome Mapping and Molecular Breeding in Plants" series edited by Dr. Kole (2007), Inoue et al. contributed an excellent review chapter on forage crops. Later on, we felt the need of a complete volume dedicated to forage crops under the new series of "Genetics, Genomics and Breeding of Crop Plants".

Here we have invited leading researchers on grasses and legume species to contribute chapters to this volume on the most important forage crops, 10 on temperate species and one on warm season grasses. In these chapters the authors have focused not only on the molecular genetics but also on the breeding efforts.

We hope this book will be useful for students and researchers with an interest in forage, turf and bioenergy crop improvements. Finally, we express our gratitude to the authors whose dedication and work have made this book possible.

February 11, 2013

Hongwei Cai, Toshihiko Yamada and
Chittaranjan Kole

Contents

List of Contributors

M.T. Abberton
International Institute of Tropical Agriculture, PMB 5320, Ibadan, Oyo State, Nigeria.
E-mail: *michael.abberton@cgiar.org*

Y. Akiyama
National Agricultural Research Organization, National Institute of Tohoku Region Agricultural Research Center, Morioka, 020-0198 Morioka, Japan.
E-mail: *akky@affrc.go.jp*

K. Amundsen
USDA-ARS, Beltsville, MD 20705, USA.
E-mail: *kamundsen2@unl.edu*

B. Boller
Agroscope Reckenholz-Taenikon Research Station ART, Reckenholzstr. 191, 8046 Zurich, Switzerland.
E-mail: *beat.boller@art.admin.ch*

E.C. Brummer
The Samuel Roberts Noble Foundation, Ardmore, OK 73401; Institute of Plant Breeding, Genetics and Genomics, The University of Georgia, Athens, GA 30602, USA.
E-mail: *ecbrummer@noble.org*

B.S. Bushman
USDA-ARS FRRL, 695 N 1100 E, Logan, UT, USA.
E-mail: *Shaun.Bushman@ars.usda.gov*

H. Cai
Department of Plant Genetics and Breeding, College of Agronomy and Biotechnology, China Agricultural University, 2, Yuanmingyuan West Road, Beijing, 100193 China; Forage Crop Research Institute, Japan Grassland Agricultural and Forage Seed Association, 388-5 Higashiakada, Nasushiobara, Tochigi 329-2742, Japan.
E-mail: *caihw@cau.edu.cn; hcai@jfsass.or.jp*

N.O.I. Cogan
Department of Environment and Primary Industries, Biosciences Research Division, AgriBio, the Centre for AgriBioscience, 5 Ring, Bundoora, Victoria 3083, Australia; Molecular Plant Breeding and Dairy Futures Cooperative Research Centres, Australia.
E-mail: *noel.cogan@depi.vic.gov.au*

M. Ebina
National Agricultural Research Organization, National Institute of Livestock and Grassland Science, Nasushiobara, 329-2793 Tochigi, Japan.
E-mail: *triticum@affrc.go.jp*

N.W. Ellison
AgResearch, Private Bag 11008, Palmerston North, New Zealand.
E-mail: *nick.ellison@agresearch.co.nz*

J.W. Forster
Department of Environment and Primary Industries, Biosciences Research Division, AgriBio, the Centre for AgriBioscience, 5 Ring, Bundoora, Victoria 3083, Australia; La Trobe University, Bundoora, Victoria 3086, Australia; Molecular Plant Breeding and Dairy Futures Cooperative Research Centres, Australia.
E-mail: *john.forster@depi.vic.gov.au*

Y. Han
The Samuel Roberts Noble Foundation, Ardmore, OK 73401, USA.
E-mail: *yhan@noble.org*

M. Hirata
Forage Crop Research Institute, Japan Grassland Agricultural and Forage Seed Association, 388-5 Higashiakada, Nasushiobara, Tochigi 329-2742, Japan.
E-mail: *hirata@jfsass.or.jp*

M. Inoue
Dept. of Plant, Soil, and Insect Sciences, 14 Stockbridge Hall, University of Massachusetts, 80 Campus Center Way, Amherst, MA 01003, USA.
E-mail: *den8mai@yahoo.co.jp*

S. Isobe
Kazusa DNA Research Institute, 2-6-7 Kazusa-Kamatari, Kisarazu, Chiba, 292-0818, Japan.
E-mail: *sisobe@kazusa.or.jp*

R. Kölliker
Agroscope Reckenholz-Taenikon Research Station ART, Reckenholzstr. 191, 8046 Zurich, Switzerland.
E-mail: *roland.koelliker@art.admin.ch*

M. Li
Department of Grassland Science, College of Animal Science and Technology, China Agricultural University, 2 Yuanmingyuan West Road, Beijing 100193, China.
E-mail: *lillian165@163.com*

M.J. Monteros
The Samuel Roberts Noble Foundation, Ardmore, OK 73401; Institute of Plant Breeding, Genetics and Genomics, The University of Georgia, Athens, GA 30602, USA.
E-mail: *mjmonteros@noble.org*

K.M. Rao
Department of Plant and Environmental Sciences/CIGENE, Norwegian University of Life Sciences, NO-1432 Ås, Norway.
E-mail: *mallikarjuna.rao.kovi@umb.no*

H. Riday
US Dairy Forage Research Center, 1925 Linden Drive West Madison, WI 53706, USA.
E-mail: *Heathcliffe.Riday@ARS.USDA.GOV*

O.A. Rognli
Department of Plant and Environmental Sciences/CIGENE, Norwegian University of Life Sciences, NO-1432 Ås, Norway.
E-mail: *odd-arne.rognli@umb.no*
Phone no: +47 64965578

M.C. Saha
Forage Improvement Division, The Samuel Roberts Noble Foundation, Ardmore, OK 73401, USA.
E-mail: *mcsaha@noble.org*

A.V. Stewart
PGG Wrightson Seeds, PO Box 175, Lincoln, Christchurch 7640, New Zealand.
E-mail: *astewart@pggwrightsonseeds.co.nz*

S. Tsuruta
National Agricultural Research Organization, National Institute of Livestock and Grassland Science, Nasushiobara, 329-2793 Tochigi, Japan.
E-mail: *stsuru@affrc.go.jp*

X. Wang
Chengdu Institute of Biology, Chinese Academy of Sciences, No.9 Section 4, Renmin Nan Road, Chengdu, Sichuan, 610041 P.R. China.
E-mail: *wangxun0104@hotmail.com*

S. Warnke
USDA-ARS, Beltsville, MD 20705, USA.
E-mail: *Scott.Warnke@ars.usda.gov*

T. Yamada
Field Science Center for Northern Biosphere, Hokkaido University, Sapporo, Japan.
E-mail: *yamada@fsc.hokudai.ac.jp*

N. Yuyama
Forage Crop Research Institute, Japan Grassland Agricultural and Forage Seed Association, 388-5 Higashiakada, Nasushiobara, Tochigi 329-2742, Japan.
E-mail: *yuyama@jfsass.or.jp*

Abbreviations

AFLP	Amplified fragment length polymorphism
AMOVA	Analysis of molecular variance
BAC	Bacterial artificial chromosome
CAPS	Cleaved amplified polymorphic sequence
CIM	Composite interval mapping
CP	Cross-pollination
cpSSR	Chloroplast microsatellite
DArT	Diversity arrays technology
DM	Dry matter
EST	Expressed sequence tag
FISH	Fluorescence *in situ* hybridization
GBS	Genotype-by-sequencing or Genotyping-by-sequencing
GCA	General combining ability
GISH	Genomic *in situ* hybridization
GRIN	Germplasm resources information network
GS	Genomic selection
GWAS	Genome-wide association studies
HRM	High-resolution melting
In-del	Insertion-deletion
ILGI	International lolium genome initiative
LD	Linkage disequilibrium
LG	Linkage group
LRR	Leucine rich repeat
MAB	Marker assisted breeding
MAS	Marker assisted selection
MITE	Miniature inverted-repeat transposable element
NBS	Nucleotide binding site
NPGS	The national plant germplasm system
OECD	Organization for economic cooperation and development
QTL	Quantitative trait loci
RAPD	Random amplified polymorphic DNA
RFLP	Restriction fragment length polymorphism
RGA	Resistance gene analog

RIL	Recombinant inbred line
SAMPL	Selectively amplified microsatellite polymorphic loci
SCA	Specific combining ability
SCAR	Sequence-characterized amplified region
SI	Self-incompatibility
SNP	Single nucleotide polymorphism
SRAP	Sequence related amplified polymorphism marker
SSR	Simple sequence repeat
TDF	Transcript-derived fragment
WGS	Whole genome selection

1

Forage Grasses and Legumes— An Introduction

Toshihiko Yamada

ABSTRACT

Grassland produces large amount of forages for ruminants, maintains soil fertility, protects and conserves soil and water resources, harbors stock of carbon in terrestrial ecosystem, creates a habitat for wildlife and provides recreational spaces for sport and leisure while contributing to the general benefits at the same time as maintaining sustainable economic outputs. In addition to food and environment, grassland biomass can be used as a renewable source of energy. Some grasses are also grown specifically for turf or amenity purposes on sports fields, golf courses, parks, lawns and roadsides. Some important grass and legume species for forage and turf have been succeeded in seed business and are sown in grasslands today. In such species breeding programs have had an impact on the improvements in both yield and quality. Many forage and turf varieties have been developed and still continue to be developed through the traditional breeding techniques of hybridization and selection. Recently forage grass and legume breeding programs have entered the biotechnology era using molecular biology tools such as transformation and DNA markers. Molecular breeding is important and will be used extensively in future in improvement of forage and turf grasses and forage legumes and also in development of energy crops as source for cellulosic biomass feedstock.

Key words: Cool-season grasses, Warm-season grasses, Cool-season legumes, Warm-season legumes, Transformation, Marker-assisted selection

Field Science Center for Northern Biosphere, Hokkaido University, Sapporo, Japan.
E-mail: *yamada@fsc.hokudai.ac.jp*

1.1 Introduction

Grassland covers 27% of the world's total land area and 71% of the world agricultural area (FAO). Grassland produces large amount of forages for ruminants, maintains soil fertility, protects and conserves soil and water resources, harbors stock of carbon in terrestrial ecosystems, creates a habitat for wildlife, and provides recreational spaces for sport and leisure while contributing to the general benefits at the same time maintaining sustainable economic outputs. In addition to food and environment, grassland biomass can be used as a renewable source of energy. Besides being used for grassland, some forage grasses are grown specifically for turf or amenity purposes on sports fields, golf courses, parks, lawns and roadsides. Turf grasses contribute considerably to the environment by adding beauty to the surroundings, providing a safe playing surface for sports and recreation and preventing erosion.

There are about 10,000 species of grasses worldwide (Watson and Dallwits 1992), making them one of the largest plant families, Poaceae. Grasses that grow best under cool and generally moist conditions typically have C_3 photosynthesis system (cool-season grasses), and those that grow best under warm and often dry condition typically have the C_4 photosynthesis system (warm-season grasses). Optimum temperature for growth of most cool-season grasses is between 20 and 25°C. Below 10°C, growth drops rapidly, but slow growth often occurs at 5°C (Moser and Hoveland 1996). Because optimum temperature for C_4 photosynthesis is in the range of 35°C to 38°C, warm-season grasses grow well at high temperatures. Warm-season grasses, with their C_4 photosynthesis system that overcomes photorespiration, use water efficiency requiring about one-third to one-half as much water to produce a unit of dry matter as do C_3 grasses (Moser et al. 2004). Most forage grasses are perennials. Some warm-season grasses considered annuals in temperate climates are actually perennial that have no freeze tolerance (Moser et al. 2004).

Most cool-season grasses are high in digestible energy and protein when utilized in the vegetative stage. In comparison to cool-season grasses, warm-season grasses have relatively larger proportions of cell-wall materials that are potentially digestible, but because of their chemical composition and anatomical structure, the rate of digestibility in warm-season grasses is relatively low (Coleman et al. 2004).

The most important cool-season grasses are members of the *Lolium-Festuca* species complex (ryegrasses and fescues), two closely related genera belonging to the Poeae tribe of subfamily Pooideae. In contrast, various species of warm-season grasses (members of the genera *Brachiaria, Paspalum,*

Eragrostis and *Pennisetum*) belong to several tribes within the subfamily Panicoideae.

The key cool-season legume species predominantly belong to the genera *Medicago* (medics, including alfalfa) and *Trifolium* (clovers, including white clover and red clover) within the Trifolieae tribe of the Fabaceae clade Hologalegina. There are also a number of agronomically important warm-season legumes, including members of the genera *Chamaecrista*, *Macroptilium, Stylosanthes, Centrosema* and *Desmodium*.

The benefits of forage legumes have been promoted firstly in terms of their contribution to the nitrogen economy of grasslands and, secondly, their superior feeding value for ruminant livestock nutrition (Frame and Laidlaw 2005). Biological N_2-fixation accounts for about 65% of the N used in world agriculture (Vance 1997), of which a high proportion is rhizobially fixed by forage legumes in grasslands. Hauck (1988) estimated that approximately 50 million tons are fixed annually by forage legumes. The efficiency of the plant-*Rhizobium* symbiosis is therefore a key factor in legumes production and N supply for soil fertility.

Table 1-1 shows the important grass and legume species for forage and turf in the world. In such species breeding programs have had impact on the improvement in both yield and quality. Many forage and turf varieties have been developed and still are continuing to be developed through the traditional breeding techniques of hybridization and selection.

1.2 Reproductive and Breeding Characteristics

Breeding systems that can be used effectively to improve a species are determined by the reproductive mode. Most forage grasses and legumes reproduce either sexually via cross-pollination or by apomixis (Sleper and Poehlman 2006). Many cool-season grasses and legume species are cross-pollinated in nature and largely self-incompatible. The grasses are pollinated by wind and legumes are pollinated by insects. Self-incompatibility is the inability of functional male and female gamates to produce normal seeds following pollination (de Nettancourt 1977). For example, in perennial ryegrass a gametophytic self-incompatibility (SI) system is controlled by two loci (*S* and *Z*). Incompatible mating occurs when the alleles at both loci in the male gametophyte (pollen grain) match one of the two alleles at each locus in the female sporophyte (Cornish et al. 1979). This mechanism ensures a very low level of self-fertilization, and also limits the level of fertility in closely related individuals such as full-sibs. White clover has a gametophytic SI system controlled by a series of alleles at a single locus (*S*) (Attwood 1940).

Apomixis, an asexual form of reproduction where seeds are produced without the fertilization of an egg, is not present in warm-season grasses,

Table 1-1 Important grasses and legumes species for forage and turf.

Cool season forage and turf

Grasses	Scientific name	Legumes	Scientific name
Perennial ryegrass	*Lolium perenne* L.	White clover	*Trifolium repens* L.
Italian ryegrass	*Lolium multiflorum* Lam.	Red clover	*Trifolium pratense* L.
Tall fescue	*Festuca arundinacea* Schreb.	Alsike clover	*Trifolium hybridum* L.
Meadow fescue	*Festuca pratensis* Huds.	Crimson clover	*Trifolium incarnatum* L.
Orchard grass	*Dactylis glomerata* L.	Subterranean clover	*Trifolium subterraneum* L.
Timothy	*Phleum pratense* L.	Alfalfa	*Medicago sativa* L.
Smooth bromegrass	*Bromus inermis* Leyss.	Birdsfoot trefoil	*Lotus corniculatus* L.
Reed canarygrass	*Phalaris arundinacea* L.	Goat's rue, Galega	*Galega orientalis* Lam.
Redtop	*Agrostis alba* L.	Sweet clover	*Melilotus alba* Desr., *Melilotus officinalis* (L.) Pall.
Kentucky bluegrass	*Poa pratensis* L.		
Creeping bentgrass	*Agrostis stolonifera* L.		
Colonial bentgrass	*Agrostis tenuis* Sibth.		

Warm season forage and turf

Grasses	Scientific name	Legumes	Scientific name
Bahiagrass	*Paspalum notatum* Flüggé	Siratro	*Macroptilium atropurpureum* (DC.) Urb.
Rhodesgrass	*Chloris gayana* Kunth	Phasey bean	*Macroptilium lathyroides* (L.) Urb.
Bermudagrass	*Cynodon dactylon* (L.) Pers.	Greenleaf desmodium	*Desmodium intortum* (Mill.) Urb.
Dallisgrass	*Paspalum dilatatum* Poir.	Silverleaf desmodium	*Desmodium uncinatum* (Jacq.) DC.
Napiergrass	*Pennisetum purpureum* Schum.	Centro	*Centrosema* (L.) Benth.
Giant stargrass	*Cynodon aethiopicus* Clayton & Harlan	Lablab bean	*Lablab purpureus* (L.) Sweet.
		Townsville stylo	*Stylosanthes humilis* H. B. K.

especially the tropical species (Moser et al. 2004). The exception of cool-season grasses is Kentucky blue grass, which is also highly apomictic. This reproductive phenomenon has impacted the breeding of many warm-season grasses because conventional breeding methods cannot be used for genetic improvement. Therefore, the reproductive behavior of suspected apomictic grasses should be clarified before a breeding program is initiated.

Characteristics on reproduction, genetics and breeding in forage grasses and legumes are summarized (Fig. 1-1). Most species have small flowers, making hand emasculation tedious and difficult. Cytoplasmic male-sterility systems have not been developed except for some species such as alfalfa. Many species are polyploids, which complicates inheritance of traits. Perennial plant species can be vegetatively propagated by stolons, rhizomes, tillers, or bud on culms. Individual plants in a population are highly heterozygous. A population consists of plants with many different genotypes, and is heterogeneous. Plants are used in thickly used seeded stands or swards as forages and turf. Individual plant selection is not possible under these conditions. Therefore, evaluation and selection are usually carried out in space-transplanted nurseries (Fig. 1-2).

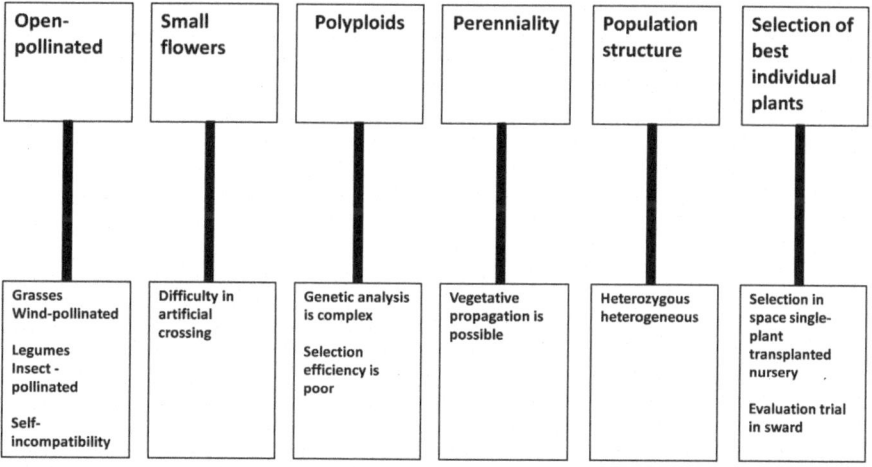

Figure 1-1 Characteristics of forage grasses and legumes—reproduction, genetics and breeding.

1.3 Breeding Systems

The most popular breeding strategy, heterosis, the superiority of a progeny to its parents, underpins many of the yield gains of maize (*Zea mays* L.) and other open-pollinated crops (Brummer 1999). A heterotic group has been defined as a collection of germplasm that tends to exhibit a higher

Figure 1-2 A space-transplanted selection nursery of timothy (Kitami Agricultural Experiment Station, Kunneppu, Hokkaido, Japan).

Color image of this figure appears in the color plate section at the end of the book.

degree of heterosis when crossed with germplasm from an external group than when crossed with a member of its own group (Lee 1995). Combining ability, the ability to produce superior progeny, is used to identify desirable combinations of genotypes to cross in the breeding of hybrid cultivars or to identify desirable parents to include in a synthetic cultivar of a forage crop (Sleper and Poehlman 2006). General combining ability (GCA) is defined as the average performance of a strain in a series of crosses and is due to mainly additive effects. Specific combining ability (SCA) is defined as the performance of specific combinations of genetic strains in crosses relating to the average performance of all combinations and is due to mainly non-additive effects. In order to exploit heterosis, a breeding approach which capitalizes on SCA using heterotic groups is more advantageous than that based on GCA. However, the primary breeding strategy in forage grasses or legumes, with perfect small flowers and without a stable cytoplasmic male sterility system, is cultivar synthesis based on GCA, using additive genetic variation. The breeding systems for cross-pollinated forage grasses and legumes are based on population genetics and recurrent selection or repeated generation breeding. The objectives are to change population means for specific traits by increasing the frequency of desirable genes for those traits. Improved populations are released as synthetic cultivars. Restricted recurrent phenotypic mass selection and between and within half-sib family selection are efficient and effective breeding programs in cross-pollinated forage species (Vogel and Pedersen 1993). These breeding methods capitalize on additive genetic variation by accumulation of desirable genes and do not exploit non-additive genetic variation, that is, heterosis.

Several breeding methods have been proposed for effective yield improvement in forage grasses or legumes with the aim of capturing heterosis and/or exploiting SCA (Rotili et al. 1996; Brummer 1999; Tamaki et al. 2007, 2009). One of the methods is a "clone and strain synthesis (CSS)" approach, which exploits both GCA and SCA and minimizes the risk of inbreeding depression. This has been used to develop strains producing significantly higher yield performance in timothy (Tamaki et al. 2009). Molecular marker data using genetic distance estimation could lead to more efficient improvement of forage grasses or legumes by exploiting heterosis or for reducing inbreeding depression (Tanaka et al. 2011a,b, 2013).

1.4 Molecular Breeding

Breeding programs in forage grasses and legumes have entered the biotechnology era using molecular biology tools. Molecular breeding is therefore a relatively new term that describes the use of transgenic and genomic approaches in a conjunction with traditional breeding programs (Bouton 2008).

Transgenesis involves the movement of specific and useful genes into the crop of choice. This approach has already shown success in introducing genes, which make many important crops resistant to insects, viruses and herbicides. The transgenic approach has also been very useful in creating unique plants that allow basic research to clarify the physiological and biochemical pathways.

Genomics research received great publicity with the successful completion of the human genome sequencing project. Following whole genome sequence of the model plant species, *Arabidopsis thaliana* and *Oryza sativa*, *Zea mays*, *Sorghum bicolor*, *Brachypodium distachyon*, *Medicago truncatula* and *Lotus corniculatus* var. *japonicus* have been well sequenced whole genome and used as a reference species for forage grasses and legumes, respectively. The sequencing data for these reference species, combined with high throughput machinery and data analysis (e.g., bioinformatics), allows more accurate determinations of species relationships and gene expression. From this understanding, new and innovative methods for improving forage grasses and legumes are evolving.

Molecular breeding is important and will be used extensively in future forage grasses and legumes improvement and development of energy crops as cellulosic biomass feedstock. Transformation techniques using *Agrobacterium* or biolistics-based method and many available molecular markers such as simple sequence repeat (SSR) markers and some functionally-associated genetic markers have been developed in many forage grasses and legumes (Yamada et al. 2010). Genetic dissection of "quantitative trait loci" referred to as QTL has become a common approach, and created a new paradigm in plant breeding. Many QTLs have been identified using linkage maps constructed from DNA markers such as SSR in forage grasses and legumes (Yamada and Spangenberg 2008). Conventional QTL analysis based on biparental mapping population is not always the best way to implement a marker-assisted selection (MAS) program in open-pollinated forage grasses and legumes, although this approach is useful for introgression of alleles from wild germplasm (Yamada and Skøt 2010). The availability of abundant markers and the reduction of cost and time of marker analyses through high-throughout genotyping system allow the development of new tools with a more accurate MAS program. For the breeding program of the complex polygenic traits that are crucial for the success of new crop varieties, genomic selection (GS) that predicts the breeding values of lines in a population by analyzing their phenotypes and high-density marker scores appear to be promising breeding schemes (Heffner et al. 2009).

We still have the remaining challenge on the successful implementation of molecular breeding in practical varietal development. Molecular breeding needs to be developed from a platform of good conventional breeding and includes supporting agronomic research and partnering with commercial industry where appropriate. With the availability of more sequencing information such expressed sequence tags (ESTs) by next generation sequencers, gene isolation has become much easier than ever before. We should focus on functional characterization of genes and their regulatory elements.

The large amounts of data involved need substantial bioinfomatics support. Data of various kinds must be integrated from an increasingly wide range of sources such as genetic resources and genes and association studies for plant populations through the transcriptome and metabolome of individual tissues with a different plant stage. The merging of data from disparate sources and multivariate data-mining across datasets can reveal novel information concerning the biology of complex forage grasses and legumes.

References

Attwood SS (1940) Genetics of cross-incompatibility among self-incompatible plants *of Trifolium repens*. J Am Soc Agron 32: 955–968.

Bouton JH (2008) Molecular breeding to improve forages for use in animal and biofuel production systems. In: Yamada T, Spangenberg G (eds) Molecular Breeding of Forage Crops. Springer, New York, USA, pp 1–13.

Brummer EC (1999) Capturing heterosis in forage crop cultivar development. Crop Sci 39: 943–954.

Coleman SW, Moore JE, Wilson JR (2004) Quality and utilization. In: Moser LE et al. (eds) Warm-season (C_4) Grasses. ASA-CSSA-SSSA, Madison pp 267–308.

Cornish MA, Hayward MD, Lawrence MJ (1979) Self-incompatibility in ryegrass. I. Genetic control in diploid *Lolium perenne* L. Heredity 43: 95–106.

de Nettancourt D (1977) Incompatibility in Angiosperms. Springer, New York, USA.

FAO http://www.fao.org/ag/agp/agpc/doc/grass_stats/grass-stats.htm.

Frame J, Laidlaw AS (2005) Prospects for temperate forage legumes. In: Reynolds SG, Frame J (eds) Grasslands: Developments Opportunities Perspectives. Food and Agriculture Organization of the United Nations, Science Publishers, Enfield, USA, pp 4–28.

Hauck RD (1988) A human ecosphere perspective of agricultural nitrogen cycling. In: Wilson JR (ed) Advances in Nitrogen Cycling in Agricultural Ecosystems. CAB International, Wallingford, UK, pp 3–19.

Heffner EL, Sorrells ME, Jannink J-L (2009) Genomic selection for crop improvement. Crop Sci 49: 1–12.

Lee M (1995) DNA markers and plant breeding programs. Adv Agron 55: 265–344.

Moser LE, Burson BL, Sollenberger LE (2004) Warm-season (C_4) grass overview. In: Moser LE et al. (eds) Warm-season (C_4) Grasses. ASA-CSSA-SSSA, Madison, pp 1–14.

Moser LE, Hoveland CS (1996) Cool-season grass overview. In: Moser LE et al. (eds) Cool-season Grasses. ASA-CSSA-SSSA, Madison, pp 1–14.

Rotili P, Busbice TH, Demarly Y (1996) Breeding and variety constitution in alfalfa: present and future. In: Parente G et al. (eds) Grassland and Land use Systems. 16th EGF meeting. EGF&ERSA, Gorizia, Italy, pp 163–180.

Sleper DA, Poehlman JM (2006) Breeding Field Crops. Fifth edn. Blackwell, Ames, IA.

Tamaki H, Sato K, Ashikaga K, Tanaka T, Yoshizawa A, Fujii H (2009) High–yield timothy (*Phleum pratense* L.) strains developed by 'clone and strain synthesis', a method for breeding perennial and self–incompatible crops. Grassland Sci 55: 57–62.

Tamaki H, Yoshizawa A, Fujii H, Sato K (2007) Modified synthetic varieties; a breeding method for forage crops to exploit specific combining ability. Plant Breed 126: 95–100.

Tanaka T, Tamaki H, Cai HW, Ashikaga K, Fujii H, Yamada T (2011a) DNA profiling of seed parents and top-cross tester and its application for yield improvement in timothy (*Phleum pratense* L.). Crop Sci 51: 612–620.

Tanaka T, Tamaki H, Ashikaga K, Fujii H, Yamada T (2011b) Proposal for shift to reciprocal recurrent selections in 'clone and strain synthesis' timothy breeding using molecular marker diversity, Crop Sci 51: 2589–2596.

Tanaka T, Tamaki H, Ashikaga K, Fujii H, Yamada T (2013) Use of molecular marker diversity to increase forage yield in timothy (*Phleum pratense* L.) Plant Breed 132: 144–148.

Vance CP (1997) Nitrogen fixation capacity. In: McKersie BD, Brown DCW (eds) Biotechnology and the Improvement of Forage Legumes. CAB International, Wallingford, UK, pp 375–407.

Vogel KP, Pedersen JF (1993) Breeding systems for cross-pollinated perennial grasses. Plant Breed Rev 11: 251–274.

Watson L, Dallwitz MJ (1992) The Grass Genera of the World. CAB International, Wallingford, UK.

Yamada T, Skøt L (2010) Allelic diversity for candidate genes and association studies: methods and results. In: Huyghe C (ed) Sustainable Use of Genetic Diversity in Forage and Turf Breeding. Springer, New York, pp 391–396.

Yamada T, Spangenberg G (eds) (2008) Molecular Breeding of Forage and Turf. Springer, New York, p 352.

Yamada T, Tamura K, Wang X, Aoyagi Y (2010) Transgensis and genomics in forage crops. In: Jain SM, Brar DS (eds) Molecular Techniques in Crop Improvement, 2nd edn, Springer, The Netherlands, Dordrecht, pp 719–744.

2

Perennial Ryegrass

Toshihiko Yamada

ABSTRACT

Perennial ryegrass (*Lolium perenne* L.) is the most widely sown perennial forage grass in temperate regions of the world, mainly in northwest Europe, New Zealand, Australia, South Africa and South America. It is a palatable, persistent grass of high tillering density that shows resistance to treading and good response to high nitrogen application. It is also widely grown in amenity turf. New cultivars, which developed through genetic improvement, have a long history of positively impacting forage and livestock systems. In general, perennial ryegrass can be grouped as diploids or tetraploids. One great achievement of perennial ryegrass breeding was to double chromosome number by colchicine treatment. The proportion of listed tetraploid varieties has increased because they have characteristics of high yielding with good sugar contents and the open sward prevent damage from snow mold. Recurrent phenotypic selection is often used in perennial ryegrass due to outbreeding property with self-incompatibility. Molecular breeding as molecular marker-assisted selection could be powerful for selection in perennial ryegrass. Many works on development of DNA markers, construction of genetic linkage maps, analyses of quantitative trait loci for agronomic important traits and trait dissection, marker-assisted selection have been reported. A more practical approach may be the use of association analysis, measuring both phenotypes and markers directly on the plants in the breeding field. Recently as the cost of genotyping has decreased genomic selection methods would have a great impact on molecular breeding in

Field Science Center for Northern Biosphere, Hokkaido University, Sapporo, Japan.
E-mail: *yamada@fsc.hokudai.ac.jp*

the future. Approach of targeted genes, such as *VRN1* gene, flowering genes and self-incompatibility genes has been studied. Transformation techniques using *Agrobacterium* or biolistics-based method have been developed in perennial ryegrass. The recent achievement of molecular breeding in perennial ryegrass is reviewed in this chapter.

Key words: Chromosome doubling, Recurrent phenotypic selection, DNA markers, Association analysis, Genomic selection, Transformation

2.1 Introduction

Perennial ryegrass (*Lolium perenne* L.) is the most widely sown perennial forage grass in temperate regions of the world. It is of interest as it is fast-growing, has high nutritional value and has good persistency under grazing. Swards containing improved perennial ryegrass varieties can produce dry matter (DM) yield approaching 15 t/ha over a grazing season with an average grass production of 60 kg/ha/day (Humphreys 2005). Perennial ryegrass also has important amenity value and provides many landscape benefits with functional (erosion control, reduction of glare, noise, heat build-up and air pollutions; stabilizing dust and soil), recreational (sport and leisure) and aesthetic impacts. To overcome the negative effect from climate changes, plants must become more resistant to many kinds of biotic and abiotic stresses such as a variety of diseases and insects, drought, and high, low or fluctuating temperatures. Other possible environmental targets include soil C-sequestration and biodiversity, bioremediation and flood mitigation.

New cultivars, which were developed through genetic improvement, have a long history of positively impacting forage and livestock systems. Recently climate changes are of global concern and also for forage breeding. Many factors for climate changes such as warmer winter, high temperature during summer, drought, sea level rise and increase in heavy precipitation, together with the side effects such as high salinity, macronutrients and micronutrients deficiency may cause stress to plants, disturb the plant growth, decrease the yield of forage, and as consequence decrease livestock production. Therefore, it is important to develop new forage crop cultivars that can adapt to climate changes. Traditional breeding methods such as phenotypic evaluation, selection and hybridization have always been, and still continue to be used. However, breeding programs have entered the biotechnological era using molecular biology tools. Molecular breeding is therefore a relatively new term that describes the use of genomic approaches in conjunction with traditional breeding programs (Bouton 2008). The recent achievements of molecular breeding in perennial ryegrass is reviewed here.

2.2 Biology of Perennial Ryegrass

2.2.1 Origin, Distribution and Taxonomy

Ryegrasses are indigenous to parts of Europe, Asia and North Africa (Jauhar 1993; Jung et al. 1996). The Mediterranean Basin is the probable origin of ryegrasses, which share a high degree of genome ancestry with cereal species of Eurasian origin, such as wheat and barley (Kellog 2011). The domestication of grasses is associated with the emergence of primitive agriculture in the Fertile Crescent of the Middle East about 10,000 years ago and its subsequent expansion into West and North Europe (Balfourier 2000). Ryegrasses probably spread as weeds of cereal crops by migrating farmers. The first records of cultivated use of perennial ryegrass came from England about 1677 (Jung et al. 1996). The spread of ryegrasses is closely associated with the development of livestock farming. Ryegrasses are widely distributed throughout temperate zones and are extensively used in Europe, Australia, New Zealand, North America and Japan (Jung et al. 1996).

Perennial ryegrass is a member of the Poaeae tribe of the Poodae super-tribe in the Pooideae subfamily of the grass and cereal family Poaceae (Soreng and Davis 1998). Members of the *Lolium* genus are diploids with a fundamental chromosome number of 7 ($2n = 2x = 14$). The genome size of perennial ryegrass has been estimated through measurements of nuclear DNA content by microdensitometry (Hutchinson et al. 1979; Seal and Rees 1982). A 2C value of 4.16 pg corresponds to a haploid genome size of c. 1.6 x 109 bp. The individual genome sizes of other *Lolium* species vary, with the inbreeding taxa such as *Lolium temulentum* exhibiting nuclear DNA contents c. 50% larger than those of the outbreeding species. In common with other Poaceae family members, the genomes of *Lolium* species contain large numbers of dispersed repetitive sequences, frequently belonging to major retroelement families (Jenkins et al. 2000).

2.2.2 Reproductive Manner

Perennial ryegrass is an obligate outbreeding species, with a gametophytic self-incompatibility (SI) system controlled by two loci (*S* and *Z*). Incompatible mating occurs when the alleles at both loci in the male gametophyte (pollen grain) match one of the two alleles at each locus in the female sporophyte (Cornish et al. 1979). This mechanism ensures a very low level of self-fertilization, and also limits the level of fertility in closely related individuals such as full-sibs. The proportions of compatible mating between related individuals depend on the degree of allelic complexity at the SI loci. Both *S* and *Z* are represented by a polyallelic series: 17 S and 17 Z alleles were discriminated in naturally occurring perennial ryegrass pasture populations (Fearon et al. 1994).

2.2.3 Economic Importance

Perennial ryegrass is a palatable, persistent grass of high tiller density that shows resistance to treading and good response to high nitrogen application. Perennial ryegrass is main forage grass sown in Northwest Europe, New Zealand, Australia, South Africa and South America. Perennial ryegrass is also widely grown in amenity turf. The importance of perennial ryegrass is reflected in the number of varieties on the Organisation for Economic Co-operation and Development (OECD) list in 2011 (1,362) and in seed production that since 2000 the EU-27 countries have produced, on average 83,660 t per year and on global scale 209,674 t per year (Humphreys et al. 2010).

2.3 Conventional Breeding

In general, perennial ryegrass can be grouped as diploids or tetraploids. One of the earliest "novel" achievements of perennial ryegrass breeding was to double chromosome number by colchicines treatment. Wit (1959) was the first in the Netherlands to treat breeding materials systematically and supply valuable tetraploid resources to commercial breeders. Until the late 1980s, confirmation of chromosome doubling was mainly done by chromosome counting using microscopes. With the introduction of flow cytometer, the detection of tetraloid plants based on DNA content per cell became much easier. Today flow cytometry is a reliable and quick technique to develop tetraploid varieties. The proportion of listed tetraploid varieties has increased because they have characteristics of high yield with good sugar contents and the open sward prevents damage from snow mold. Non-suitable climatic areas such as Japan have shown good performance for perennial ryegrass cultivation in only tetraploid varieties.

Due to outbreeding property, recurrent phenotypic selection is often used in perennial ryegrass. Wilkins and Humphreys (2003) reviewed how four generations of combined phenotypic and half-sib family selection over 12 yr was successful in simultaneously improving dry mater (DM) yield and water soluble carbohydrate (WSC) content in perennial ryegrass. Current breeding methods and target breeding traits in perennial ryegrass were reviewed by Humphreys et al. (2010).

The ability to produce enough seed yield is essential to ensure that new varieties are sown in grassland in practice. The breeding program for high seed productivity is also important.

2.4 Genomics

Molecular breeding as molecular marker-assisted selection could be useful for the selection. In order to establish molecular breeding, several research studies on genome resources, DNA marker, genetic linkage map, quantitative trait loci (QTL) analysis and association analysis have been carried out in perennial ryegrass. Herbage yield and forage quality are important breeding objectives in a combination with several traits such as abiotic stresses such as drought, heat and cold tolerance as well as biotic stresses as disease and insect resistance. DNA marker selection approaches are expected to accelerate the conventional breeding approaches because agriculturally important traits showing continuous phenotypic variation are controlled by a variable number of QTLs.

2.4.1 Genome Resources

2.4.1.1 ESTs

Gene discovery by expressed sequence tag (EST) sequencing has generated substantial genomic resources. In an Australian group, a collection of 44,534 perennial ryegrass ESTs was generated from single pass sequencing of randomly selected clones from 29 cDNA libraries that represent a range of plant organs, developmental stages and environmental conditions (Sawbridge et al. 2003). The sequences were annotated by comparison to GenBank and SwissProt public sequence databases and automated intermediate Gene Ontology (GO) annotation was obtained (Spangenberg et al. 2005). All sequences and annotation are maintained within ASTRA format MySQL databases, with web-based access for text searching, BLAST sequence comparison and GO hierarchical tree browsing. Each ryegrass sequence was mapped onto an EnsEMBL genome viewer for comparison with the complete genome sequence of rice and expressed sequences from related species.

2.4.1.2 BAC Libraries

Complementing the EST resources, large insert DNA libraries have been generated using bacterial artificial chromosome (BAC) vectors. Two BAC libraries were constructed for perennial ryegrass (Farrar et al. 2007). The libraries consisted of 98,304 and 101,376 BAC clones for perennial ryegrass genotypes LTS18 and NV#20F1-30, respectively. The estimated average insert size of both libraries was approximately 100 Kb. BAC libraries of perennial ryegrass (50,304 BAC clones with 113 kb average insert size, corresponding to 3.4 genome equivalents and 97% genome coverage) have been established (Spangenberg et al. 2005).

2.4.2 Transcriptomics

High-density cDNA microarrays representing approximately 15,000 unique genes for perennial ryegrass (Sawbridge et al. 2003) have been developed. The forage crops-derived microarrays have been applied in hybridization with labeled total RNA isolated from a variety of genotypes, plant organs, developmental stages and growth conditions. Results from these studies have enabled validation of functions predicted through comparative sequence annotation, and also suggested roles for novel genes which lack comparative sequence annotation (Spangenberg et al. 2005). Unannotated genes which are co-regulated with the perennial ryegrass *LpCAD1* and *LpCAD3* (cinnamyl alcohol dehydrogenase) lignin biosynthesis genes have been identified and mapped as candidate gene-based markers to regions of the genome associated with herbage QTL (Cogan et al. 2006). Microarrays may also be used for gene and promoter discovery when used in concert with the BAC libraries established for each species (Spangenberg et al. 2005).

Tamura and Yonemaru (2010) developed an effective method for comparing transcriptome data during cold acclimation using next-generation sequencing technique as 454 sequencing. They obtained the contigs and singletons into 661 (perennial ryegrass) and 1082 (meadow fescue) clusters based on sequence similarity.

2.4.3 DNA Markers

2.4.3.1 SSR Markers

Simple sequence repeats (SSR) markers provide the current marker system of choice due to their abundance, ubiquitous distribution in plant genomes, high level of reproducibility, ease of PCR-based analysis and detection of co-dominant multiallelic loci. Thus, SSR markers have been developed from perennial ryegrass (Jones et al. 2001; Kubik et al. 2001; Asp et al. 2007; Jensen et al. 2007; Studer et al. 2008a; King et al. 2008). Also the high degree of cereal EST-SSR can be used for perennial ryegrass but sequencing analysis is essential before transferring genetic information using comparative analysis (Sim et al. 2009). For example use of SSR markers, linkage maps have been constructed using SSR markers in perennial ryegrass (Jones et al. 2002b; Gill et al. 2006).

2.4.3.2 DNA Markers from Rice Genome Information

Intronic regions generally include more polymorphisms than exonic regions, a high frequency of polymorphic markers is expected using this system.

Novel PCR-based EST markers were developed all designed around intronic regions which show higher polymorphism than exonic regions. Intronic regions of the grass genes were speculated using rice genomic information. Two hundred and nine primer sets were designed from *Lolium/Festuca* ESTs showing high similarity to unique rice genes dispersed uniformly throughout the rice genome. Sixty one of these primer sets as insertion-deletion (indel)-type markers and 82 primer sets as cleaved amplified polymorphic sequence (CAPS) markers to distinguish between perennial ryegrass and meadow fescue (*Festuca pratensis* Huds) (Tamura et al. 2009). Many indel-type markers and CAPS markers had high species specificity to *L. perenne* and *F. pratensis*. Chromosome mapping of these markers using *Lolium/Festuca* substitution plants revealed syntenic relationships between *Lolium/Festuca* and rice largely consistent with previous reports (Tamura et al. 2009). This marker system based on intron polymorphisms with a high frequency between species and high species specificity could consequently be a useful tool for breeding of *Festulolium*, intergeneric hybrid between *Lolium* and *Festuca*.

2.4.3.3 Genetic Variation

Assignment or exclusion of an individual to specific populations or cultivars based on molecular genetic markers provides an attractive approach for varietal identification at the individual level in cross-pollinated plant species such as perennial ryegrass. The studies on genotyping eight populations comprising 48 individual plants per population using 29 SSR marker loci by Wang et al. (2009) indicated that SSR marker profiles can be effectively used to assign individuals for outbreeding populations such as perennial ryegrass.

2.4.4 Genetic Map

An enhanced molecular marker-based genetic linkage map of perennial ryegrass has been constructed through the activities of the International Lolium Genome Initiative (ILGI) (Forster et al. 2001), using the p150/112 one-way pseudo-testcross mapping population. The map contains 109 restriction fragment length polymorphism (RFLP) loci detected by heterologous probes from wheat, barley, oat and rice. Comparative genetic mapping has allowed the alignment of the perennial ryegrass genetic map with those of wheat, rice and oat, revealing substantial conserved synteny with the genomes of Poaceae species (Jones et al. 2002a). At the macrosyntenic level, each of the seven linkage groups (LGs) of perennial ryegrass chiefly corresponds to one of the seven basic homeologous chromosome groups of the Triticeae cereals, and they have been numbered accordingly. Seven LGs of perennial ryegrass also correspond to 12 LGs of rice.

Two genetic mapping populations of perennial ryegrass (F_2 (Aurora x Perma) and F_1 (North African6 (NA_6) x Aurora6 (AU_6)) have been independently developed as successors to the p150/112 population, and have been aligned to the reference map using common markers (Armstead et al. 2004; Faville et al. 2004). These genetic maps contain functionally-associated molecular marker information through the inclusion of gene-associated CAPS markers, and both RFLP and SSR markers from ESTs, respectively. High-density molecular marker-based genetic linkage maps have also been constructed for interspecific hybrid populations based on crosses between *L. multiflorum* and *L. perenne* (Warnke et al. 2004; Sim et al. 2005; Wang et al. 2011). Genetic maps are consequently available for detailed dissection of complex phenotypes to resolve the locations of pleiotropic and interacting genetic factors.

Recently Studer et al. (2010) reported the construction of a consensus linkage map in perennial ryegrass. A set of 204 EST-derived SSR markers has been assigned to map positions using eight different mapping populations of perennial ryegrass. Marker properties of a subset of 64 EST-SSRs were assessed in six to eight individuals of each mapping population and revealed 83% of the markers to be polymorphic in at least one population and an average number of alleles of 4.88. EST-SSR markers polymorphic in multiple populations served as anchor markers and allowed the construction of the first comprehensive consensus map for ryegrass. The integrated map was complemented with 97 SSRs from previously published linkage maps and finally contained 284 EST-derived and genomic SSR markers. The total map length was 742 cM, ranging for individual chromosomes from 70 cM of LG 6 to 171 cM of LG 2. The consensus linkage map for perennial ryegrass based on eight mapping populations and constructed using a large set of publicly available *Lolium* EST-SSRs mapped for the first time together with previously mapped SSR markers will allow for consolidating existing mapping and QTL information in perennial ryegrass (Studer et al. 2010).

2.4.5 Trait Dissection

The application of genetic marker analysis to trait dissection has been reviewed by Yamada and Forster (2005) and Yamada et al. (2005). Trait dissection for perennial ryegrass has been performed in multiple populations to allow QTL analysis. The p150/112 population has been analyzed for traits such as vegetative and reproductive morphogenesis, reproductive development and winterhardiness, herbage quality and gametophytic self-incompatibility (Thorogood et al. 2002; Yamada et al. 2004; Cogan et al. 2005; Shinozuka et al. 2006), while the F_1 (NA_6 x AU_6) population has been studied for a range of root and shoot morphogenesis, photosynthetic efficiency, pseudostem water soluble carbohydrate content, and crown rust resistance characters (Forster et al. 2004). Other perennial ryegrass populations have been analyzed to detect genetic control of crown rust resistance (Dumsday et al. 2003; Muylle et al. 2005a, b), vernalization response (Jensen et al. 2005), flowering time variation (Armstead et al. 2004). Further QTLs for seed yield and fertility traits (Studer et al. 2008b) and resistance to crown rust (Schejbel et al. 2007) and powdery mildew (Schejbel et al. 2008) have been identified in mapping population of perennial ryegrass. Some QTLs for morphological traits influencing waterlogging tolerance have been identified (Pearson et al. 2011). Interspecific hybrid populations between *L. multiflorum* and *L. perenne* have been used to identify QTLs for flowering time variation (Warnke et al. 2004), fiber components and crude protein content (Xiong et al. 2007a), winter hardiness (Xiong et al. 2007b), gray leaf spot resistance (Curley et al. 2005), crown rust resistance (Sim et al. 2007) and stem and leaf spot resistance (Jo et al. 2008). Jo et al. (2008) compared locations of QTLs for four fungal diseases resistance with equivalent genomic region in cereal crops for comparative genomic analyses.

Biomass yield is a complex trait influenced by several interacting morphological traits controlled a lot of genes with genotype x environment interaction. Hence, a genetic analysis for biomass yield is a major challenge. QTLs for some traits linked to biomass production have been identified; leaf length (Yamada et al. 2004; Armstead et al. 2008; Barre et al. 2009; Kobayashi et al. 2011; Sartie et al. 2011), plant height (Yamada et al. 2004; Turner et al. 2008; Studer et al. 2008b; Barre et al. 2009; Kobayashi et al. 2011), leaf elongation rate (Turner et al. 2008; Barre et al. 2009; Sartie et al. 2011), fresh weight (Yamada et al. 2004; Turner et al. 2008; Anhalt et al. 2009), dry weight (Anhalt et al. 2009; Sartie et al. 2011) and DM (%) (Turner et al. 2008; Anhalt et al. 2008). QTL information identified by these studies will require validation across populations and environment for confirmation of association with trait performance.

2.4.6 Marker-assisted Selection

Genetic gain from phenotypic selection in open-pollinated forage crops is constrained by the inability to accurately use phenotype to estimate genotype, prior to parent selection for polycrossing. The use of marker-assisted selection (MAS) offers the potential to accelerate genetic gain by overcoming this constraint. Identifying markers based on biparental mapping populations is probably not the best way to implement a MAS program, although this approach is useful to introgress alleles from wild germplasm (Brummer and Casler 2008). MAS was tested for water-soluble carbohydrate QTLs identified F_2 mapping population of perennial ryegrass (Turner et al. 2010). The study revealed some the challenges and opportunities ahead for the use of MAS in breeding programs handling complex traits in outbreeding species. Instead, a more practical approach may be the use of association analysis, measuring both phenotypes and markers directly on the plants in the breeding nursery.

2.4.7 Association Analysis

The marker-trait gene linkage analysis method is not used often for outbreeding forage crops. Functionally-associated candidate genes have been developed in perennial ryegrass. Forage crops such as perennial ryegrass would have the potential to dispose limited linkage disequilibrium (LD), extending over relatively short molecular distances (Forster et al. 2004; Dobrowolski and Forster 2007). Forage crops contain many accessions with long-established populations derived from a large number of founding parents, as expected for ecotypes and long-established varieties, in which many rounds of recombination have occurred. These accessions allow the use of candidate gene-based functionally-associated marker analyses (Andersen and Lübberstedt 2003). Nucleotide variation in qualified candidate genes will be closely associated with the casual mutations that generate variability for key agronomic traits, and may be used as diagnostics for such variations. Successful correlation of gene haplotype structure and phenotypic variation will provide the basis for a new paradigm in forage crop molecular breeding based on direct selection of superior allele content at target genetic loci, allowing highly effective exploitation of germplasm collections for identification of potential parental genotypes (Forster et al. 2004; Spangenberg et al. 2005).

LD was investigated in the gibberelic acid insensitive gene region in three synthetic varieties of perennial ryegrass chosen for their contrasting number of parents in the initial poly-cross (Auzanneau et al. 2007). Significant LD was observed up to 1.6 Mb in a variety that originated from six related parents and not above 174 kb in a variety from 336 parents, suggesting that a genome-wide association approach may be possible in synthetic populations with few parents.

Single nucleotide polymorphisms (SNPs) are the most abundant class of genetic markers. Efficient methods for discovery of SNPs and characterization of SNP haplotype structure have been described for perennial ryegrass (Spangenberg et al. 2005; Cogan et al. 2006; Ponting et al. 2007). Multiple SNPs at regular intervals across an amplicon were detected within and between the heterozygous parents and validated in the progeny of the F_1 (NA$_6$ x AU$_6$) genetic mapping family. Decay of LD to r^2 values of c. 0.2 typically occurs over 500–3,000 bp, comparable with gene length and with little apparent variation between diverse, ecotypic and varietal population sub-groups (Smith et al. 2008).

A high level of genetic diversity was observed in a set of 380 perennial ryegrass European elite genotypes when genotyped with 40 SSRs and 2 STS markers (Brazauskas et al. 2011). One subpopulation consisted mainly of genotypes from the UK, while germplasm mostly from continental Europe was grouped into the second subpopulation. LD (r^2) decay was rapid and occurred within 0.4 cM across European varieties, when population structure was taken into consideration. However, an extended LD of up to 6.6 cM was detected within the British variety Aberdart. High genetic diversity and rapid LD decay provide means for high resolution association mapping in elite materials of perennial ryegrass (Brazauskas et al. 2011).

A candidate gene approach for associating SNPs with variation in flowering time and water-soluble carbohydrate content (WSC) and other traits has been described for perennial ryegrass (Skøt et al. 2007). Perennial ryegrass ortholog of *Arabidopsis thaliana FLOWERING LOCUS T (FT)* gene, designated *LpFT3*, was assessed for the genomic and phenotypic variations in a diverse collection of nine European germplasm populations (Skøt et al. 2011). Genotyping assays designed to detect genomic variation showed that three haplotypes were present in approximately equal proportions and represented 84% of the total, with a fourth representing a further 11%. Of the three major haplotypes, two were predicted to code for identical protein products and the third contained two amino acid substitutions. Association analysis using either a mixed model with a relationship matrix to correct for population structure and relatedness or structured association with further correction using genomic control indicated significant associations between *LpFT3* and variation in flowering time (Skøt et al. 2011).

Eleven expressed disease resistance candidate (R) genes including six nucleotide binding site and leucine rich repeat (NBS-LRR) like genes and five non-NBS-LRR genes were analyzed by allelic diversity (Xing et al. 2008). NBS-LRR like gene fragments showed a high degree of nucleotide diversity. Substantial LD decay was found within 500 bp for most resistance candidate genes. An approach based on *in vitro* SNP discovery in candidate defence response (DR) genes has been used to develop potential diagnostic genetic markers in perennial ryegrass (Dracatos et al. 2008). SNPs were predicted, validated and mapped for representatives of the pathogenesis-related (PR) protein-encoding and reactive oxygen species (ROS)-generating gene classes.

2.4.8 Genomic Selection

As the cost of genotyping has decreased, the interest in genomic selection (GS) methods has increased. It represents an alternative use of MAS compared to the introgression and backcross method, which is unsuitable for complex traits controlled by many QTLs with small effects. GS is related to genome-wide association mapping in which populations of unknown pedigree are used for associating genotype to phenotype, usually on a marker by marker basis, and very a stringent significance criteria. GS uses all marker locus data across a genome in a joint analysis to capture all locus and haplotype effects. The genetic variance obtained by summing individual marker effects can be used to estimate the breeding value of individual genotypes (the average genotypic value of its progeny) (Heffner et al. 2009). GS is particularly useful in breeding programs that want to utilize small-effect QTLs or gene pyramiding.

2.4.9 Gene Approach

2.4.9.1 VRN1 Gene and Flowering Genes

Vernalization, a period of low temperature to induce transition from vegetative to reproductive state, is an important environment stimulus for cool season grasses such as perennial ryegrass. The major QTL explaining 28% of the phenotypic variation for vernalization response in the mapping population was identified on LG 4 and found to co-localize with the *LpVRN1* gene, a putative ortholog of the *VRN1* gene from *T. monococcum* (Jensen et al. 2005). Asp et al. (2011) identified, sequenced and characterized *VRN1* locus by the analysis using BAC clones covering the *VRN1* locus from two genotypes with contrasting vernalization requirements and compared *VRN1* locus with cereals. Analysis of the allelic sequences identified an 8.6-kb deletion in the first intron of the *VRN1* gene in the genotype with low vernalization requirement and this deletion region if the first intron of the *VRN1* gene is an important regulatory region for vernalization response in perennial ryegrass (Asp et al. 2011).

A differential gene expression study was performed between two full sibling lines of perennial ryegrass with contrasting flowering time using suppression subtractive hybridization (Byrne et al. 2009). Five genes with greater difference in expression between the lines during floral induction were identified. Furthermore, putative methyl binding domain protein and bHLH transcription factor were identified, which show clear differential expression patterns through floral induction and may act as potential enhancers of flowering in perennial ryegrass.

2.4.9.2 Self-incompatibility Genes

The self-incompatibility (SI) system has provided a major obstacle to targeted varietal development and enhanced knowledge is expected to support more efficient breeding strategies. Therefore genetic analysis of the loci of SI system has been carried out. Genetic mapping using molecular markers has been used to pinpoint the locations of the S and Z loci involved in the SI response. The isozyme phosphoglycoisomerase (PGI-2) was found to be linked to the S locus in *L. perenne* (Cornish et al. 1980). S and Z loci have been assigned to LGs 1 and 2, respectively, in accordance with the Triticeae consensus map (Thorogood et al. 2002). These regions show synteny to regions of rice chromosomes 5 (R5) and R4, respectively. Additionally, a self-fertile locus has been mapped on LG 5 (Thorogood et al. 2005).

Van Daele et al. (2008) generated SI-related transcript-derived fragment (TDF) markers via cDNA-AFLP from pistil mRNA in *Lolium*. A total of 169 TDFs were expressed in an allele specific way for both the S and the Z haplotypes. Sequence analysis identified gene functions of general cell metabolism, but also gene functions known to be involved in SI in other systems, such as ubiquitinization and receptor kinases. Transcripts were identified from the SI libraries that were orthologous to sequences on rice R4 and R5. These represent potential SI candidate genes. Altogether 10 expressed SI candidate genes were identified. A rapid increase in gene expression within 2 min after pollen-stigma contact was revealed, reaching a maximum between 2 and 10 min (Yang et al. 2009).

Recently Shinozuka et al. (2010) studied comparative genetics and physiological mapping approaches to screen a representative BAC library from perennial ryegrass. For the generation of a fine-scale map around the S and Z loci, Shinozuka et al. (2010) used comparative genomic information to develop cDNA and single sequence repeat-derived markers based on public sequence information of genes and sequences that were known to map close to the S and Z loci in other grasses. The alignment of orthologous sequences facilitated the identification of conserved sequences suitable for the design of PCR primers that were able to amplify intervening corresponding *Lolium* genomic targets. The new markers were used for genotyping F_1 (NA$_6$ x AU$_6$) two-way pseudo-testcross mapping population. Although marker order was largely conserved between the grasses, several markers were reversed against the model species rye and rice. In an effort to isolate the Z locus physically, Shinozuka et al. (2010) used three of the markers close to the Z locus to screen a representative BAC library from perennial ryegrass DNA. Three BAC clones containing Z-related DNA were identified and sequenced, and putative genes were analyzed by BLAST database searches. Nine gene-like sequences were identified. Four were expressed in the anther and pistil; two of these had some degree of nucleotide diversity expected of a Z gene, but the true identity of Z gene(s) will remain elusive until further functional studies are carried out, and mutations directly correlated to particular Z phenotypes are identified.

2.4.9.4 CBF Genes

Molecular mechanisms involved in cold acclimation are largely unknown but information from model species whose genomes have recently been sequenced such as *Arabidopsis* and rice and the development of microarray technologies are providing insight into the complexity of the processes (Yamaguchi-Shinozaki and Shinozaki 2008). The CBF (C-repeat binding factor)/DREB1 (dehydration-responsive element-binding protein 1) regulon is the most important transcription unit involved in cold acclimation in plants. cDNAs encoding CBF have been isolated from cold-treated ryegrass plants. Ten novel putative CBF cDNAs have been isolated from cold-treated leaf tissue of perennial ryegrass (Tamura and Yamada 2007). Their primary structures contain some conserved motifs characteristic of the gene class. Phylogenetic analysis revealed that *LpCBF* genes were attributable to the *HvCBF3*-, and *HvCBF4*-subgroups following the previously proposed classification of barley *CBF* genes (Skinner et al. 2005). Four *LpCBF* genes were mapped on LG 5 forming a cluster within 2.2 cM, while one *LpCBF* gene on LG 1. Based on comparative genetic studies, conserved synteny for *CBF* gene family was observed between the Triticeae cereals and perennial ryegrass (Tamura and Yamada 2007). *CBF* genes cluster is positioned at the frost resistance locus, *Fr-H2* in barley (Francia et al. 2007). One *CBF* gene (*FpCBF6*) was identified in meadow fescue (Alm et al. 2011). This gene was mapped to chromosome 5F which is an orthologous of *Fr-H2* (Alm et al. 2011). Interestingly, while meadow fescue had two QTLs on chromosome 5F which corresponded to *FR-A1* and *FR-A2*, its vernalization gene *FpVRN1* was located at chromosome 4F as perennial ryegrass, instead of chromosome 5F as in wheat and barley (Alm et al. 2011).

2.4.9.5 Fructosyltransferase Genes

Fructans are linear or branched forms of fructose polymers, which are derived from sucrose. Fructans are present in 15% of the angiosperm flora, and are particularly widespread in grasses (Chatterton et al. 1989; Hendry 1993). Fructans accumulate in plant cells as a carbohydrate reserve in addition to or instead of starch (Hendry 1993), and are also thought to be involved in the maintenance of osmotic potentials (Pavis et al. 2001). Accumulation of fructans in plants has been found to be associated with tolerance to cold and drought, particularly in the development of freezing tolerance (Yoshida et al. 1998; Kawakami and Yoshida 2002). Fructan is synthesized by a combination of multiple fructosyltransferases (FTs). Two FT genes have been identified in perennial ryegrass: 1-SST (Chalmers et al. 2003) and 6G-FFT (Lasseur et al. 2006). FT genes have been mapped to perennial ryegrass LGs using a mapping population derived from a pair-cross between North African6 (NA$_6$) and Aurora6 (AU$_6$) plants (Chalmers et al. 2005). The 1-SST gene (Lp1-SST; marker xlp1-sst) has been mapped to a distal region of LG7 of NA6.

Six cDNAs encoding FTs (*prft1-prft6*) have been isolated from cold-treated perennial ryegrass plants (Hisano et al. 2008). The *prft1* and *prft4* genes were both located near a gene for soluble invertase in the distal part of the LG 7 using F$_2$ (Aurora x Perma) mapping population. This population was derived from a cross between high WSC and low WSC. The *prft3* gene was located in the distal part of the LG 3. Functional characterization using *Pichia pastoris* revealed that the *prft4* encodes sucrose-sucrose 1-fructosyltransferase (1-SST), and the *prft3* and *prft5* encode fructan-fructan 6G-fructosyltransferase (6G-FFT). Protein sequences for the other genes (*prfts 1, 2,* and *6*) were similar to sucrose-fructan 6-fructosyltransferase (6-SFT). The mRNA levels of *prft1* and *prft2* gradually increased during cold treatment while those of the 1-SST and 6G-FFT genes first increased but then decreased before increasing again during a longer period of cold treatment. At least two different patterns of expression of FT genes appear to have developed during the evolution of functionally diverse FT genes which are associated in a coordinated way with fructan synthesis in a cold environment (Hisano et al. 2008).

QTLs for fructans and the other component of WSC (sucrose, glucose and fructose) in leaves and tiller bases have been mapped using F$_2$ (Aurora x Perma) mapping population (Turner et al. 2006). Fructan QTLs were identified on chromosome 1, 2, 5 and 6. Also many QTLs for growth-trait and drought-stress response were identified on all chromosomes except for chromosome 7 (Turner et al. 2008). Therefore there is currently little evidence for FT involvement in these fructan QTLs. Further isolation of FT genes and mapping may show a closer association. Alternatively the QTLs could result from variation in fructan hydrolases or from the activity of regulatory genes (Turner et al. 2006).

2.4.9.6 Other Genes

The α-subunit of the casein protein kinase CK2 has been implicated in both light-regulated and circadian rhythm-controlled plant gene expression, including control of flowering time. Two putative CK2 α genes of perennial ryegrass (*Lpck2a-1* and *Lpck2a-2*) have been obtained from a cDNA library constructed with mRNA isolated from cold-acclimated crown tissue (Shinozuka et al. 2005). The Lpck2a-1 CAPS marker was assigned to perennial ryegrass LG 4 and the Lpck2a-2 CAPS marker was assigned to

LG 2. Allelic variation at the Lpck2a-1 and Lpck2a-2 gene loci was correlated with phenotypic variation for heading time and winter survival, respectively (Shinozuka et al. 2005). The gene for a putative glycine-rich RNA binding protein, LpGRP1, was isolated from a cDNA library constructed from crown tissues of cold-treated perennial ryegrass plants (Shinozuka et al. 2006). An RFLP locus detected by the LpGRP1 cDNA probe was mapped to a distal location on LG 2 in the p150/112 population. A significant increase in the mRNA level of LpGRP1 was detected in root, crown and leaf tissues during the treatment of plants at 4°C, through which freezing tolerance is attained. LpGRP1 protein could play an important role for adaptation to cold environments.

2.5 Transgenesis

Conventional forage grass breeding has been based on the use of natural genetic variation as found between and within ecotypes and cultivars or created through sexual recombination. Gene technology and the production of transgenic plants offer the opportunity to generate unique genetic variation. Application of transgenesis to forage plant improvement has been focused on the development of transformation events with unique genetic variation and in studies on the molecular dissection of plant biosynthetic pathways and developmental processes of high relevance for forage production (Spangenberg et al. 2001).

2.5.1 Transformation Methods

Biolistic methodology, based on particle bombardment used high-velocity gold or tungsten particles to deliver DNA into living cells for stable transformation. Because biolistic methodology is a physical process that involves only one biological system, it is a fairly reproducible method that can be easily adapted from one laboratory to another laboratory. Transgenic forage plants have been obtained by particle bombardment of embryogenic cell in perennial ryegrass (Spangenberg et al. 1995; Dalton et al. 1999; Altpeter et al. 2000; Xu et al. 2001; Petrovska et al. 2004; Chen et al. 2005).

Agrobacterium-mediated transformation has the advantage of allowing for low copy number integration of the transgenes into the plant genome. In recent years, significant progress has been made in developing transformation protocols using *Agrobacterium tumefaciens* as a vector. Transgenics have been obtained by *Agrobacterium*-mediated transformation perennial ryegrass (Altpeter et al. 2004; Wu et al. 2005; Sato and Takamizo 2006; Bajaj et al. 2006) and *Festulolium* (*Lolium/Festuca* hybrids) (Guo et al. 2009).

2.5.2 Manipulation of Fructan Biosynthesis

Fructans, a polymer of fructose and a major component of nonstructural carbohydrates is accumulated in temperate grasses. The increased level of soluble carbohydrates appears to improve the nutritional value of grasses, particularly during summer when grasses suffer a major decline in digestibility. Gadegaard et al. (2008) developed perennial ryegrass lines expressing sucrose:sucrose 1-fructosyltransferase and fructan:fructan 6G-fructosyltransferase genes from onion (*Allium cepa* L.) which exhibited up to a 3-fold increased fructan content. Fructan accumulation is also associated with winter hardiness. Transgenic perennial ryegrass plants that over-expressed the wheat fructosyltransferase genes, *wft1* and *wft2*, which encode sucrose-fructan 6-fructosyltransferase (6-SFT) and sucrose-sucrose 1-fructosyltransferase (1-SST), respectively, under the control of CaMV 35S promoter have been produced using a biolistic transformation (Hisano et al. 2004). Transgenic plants that accumulated a greater amount of fructan than non-transgenic plants showed increased tolerance to cellular freezing. The results suggest that the over-expression of the genes involved in fructan synthesis serves as a novel strategy to produce freezing-tolerant grasses (Hisano et al. 2004).

2.5.3 Manipulation of Lignin Biosynthesis

Forage digestibility is a limiting factor for animal productivity. Lignification of plant cell walls has been identified as the major factor responsible for lowering digestibility of forage tissues. Molecular breeding for improved digestibility by down-regulating monolignol biosynthetic enzymes through transgenesis has been explored. Cinnamoyl CoA-reductase (CCR) and caffeic acid O-methyltransferase (COMT) catalyze key steps in the biosynthesis of monolignols, which serve as building blocks in the formation of plant lignin. Down- regulation of *CCR1* and *caffeic acid O-methyltransferase 1* (*OMT1*) using an RNA interference–mediated silencing strategy caused dramatic changes in lignin level and composition in transgenic perennial ryegrass plants grown under both glasshouse and field conditions. Both field-grown *CCR1*-deficient and *OMT1*-deficient perennial ryegrass plants showed enhanced digestibility without obvious detrimental effects on either plant fitness or biomass production (Tu et al. 2010).

2.6 Conclusion

Molecular breeding is important and will be used extensively in future perennial ryegrass improvement. Transformation techniques using *Agrobacterium* or biolistics-based method and many available molecular markers such as SSR markers and some functionally-associated genetic markers have been developed in perennial ryegrass. However the challenge on the successful implementation of molecular breeding in practical varietal development still remains. Molecular breeding needs to develop from a platform of good conventional breeding and includes supporting agronomic research and partnering with the commercial industry wherever appropriate. With the advance in genome sequencing technologies using next-generation sequencer, QTL/gene mapping and identification have accelerated and it becomes possible to perform QTL analysis, which often has limited molecular markers. In addition, the breeding technique will be also improved by new techniques such as GS.

With a widening range of traits, techniques for more accurate, rapid and non-invasive phenotyping and genotyping become increasingly important. The large amounts of data involved require good bioinfomatics support. Data of various kinds must be integrated from an increasingly wide range of sources such as genetic resources and mapping information for plant populations through to the transcriptome and metabolome of individual tissues. The merging of data from disparate sources and multivariate data-mining across datasets can reveal novel information concerning the complexity of biology.

References

Alm V, Busso CS, Ergon A, Rudi H, Larsen A, Humphreys MW, Rognli OA (2011) QTL analyses and comparative genetic mapping of frost tolerance, winter survival and drought tolerance in meadow fescue (*Festuca pratensis* Huds.). Theor Appl Genet 123: 369–382.

Altpeter F, Xu JP, Ahmed S (2000) Generation of large numbers of independently transformed fertile perennial ryegrass (*Lolium perenne* L.) plants of forage- and turf-type cultivars. Mol Breed 6: 519–528.

Altpeter F, Fang YD, Xu JP, Ma XR (2004) Comparison of transgene expression stability after Agrobacterium-mediated or biolistic gene transfer into perennial ryegrass (*Lolium perenne* L.). In: Hopkins A, Wang Z-Y, Mian R, Sledge M, Barker RE (eds) Molecular Breeding of Forage and Turf. Kluwer Academic Publishers, Dordrecht, The Netherlands pp 255–260.

Andersen JR, Lübberstedt T (2003) Functional markers in plants. Trends Plant Sci 8: 554–560.

Anhalt UCM, Heslop-Harrison J, Byrne S, Guillard A, Barth S (2008) Segregation distortion in *Lolium*: evidence for genetic effects. Theor Appl Genet 117: 297–306.

Anhalt UCM, Heslop-Harrison J, Piepho H, Byrne S, Barth S (2009) Quantitative trait loci mapping for biomass yield traits in a *Lolium* inbred line derived F$_2$ population. Euphytica 170: 99–107.

Armstead IP, Turner LB, Farrell M, Skøt L, Gomez P, Montoya T, Donnison IS, King IP, Humphreys MO (2004). Synteny between a major heading-date QTL in perennial ryegrass (*Lolium perenne* L.) and the *Hd3* heading-date locus in rice. Theor Appl Genet 108: 822–828.

Armstead IP, Turner LB, Marshall AH, Humphreys MO, King IP, Thorogood D (2008) Identifying genetic components controlling fertility in the outcrossing grass species perennial ryegrass (*Lolium perenne*) by quantitative trait loci analysis and comparative genetics. New Phytol 178: 559–571.

Asp T, Frei UK, Didion T, Nielsen KK, Lübberstedt T (2007) Frequency, type, and distribution of EST-SSRs from three genotypes of *Lolium perenne*, and their conservation across orthologous sequences of *Festuca arundinacea, Brachypodium distachyon*, and *Oryza sativa*. BMC Plant Biol 7: 36.

Asp T, Byrne S, Gundlach H, Bruggmann R, Mayer KFX, Andersen JR, Xu M, Greve M, Lenk I, Lübberstedt T (2011) Comparative sequence analysis of *VRN1* alleles of *Lolium perenne* with the co-linear regions in barley, wheat, and rice. Theor Appl Genet 286: 433–447.

Auzanneau J, Huyghe C, Julier B, Barre P (2007) Linkage disequilibrium in synthetic varieties of perennial ryegrass. Theor Appl Genet 115: 837–847.

Bajaj S, Ran Y, Phillips J, Kularajathevan G, Pal S, Cohen D, Elborough K, Puthigae S (2006) A high throughput Agrobacterium tumefaciens-mediated transformation method for functional genomics of perennial ryegrass (*Lolium perenne* L.). Plant Cell Rep 25: 651–659.

Balfourier, F (2000) Evidence for phylogeographic structure in *Lolium* species related to the spread of agriculture in Europe. A cpDNA study. Theor. Appl Genet 101: 131–138.

Barre P, Moreau L, Mi F, Turner L, Gastal F, Julier B, Ghesquiere M (2009) Quantitative trait loci for leaf length in perennial ryegrass (*Lolium perenne* L.). Grass Forage Sci 64: 310–321.

Bouton JH (2008) Molecular breeding to improve forages for use in animal and biofuel production systems. In: Yamada T, Spangenberg G (eds) Molecular Breeding of Forage Crops. Springer, New York, USA, pp 1–13.

Brazauskas G, Lenk I, Pedersen MG, Studer B, Lübberstedt T (2011) Genetic variation, population structure, and linkage disequilibrium in European elite germplasm of perennial ryegrass. Plant Sci 181: 412–420.

Brummer EC, Casler MD (2008) Improving selection in forage, turf, and biomass crops using molecular markers. In: Yamada T, Spangenberg G (eds) Molecular Breeding of Forage Crops. Springer, New York, USA, pp 193–209.

Byrne SL, Guiney E, Donnison IS, Mur LAJ, Milbourne D, Barth S (2009) Identification of genes involved in the floral transition at the shoot apical meristem of *Lolium perenne* L. by use of suppression subtractive hybridization. Plant Growth Regul 59: 215–225.

Chalmers J, Johnson X, Lidgett A, Spangenberg G (2003) Isolation and characterisation of a sucrose:sucrose 1-fructosyltransferase gene from perennial ryegrass (*Lolium perenne* L.). J Plant Physiol 160: 1385–1391.

Chalmers J, Lidgett A, Cummings N, Cao Y, Forster J, Spangenberg G (2005) Molecular genetics of fructan metabolism in perennial ryegrass. Plant Biotechnol J 3: 459–474.

Chatterton NJ, Harrison PA, Bennett JH, Asay KH (1989) Carbohydrate partitioning in 185 accessions of Gramineae under warm and cool temperatures. J Plant Physiol 134: 169–179.

Chen X, Yang WQ, Sivamani E, Bruneau AH, Wang BH, Qu RD (2005) Selective elimination of perennial ryegrass by activation of a pro-herbicide through engineering *E. coli argE* gene. Mol Breed 15: 339–347.

Cogan NOI, Smith KF, Yamada T, Francki MG, Vecchies AC, Jones ES, Spangenberg GC, Forster JW (2005) QTL analysis and comparative genomics of herbage quality traits in perennial ryegrass (*Lolium perenne* L.). Theor Appl Genet 110: 364–380.

Cogan NOI, Ponting RC, Vecchies AC, Drayton MC, George J, Dobrowolski MP, Sawbridge TI, Spangenberg GC, Smith KF, Forster JW (2006) Gene-associated single nucleotide polymorphism (SNP) discovery in perennial ryegrass (*Lolium perenne* L.). Mol Genet Genom 276: 101–112.

Cornish MA, Hayward MD, Lawrence MJ (1979) Self-incompatibility in ryegrass. I. Genetic control in diploid *Lolium perenne* L. Heredity 43: 95–106.

Cornish MA, Hayward MD, Lawrence MJ (1980) Self-incompatibility inryegrass III. The joint segregation of S and PGI-2 in *Lolium perenne* L. Heredity 44: 55–62.

Curley J, Sim SC, Warnke S, Leong S, Barker R, Jung G (2005) QTL mapping of resistance to grey leaf spot in ryegrass. Theor Appl Genet 111: 1107–1117.

Dalton SJ, Bettany AJE, Timms E, Morris P (1999) Co-transformed, diploid *Lolium perenne* (Perennial ryegrass), *Lolium multiflorum* (Italian ryegrass) and *Lolium temulentum* (Darnel) plants produced by microprojectile bombardment. Plant Cell Rep 18: 721–726.

Dobrowolski MP, Forster JW (2007) Linkage disequilibrium-based association mapping in forage species. In: Oraguzie NC, Rikkerink EHA, Gardiner SE, Silva HND (eds) Association Mapping in Plants. Springer, New York, USA, pp 197–209.

Dracatos PM, Cogan NOI, Dobrowolski MP, Sawbridge TI, Spangenberg GC, Smith KF, Forster W (2008) Discovery and genetic mapping of single nucleotide polymorphisms in candidate genes for pathogen defence response in perennial ryegrass (*Lolium perenne* L.). Theor Appl Genet 117: 203–219.

Dumsday JL, Smith KF, Forster JW, Jones ES (2003) SSR-based genetic linkage analysis of resistance to crown rust (*Puccinia coronata* Corda f. sp. *lolii*) in perennial ryegrass (*Lolium perenne* L.). Plant Pathol 52: 628–637.

Farrar K, Asp T, Lübberstedt T, Xu M, Thomas AM, Christiansen C, Humphreys MO, Donnison IS (2007) Construction of two *Lolium perenne* BAC libraries and identification of BACs containing candidate genes for disease resistance and forage quality. Mol Breed 19: 15–23.

Faville M, Vecchies AC, Schreiber M, Drayton MC, Hughes LJ, Jones ES, Guthridge KM, Smith KF, Sawbridge T, Spangenberg GC, Bryan GT, Forster JW (2004) Functionally-associated molecular genetic marker map construction in perennial ryegrass (*Lolium perenne* L.). Theor Appl Genet 110: 12–32.

Fearon CH, Cornish MA, Hayward MD, Lawrence MJ (1994) Self-incompatibility in ryegrass. X. Number and frequency of alleles in a natural population of *Lolium perenne* L. Heredity 73: 254–261.

Forster JW, Jones ES, Kölliker R, Drayton MC, Dumsday J, Dupal MP, Guthridge KM, Mahoney NL, van Zijll de Jong E, Smith KF (2001) Development and Implementation of Molecular Markers for Forage Crop Improvement. In: Spangenberg G (ed) Molecular Breeding of Forage Crops. Kluwer Academic Press, Dordrecht, The Netherlands, pp 101–133.

Forster JW, Jones ES, Batley J, Smith KF (2004) Molecular marker-based genetic analysis of pasture and turf grasses. In: Hopkins A, Wang Z-Y, Mian R, Sledge M, Barker RE (eds). Molecular Breeding of Forage and Turf. Kluwer Academic Press, Dordrecht, The Netherlands, pp 197–239.

Francia E, Barabaschi D, Tondelli A, Laidò G, Rizza F, Stanca AM, Busconi M, Fogher C, Stockinger EJ, Pecchioni N (2007) Fine mapping of a *Hv CBF* gene cluster at the frost resistance locus *Fr-H2* in barley. Theor Appl Genet 115: 1083–1091.

Gadegaard G, Didion T, Folling M, Storgaard M, Andersen CH, Nielsen KK (2008) Improved fructan accumulation in perennial ryegrass transformed with the onion fructosyltransferase genes 1-SST and 6G-FFT. J Plant Physiol 165: 1214–1225.

Gill G P, Wilcox PL, Whittaker DJ, Winz RA, Bickerstaff P, Echt CE, Kent J, Humphreys MO, Elborough KM, Gardner RC (2006) A framework linkage map of perennial ryegrass based on SSR markers. Genome 49: 354–364.

Guo Y-D, Hisano H, Shimamoto Y, Yamada T (2009) Transformation of androgenic-derived *Festulolium* plants (*Lolium perenne* L. x *Festuca pratensis* Huds.) by *Agrobacterium tumefaciens*. Plant Cell Tiss Org Cult 96: 219–227.

Heffner EL, Sorrells ME, Jannink J-L (2009) Genomic selection for crop improvement. Crop Sci 49: 1–12.

Hendry GAF (1993) Evolutionary origins and natural functions of fructans—a climatological, biogeographic and mechanistic appraisal. New Phytol 123: 3–14.

Hisano H, Kanazawa A, Kawakami A, Yoshida M, Shimamoto Y, Yamada T (2004) Transgenic perennial ryegrass plants expressing wheat fructosyltransferase genes accumulate increased amounts of fructan and acquire increased tolerance on a cellular level to freezing. Plant Sci 167: 861–868.

Hisano H, Kanazawa A, Yoshida M, Humphreys MO, Iizuka M, Kitamura K, Yamada T (2008) Coordinated expression of functionally diverse fructosyltransferase genes is associated with fructan accumulation in response to low temperature in perennial ryegrass. New Phytol 178: 766–780.

Hutchinson J, Rees H, Seal AG (1979) An assay of the activity of supplementary DNA in *Lolium*. Heredity 43: 411–421.

Humphreys MO (2005) Genetic improvement of forage crops—past, present and future. J Agri Sci 143: 441–448.

Humphreys M, Feurstein U, Vandewalle M, Baert J (2010) Ryegrasses. In: Boller B, Posselt UK, Veronesi F (eds) Fodder Crops and Amenity Grasses, Handbook of Plant breeding 5. Springer, New York, USA, pp 211–260.

Jauhar PP (1993) Cytogenetics of the *Festuca-Lolium* complex: Relevance to Breeding, Springer, Berlin, Germany.

Jenkins G, Head J, Foster JW (2000) Probing meiosis in hybrids of *Lolium* (Poaceae) with a discriminatory repetitive genomic sequence. Chromosoma 109: 280–286.

Jensen LB, Andersen JR, Frei U, Xing Y, Taylor C, Holm PB, Lübberstedt T (2005) QTL mapping of vernalisation response in perennial ryegrass (*Lolium perenne* L.) reveals co-location with an orthologue of wheat *VRN1*. Theor Appl Genet 110: 527–536.

Jensen LB, Holm PB, Lübberstedt T (2007) Cross-species amplification of 105 *Lolium perenne* SSR loci in 23 species within the Poaceae. Mol Ecol Notes 7: 1155–1161.

Jo Y, Barker R, Pfender W, Warnke S, Sim S, Jung G (2008) Comparative analysis of multiple disease resistance in ryegrass and cereal crops. Theor Appl Genet 117: 531–543.

Jones ES, Dupal MP, Kölliker R, Drayton MC, Forster JW (2001) Development and characterisation of simple sequence repeat (SSR) markers for perennial ryegrass (*Lolium perenne* L.). Theor Appl Genet 102: 405–415.

Jones ES, Mahoney NL, Hayward MD, Armstead IP, Jones JG, Humphreys MO, King IP, Kishida T, Yamada T, Balfourier F, Charmet C, Forster JW (2002a) An enhanced molecular marker-based map of perennial ryegrass (*Lolium perenne* L.) reveals comparative relationships with other Poaceae species. Genome 45: 282–295.

Jones ES, Dupal MD, Dumsday JL, Hughes LJ, Forster JW (2002b) An SSR-based genetic linkage map for perennial ryegrass (*Lolium perenne* L.). Theor Appl Genet 105: 577–584.

Jung GA, Van Wijk AJP, Hunt WF, Watson CE (1996) Ryegrasses. In: Moser LE et al. (eds) Cool-season forage grasses. ASA-CSSA-SSSA, Madison, USA, pp 605–641.

Kawakami A, Yoshida M (2002) Molecular characterization of sucrose:sucrose 1-fructosyltransferase and sucrose:fructan 6-fructosyltransferase associated with fructan accumulation in winter wheat during cold hardening. Biosci Biotechnol Biochem 66: 2297–2305.

Kellog, EA (2001) Evolutionary history of the grasses. Plant Physiol 125: 1198–1205.

King, J, Thorogood D, Edwards KJ, Armstead IP, Roberts L, Skøt K, Hanley Z, King IP (2008) Development of a genomic microsatellite library in perennial ryegrass (*Lolium perenne*) and its use in trait mapping. Ann Bot 101: 845–853.

Kobayashi S, Humphreys MO, Tase K, Sanada Y, Yamada T (2011) Molecular marker dissection of ryegrass plant development and its response to growth environments and foliage cuts. Crop Sci 51: 600–611.

Kubik C, Sawkins M, Meyer WA, Gaut BS (2001) Genetic diversity in seven perennial ryegrass (*Lolium perenne* L.) cultivars based on SSR markers. Crop Sci 41: 1565–1572.

Lasseur B, Lothier J, Djoumad A, De Coninck B, Smeekens S, Van Laere A, Morvan-Bertrand A, Van den Ende W, Prud'homme MP (2006) Molecular and functional characterization of a cDNA encoding fructan: fructan 6G-fructosyltransferase (6G-FFT)/ fructan: fructan 1-fructosyltransferase (1-FFT) from perennial ryegrass (*Lolium perenne* L.). J Exp Bot 57: 2719–2734.

Muylle H, Baert J, Van Bockstaele E, Petijs J, Roldán-Ruiz I (2005a) Four QTLs determine crown rust (*Puccinia coronata* f. sp. *lolii*) resistance in a perennial ryegrass (*Lolium perenne*) population. Heredity 95: 348–357.

Muylle H, Baert J, Van Bockstaele E, Moerkerke B, Goetghebeur E, Roldán-Ruiz I (2005b) Identification of molecular markers linked with crown rust (*Puccinia coronata* f.sp. *lolii*) resistance in perennial ryegrass (*Lolium perenne*) using AFLP markers and a bulked segregant approach. Euphytica 143: 135–144.

Pavis N, Boucaud J, Prud'homme MP (2001) Fructans and fructan-metabolizing enzymes in leaves of *Lolium perenne*. New Phytol 150: 97–109.

Pearson A, Cogan NOI, Baillie RC, Hand ML, Bandaranayake CK, Erb S, Wang J, Kearney GA, Gendall AR, Smith KF, Forster JW (2011) Identification of QTLs for morphological traits influencing waterlogging tolerance in perennial ryegrass (*Lolium perenne* L.). Theor Appl Genet 122: 609–622.

Petrovska N, Wu X, Donato R, Wang Z-Y, Ong E-K, Jones E, Forster J, Emmerling M, Sidoli A, O'Hehir R, Spangenberg G (2004) Transgenic ryegrasses (*Lolium* spp.) with down-regulation of main pollen allergens. Mol Breed 14: 489–501.

Ponting RC, Drayton MC, Cogan NOI, Dobrowolski MP, Spangenberg GC, Smith KF, Forster JW (2007) SNP discovery, validation, haplotype structure and linkage disequilibrium in full-length herbage nutritive quality genes of perennial ryegrass (*Lolium perenne* L.). Mol Genet Genom 278: 585–597.

Sartie AM, Matthew C, Easton HS, Faville MJ (2011) Phenotypic and QTL analyses of herbage production-related traits in perennial ryegrass (*Lolium perenne* L.). Euphytica 182: 295–315.

Sato H, Takamizo T (2006) *Agrobacterium* tumefaciens-mediated transformation of forage-type perennial ryegrass (*Lolium perenne* L.). Grassland Sci 52: 95–98.

Sawbridge T, Ong E-K, Binnion C, Emmerling M, McInnes R, Meath K, Nguyen N, Nunan K, O'Neill M, O'Toole F, Rhodes C, Simmonds J, Tian P, Wearne K, Webster T, Winkworth A, Spangenberg G (2003) Generation and analysis of expressed sequence tags in perennial ryegrass (*Lolium perenne* L.). Plant Sci 165: 1089–1100.

Schejbel B, Jensen LB, Asp T, Xing Y, Lübberstedt T (2007) QTL analysis of crown rust resistance in perennial ryegrass under conditions of natural and artificial infection. Plant Breed 126: 347–352.

Schejbel B, Jensen LB, Asp T, Xing Y, Lübberstedt T (2008) Mapping of QTL for resistance to powdery mildew and resistance gene analogues in perennial ryegrass. Plant Breed 127: 368–375.

Seal AG, Rees H (1982) The distribution of quantitative DNA changes associated with the evolution of diploid Festuceae. Heredity 49: 179–190.

Shinozuka H, Hisano H, Ponting RC, Jones ES, Cogan NOI, Forster JW, Yamada T (2005) Molecular cloning and genetic mapping of perennial ryegrass protein kinase CK2α-subunit genes. Theor Appl Genet 112: 167–177.

Shinozuka H, Hisano H, Yoneyama S, Shimamoto Y, Jones ES, Forster JW, Yamada T, Kanazawa A (2006) Cold-responsive gene expression, linkage mapping and protein localization in the nucleus of glycine-rich RNA-binding protein gene from perennial ryegrass suggest a role for cold adaptation. Mol Genet Genom 275: 399–408.

Shinozuka H, Cogan NOI, Smith KF, Spangenberg GC, Forster JW (2010) Fine-scale comparative genetic and physical mapping supports map-based cloning strategies for the self-incompatibility loci of perennial ryegrass (*Lolium perenne* L.). Plant Mol Biol 72: 343–355.

Sim S, Chang T, Curley J, Warnke SE, Barker R, Jung G (2005) Chromosomal rearrangements differentiating the ryegrass genome from the Triticeae, oat and rice genomes using common heterologous RFLP probes. Theor Appl Genet 110: 1011–1019.

Sim S, Diesburg K, Casler M, Jung G (2007) Mapping and comparative analysis of QTL for crown rust resistance in an Italian x perennial ryegrass population. Phytopathology 97: 767–776.

Sim S, Yu J, Jo Y, Sorrells ME, Jung G (2009) Transferability of cereal EST-SSR markers to ryegrass. Genome 52: 431–437.

Skinner JS, Zitzewitz von J, Szűcs P, Marquez-Cedillo L, Filichkin T, Amundsen K, Stockinger EJ, Thomashow MF, Chen THH, Hayes PM (2005) Structural, functional, and phylogenetic characterization of a large CBF gene family in barley. Plant Mol Biol 59: 533–551.

Skøt L, Humphreys J, Humphreys MO, Thorogood D, Gallagher J, Sanderson R, Armstead IP, Thomas ID (2007) Association of candidate genes with flowering time and water-soluble carbohydrate content in *Lolium perenne* L. Genetics 177: 535–547.

Skøt L, Sanderson R, Thomas A, Skøt K, Thorogood D, Latypova G, Asp T, Armstead I (2011) Allelic variation in the perennial ryegrass *FLOWERING LOCUS T* gene is associated with changes in flowering time across a range of populations. Plant Physiol 155: 1013–1022.

Smith KF, Dobrowolski MP, Cogan NOI, Spangenberg GC, Forster JW (2008) Utilizing linkage disequilibrium and association mapping to implement candidate gene based markers in perennial ryegrass breeding In: Yamada T, Spangenberg G(eds) Molecular Breeding of Forage and Turf. Springer, New York, USA, pp 335–340.

Soreng RJ, Davis JI (1998) Phylogenetics and character evolution in the grass family (Poaceae): simultaneous analysis of morphological and chloroplast DNA restriction site characters. Bot Rev 64: 1–85.

Spangenberg G, Wang Z-Y, Wu XL, Nagel J, Potrykus I (1995) Transgenic perennial ryegrass (*Lolium perenne*) plants from microprojectile bombardment of embryogenic suspension cells. Plant Sci 108: 209–217.

Spangenberg G, Kalla R, Lidgett A, Sawbridge T, Ong EK, John U (2001) Breeding forage plants in the genome era. In: Spangenberg G (ed) Molecular Breeding of Forage Crops. Kluwer Academic Publishers, Dordrecht, The Netherlands, pp 1–39.

Spangenberg GS, Forster JW, Edwards D, John U, Mouradov A, Emmerling M, Batley J, Felitti S, Cogan NOI, Smith KF, Dobrowolski MP (2005) Future directions in the molecular breeding of forage and turf. In: Humphreys MO (ed) Molecular Breeding for the Genetic Improvement of Forage Crops and Turf. Wageningen Academic Publishers, Wageningen, The Netherlands, pp 83–97.

Studer B, Asp T, Frei U, Hentrup S, Meally H, Guillard A, Barth S, Muylle H, Rolda´n-Ruiz I, Barre P, Koning-Boucoiran C, Uenk-Stunnenberg G, Dolstra O, Skøt L, Skøt KP, Turner LB, Humphreys MO, Kölliker R, Roulund N, Nielsen KK, Lübberstedt T (2008a) Expressed sequence tag-derived microsatellite markers of perennial ryegrass (*Lolium perenne* L.) Mol Breed 21: 533–548.

Studer B, Jensen LB, Hentrup S, Brazauskas G, Kölliker R, Lübberstedt T (2008b) Genetic characterisation of seed yield and fertility traits in perennial ryegrass (*Lolium perenne* L.). Theor Appl Genet 117: 781–791.

Studer B, Kölliker R, Muylle H, Asp T, Frei U, Roldán-Ruiz I, Barre P, Tomaszewski C, Meally H, Barth S, Skøt L, Armstead IP, Dolstra O, Lübberstedt T (2010) EST-derived SSR markers used as anchor loci for the construction of a consensus linkage map in ryegrass (*Lolium* spp.). BMC Plant Biol 10: 177.

Tamura K, Yamada T (2007) A perennial ryegrass *CBF* gene cluster is located in a region predicted by conserved synteny between Poaceae species. Theor Appl Genet 114: 273–283.

Tamura K, Yonemaru J, Hisano H, Kanamori H, King J, King IP, Tase K, Sanada Y, Komatsu T, Yamada T (2009) Development of the intron-flanking EST markers for the *Lolium-Festuca* complex using rice genomic information. Theor Appl Genet 118: 1549–1560.

Tamura K, Yonemaru J (2010) Next-generation sequencing for comparative transcriptomics of perennial ryegrass (*Lolium perenne* L.) and meadow fescue (*Festuca pratensis* Huds.) during cold acclimation. Grassland Sci 56: 230–239.

Thorogood D, Kaiser WJ, Jones JG, Armstead I (2002) Self-incompatibility in ryegrass 12: Genotyping and mapping the S and Z loci of *Lolium perenne* L. Heredity 88: 385–390.

Thorogood D, Armstead IP, Turner LB, Humphreys MO, Hayward MD (2005) Identification and mode of action of self-compatibility loci in *Lolium perenne* L. Heredity 94: 356–363.

Tu Y, Rochfort S, Liu Z, Ran Y, Griffith M, Badenhorst P, Louie GV, Bowman ME, Smith KF, Noel JP, Mouradov A, Spangenberg G (2010) Functional analyses of *caffeic acid O-methyltransferase* and *cinnamoyl-CoA-reductase* genes from perennial ryegrass (*Lolium perenne*). Plant Cell 22: 3357–3373.

Turner LB, Cairns AJ, Armstead IP, Ashton J, Skøt K, Whittaker D, Humphreys MO (2006) Dissecting the regulation of fructan metabolism in perennial ryegrass (*Lolium perenne*) with quantitative trait locus mapping. New Phytol 169: 45–58.

Turner LB, Cairns AJ, Armstead IP, Thomas H, Humphreys MW, Humphreys MO (2008) Does fructan have a functional role in physiological traits? Investigation by quantitative trait locus mapping. New Phytol 179: 765–775.

Turner LB, Farrell M, Humphreys MO, Dolstra O (2010) Testing water-soluble carbohydrate QTL effect in perennial ryegrass (*Lolium perenne* L.) by marker selection. Theor Appl Genet 121: 1405–1417.

Van Daele I, Van Bockstaele E, Martens C, Roldán-Ruiz I (2008) Identification of transcribed derived fragments involved in self-incompatibility in perennial ryegrass (*Lolium perenne* L.) using cDNA-AFLP. Euphytica 163: 67–80.

Wang J, Dobrowolski MP, Cogan NOI, Forster JW, Smith KF (2009) Assignment of individual genotypes to specific forage cultivars of perennial ryegrass based on SSR markers. Crop Sci 49: 49–58.

Wang J, Baillie RC, Cogan NOI, McFarlane NM, Dupal MP, Smith KF, Forster JW (2011) Molecular genetic marker-based analysis of species differentiated phenotypic characters in an interspecific ryegrass mapping population. Crop Pasture Sci 62: 892–902.

Warnke SE, Barker RE, Jung G, Rouf Mian MA, Saha MC, Brilman LA, Dupal MD, Forster JW (2004) Genetic linkage mapping of an annual x perennial ryegrass population. Theor Appl Genet 109: 294–304.

Wilkins PW, Humphreys MO (2003) Progress in breeding perennial forage grasses for temperature agriculture. J Agri Sci 140: 129–150.

Wit, F (1959) Chromosome doubling and the improvement of grasses. Genet Agraria 9: 97–115.

Wu YY, Chen QJ, Chen M, Chen J, Wang XC (2005) Salt-tolerant transgenic perennial ryegrass (*Lolium perenne* L.) obtained by *Agrobacterium tumefaciens*-mediated transformation of the vacuolar Na+/H+ antiporter gene. Plant Sci 169: 65–73.

Xing Y, Frei U, Schejbel B, Asp T, Lübberstedt T (2008) Nucleotide diversity and linkage disequilibrium in 11 expressed resistance candidate genes in *Lolium perenne*. BMC Plant Biol 7: 43.

Xiong Y, Fei S-Z, Brummer EC, Moore KJ, Barker RE, Jung G, Curley J, Warnke SE (2007a) QTL analyses of fiber components and crude protein in an annual x perennial ryegrass interspecific hybrid population. Mol Breed 18: 327–340.

Xiong Y, Fei S-Z, Arora R, Brummer EC, Barker RE, Jung G, Warnke SE (2007b) Identification of quantitative trait loci controlling winter hardiness in an annual x perennial ryegrass interspecific hybrid population. Mol Breed 19: 125–136.

Xu JP, Schubert J, Altpeter F (2001) Dissection of RNA-mediated ryegrass mosaic virus resistance in fertile transgenic perennial ryegrass (*Lolium perenne* L.). Plant J 26: 265–274.

Yamada T, Jones ES, Cogan NOI, Vecchies AC, Nomura T, Hisano H, Shimamoto Y, Smith KF, Forster JW (2004) QTL analysis of morphological, developmental and winter hardiness-associated traits in perennial ryegrass (*Lolium perenne* L.). Crop Sci 44: 925–935.

Yamada T, Forster JW (2005) QTL analysis and trait dissection in ryegrasses (*Lolium* spp.). In: Humphreys MO (ed) Molecular Breeding for the Genetic Improvement of Forage Crops and Turf. Wageningen Academic Publishers, Wageningen, The Netherlands, pp 43–53.

Yamada T, Forster JW, Humphreys MW, Takamizo T (2005) Genetics and molecular breeding in *Lolium/Festuca* grass species complex. Grassland Sci 51: 89–106.

Yamaguchi-Shinozaki K, Shinozaki K (2008) DREB regulons in abiotic-stress-responsive gene expression in plants. In: Yamada T, Spangenberg G (eds) Molecular Breeding of Forage Crops. Springer New York, USA, pp 15–27.

Yang B, Thorogood D, Armstead I, Franklin FCH, Barth S (2009) Identification of genes expressed during the self-incompatibility (SI) response in perennial ryegrass (*Lolium perenne* L.). Plant Mol Biol 70: 709–723.

Yoshida M, Abe J, Moriyama M, Kuwabara T (1998) Carbohydrate levels among wheat cultivars varying in freezing tolerance and snow mold resistance during autumn and winter. Physiol Plant 103: 8–16.

Italian Ryegrass

Maiko Inoue,[1] Alan V. Stewart[2] and Hongwei Cai[3,4],*

ABSTRACT

Italian ryegrass (*Lolium multiflorum* Lam.) is the most widely used forage species in the world. The species is used for turf, forage and quick cover in erosion. Italian ryegrass is an outcrossing species and easy to cross among other closely related species, ryegrass species as well as some fescues. First a brief history of the crop and classical breeding efforts are discussed, and then a detailed account of the results for diversity analysis based on both phenotype and molecular markers, molecular genetic map construction and QTL analysis for agricultural important traits are described. Finally, molecular breeding, transgenic study and more recent results for structural and functional genomics are also reviewed.

Key words: Italian ryegrass, Breeding, Genetic resource, Molecular markers, Genomics

[1]Dept. of Plant, Soil, and Insect Sciences, 14 Stockbridge Hall, University of Massachusetts, 80 Campus Center Way, Amherst, MA 01003, USA.
[2]PGG Wrightson Seeds, PO Box 175, Lincoln, Christchurch 7640, New Zealand.
[3]Dept. of Plant Genetics and Breeding, College of Agronomy and Biotechnology, China Agricultural University, 2, Yuanmingyuan West Road, Beijing, 100193 China.
[4]Forage Crop Research Institute, Japan Grassland Agricultural and Forage Seed Association, 388-5 Higashiakada, Nasushiobara, Tochigi 329-2742, Japan.
*Corresponding author: *caihw@cau.edu.cn or hcai@jfsass.or.jp*

3.1 Basic Information on the Plant

3.1.1 Botanical Descriptions

Lolium multiflorum Lam. is the most widely used forage species in the world. The species contain annual forms known as annual or westerwolds ryegrass as well as biennial forms known as Italian ryegrass. They are widely used throughout the world in winter or summer growing conditions which are not subject to severe cold, heat or drought during the relevant growth period. In terms of soil conditions, they require good fertile and soil moisture, however, they can withstand wet conditions and low pH soils. The species are used for turf, forage and quick cover against erosion because of their high tolerance of frequent mowing, high digestibility, palatability and rapid growth. The ryegrasses are often used in mixtures with legumes to provide nitrogen and to improve the quality of forage produced.

They grow vigorously in mild winters and early spring, from 2 to 3 feet (0.6–0.9 m) tall. The plants have a bunchy form. The surface of leaves are bright, glossy and smooth. The leaves range from 1/10 to 2/5 of an inch (3–10 mm) in width (Fig. 3-1).

Figure 3-1 Illustration of *L. multiflorum* (right) and *L. perenne* (left). From Otto Wilhelm Thomé (1885): Flora von Deutschland Österreich und der Schweiz (Kurt Stueber. Kurt Stübers Online Library, http://www.biolib.de).

Color image of this figure appears in the color plate section at the end of the book.

Italian ryegrass is used as a model plant of forage grasses in genetics, cytogenetics, breeding and genomics works. The basic chromosome number is seven and genome size is 1C = 2000 Mb (Hutchinson et al. 1979). Italian ryegrass is an outcrossing species and easy to cross with other closely related species, ryegrass species as well as some fescues. The ploidy level in nature is restricted to diploid (2x) but breeders have developed numerous autotetraploid (4x) cultivars. The reproductive pattern is allogamous. Therefore the genetic construction is highly heterozygous, each individual in an open-pollinated population represents a different genotype. Inbreeding leads to depression of vigor and loss of fertility for self-incompatibility habit. Genotypes of Italian ryegrass may be propagated vegetatively as clones.

3.1.2 Brief History of the Crop

Italian ryegrass is also called annual ryegrass is native to southern Europe and, in some parts of the world, may be called "Westerwold" ryegrass. This is from a variety or varieties that originated in the Westerwolde area in the province of Groningen, the Netherlands from the practice of spring sowing (De Haan 1955). The first reported annual ryegrass that was cultivated was in northern Italy. It was reported in France in 1818, in Switzerland in 1820, and in England in 1831 (Lacefield et al. 2003).

3.2 Classical Genetics and Traditional Breeding

3.2.1 Description of germplasm

There are total of 2660 of genetic resource collections of Italian ryegrass throughout the world. The three main international collections of Italian ryegrass are in the USA with the National Plant Germplasm System of the USDA-ARS (*http://www.ars-grin.gov/npgs/index.html*), in Europe in the Eurisco system (*http://eurisco.ecpgr.org/*) and in the Margot Forde Germplasm Centre in New Zealand (*http://www.agresearch.co.nz/seeds*).

3.2.2 Classical Mapping Efforts

Linkage map construction using genetic markers and quantitative trait loci (QTL) analysis in *Lolium* started in the 1990s (Hayward et al. 1994; Hayward et al. 1998). A genetic linkage map using a segregating family derived from an F_1 hybrid plant of *L. perenne* x *L. multiflorum* provenance, crossed on to a doubled haploid *L. perenne* has been produced using isozyme, restriction fragment length polymorphism (RFLP) and random amplified polymorphic DNA (RAPD) markers (Hayward et al. 1998). A total of 106 markers have been assigned to seven linkage groups covering a map distance of 692 cM. At that time, it was presumed that some markers showed severe segregation distortion which could be accounted for by the use of an interspecific hybrid between two species of differing genome size, with consequent cytological imbalance.

3.2.3 Limitations of Classical Endeavors and Utility of Molecular Mapping

Classical markers such as morphological markers, isozyme and protein markers have limitations on detection and available numbers, for example, isozyme has a limited number per species, and this limits the number of polymorphisms that may be detected. Morphological markers derived from quantitative traits are affected by environmental conditions in the field. However, new molecular marker techniques such as RFLP, AFLP (amplified fragment length polymorphism), RAPD, SSR (simple sequence repeat), etc. provide unlimited number of polymorphism, and molecular markers are stable regardless of environmental conditions. In addition, it is possible to use these markers for comparative mapping among maps if the maps utilize common markers. Marker data can indicate more detailed information on position such as flanking of genes, upstream, downstream and approximate distance.

3.2.4 Breeding Objectives

Generally, the breeding objectives in Italian ryegrass are improvement of forage and seed yield, persistency, disease resistance, nutritional value, nitrogen use efficiency, extended seasonal growth, digestibility, higher water-soluble carbohydrate (WSC) content, and abiotic stress tolerance such as winter hardiness.

3.2.5 Classical Breeding Achievements

As described already, Italian ryegrass is an outcrossing species that possess the characteristics of heterozygosity and self-incompatibility that differ from those of cereal crops. Therefore, the methods for breeding Italian ryegrass are also different. The methods of mass selection and synthetic cultivars are the most widely used breeding methodologies. Examples of achievements in breeding are many, but include the variety "Tribune" where digestibility and persistency were improved which resulted in producing 6% more milk when fed as silage (Wilman et al. 1992), and this was one of the first demonstrations of the impact that breeding for a nutritive value trait could have on livestock production (Humphreys 2005b). Other achievements include crown rust resistance which was transferred from meadow fescue into diploid Italian ryegrass by backcrossing (Oertel and Matzk 1999).

So far, a total of 489 Italian ryegrass cultivars have been included in the Organization for Economic Cooperation and Development's list of varieties eligible for seed certification (OECD 2009).

3.2.6 Limitations of Traditional Breeding and Rationale for Molecular Breeding

Screening breeding materials only for morphological traits is difficult because morphological traits are usually affected by environmental conditions and also many genotypes with heterozygous alleles expressing intermediate traits are often overlooked as resources for breeding in field evaluations. Investigation of targeting traits needs time, labor and excellent field experimentation techniques. Screening with molecular markers which link the traits of interest will allow us to screen plants at the early stage and decrease time and labor. Use DNA markers can also overcome environmental variation. However, construction of the linkage map is necessary to get markers that link the traits of interest. One has to keep in mind the possibility of recombination between the marker and gene that could occur, so it is still necessary when phenotyping plants in the field to characterize markers for molecular breeding.

3.3 Diversity Analysis

In Europe, Italian ryegrass is one of the most important forage grasses used. It is also one of the oldest as for example, in Switzerland, early in the 19th century Italian ryegrass was introduced in agriculture (Stebler and Volkart 1913). Nowadays, the use is not only for forage but also amenity use and as a protection against erosion (Lübbersted et al. 2003). Many researchers recognize the value of ecotypes as resources of breeding materials and many diversity analyses have been performed on ecotypes (Peter-Schmid et al. 2008a; Peter-Schmid et al. 2008b; Boller et al. 2009; Lopes et al. 2009).

3.3.1 Phenotype-based Diversity Analysis

Nineteen populations of Italian ryegrass were collected in Switzerland and 16 morpho-physiological characters were assessed. In terms of morphology, no clear separation of Italian ryegrass cultivars from ecotype populations was observed. Eight out of 16 characteristics did not differ significantly among regional groups of the ecotype populations or the group of cultivars. However, individuals within ecotypes from the northern foothills of the Alps were consistently shorter than individuals of populations from the Swiss plateau and of the cultivars. Ecotype populations of both geographic regions headed significantly earlier than the cultivars (Peter-Schmid et al. 2008b).

A total of 20 Italian ryegrass ecotypes were collected in Switzerland and investigated for performance characteristics (yield, vigor and disease resistance) compared to four locally recommended diploid cultivars (Boller et al. 2009). The high agronomic potential of an ecotype collected from permanent grassland was revealed in this study. The five best ecotypes originated from a relatively small region in East Central Switzerland. Ecotypes showed a better resistance against snow mold than did the cultivars, however, they were less resistant against leaf diseases than cultivars. This was particularly striking for crown rust. Boller et al. (2009) also suggested that if these ecotypes were to be used as breeding materials, employing disease resistance molecular markers could help to screen and produce new cultivars, and at that point, the possibility of variation between rust inoculation results from the field and greenhouse needed to be considered. Lopes et al. (2009) evaluated the morphological variability between Italian ryegrass landraces, to analyze the existence of duplicates in the germplasm collection in Portugal and the "farmer" classification. Seventeen morphological traits were evaluated on 40 to 70 plants per accession. Of those, the most discriminated traits were the heading stage, mean height of plant (MHPL), length and width leaf (LL, WL), maximum height of plant (MxHPL) and heading date (HDATE). In the results, five traits (MHPL, WL, LL, MxHPL and HDATE) explained 89% of total variation of morphological traits. All landraces showed inter-population variability.

3.3.2 Genotype-based Diversity Analysis, Molecular Markers Applied

Several authors have reported the results of genetic diversity studies on *Lolium* species using molecular markers (Cresswell et al. 2001; McGrath et al. 2007; Peter-Schmid et al. 2008a). Cresswell et al. (2001) evaluated the use of AFLP markers to distinguish genotypes, populations, and species of *Lolium* collected from Portugal. The principal coordinate analysis (PCA) of similarities between 127 plants based on 765 polymorphic bands generated from three primer pairs showed high dimensionality in the data. Axes 1–3 were associated primarily with species differences; axes 4–14 with population differences within species; and axis 15 onwards with differences within populations. The unweighted pair group method with arithmetic mean (UPGMA) analysis confirmed the groupings. Analyses of individual bands showed that every interspecific and intraspecific contrast involved a different set of bands, again confirming the high dimensionality of the data. No single band was strictly diagnostic of any population or species.

McGrath et al. (2007) described genetic diversity of *Lolium* species, mostly *Lolium perenne* but included *Lolium multiflorum*, *Festuca* species and x *Festulolium* cultivars using chloroplast microsatellite (cpSSR). All 10 cpSSR marker loci used in the analysis of 1575 individuals were found to be polymorphic, locus TeaCpSSR3 was extremely rich in alleles, including 11 alleles for the *Lolium* species. Of them three alleles were unique for non-*L. perenne Lolium* species (at loci TeaCpSSR3 and TeaCpSSR8). The UPGMA dendrogram showing similarities between 11 groups of accessions was constructed to support the AMOVA analysis. The group of *Festuca* species outlay all other groups. The rest of the dendrogram was split into two major groups. One group contained the Irish *L. perenne* ecotypes, the *L. perenne* cultivars, other *Lolium* species and the x *Festulolium* cultivars.

Twenty-four ecotype populations of either Italian ryegrass or *F. pratensis* were sampled in Switzerland and evaluated for their diversity using Expressed sequence repeat (EST)-SSR and genomic SSR markers derived from *L. multiflorum, L. perenne* and *F. arundinacea* (Peter-Schmid et al. 2008a). Analysis of molecular variance (AMOVA) revealed that most of the variation was due to change within populations. Analysis of ecotype populations and cultivars showed higher molecular variance within the ecotype populations when compared to cultivars for both species. Only a small amount of variation (0.6%) was due to changes between the groups of cultivars and ecotype populations for Italian ryegrass. In this study no clear grouping of populations were observed in relation to the geographical origin.

3.3.3 Extent of Genetic Diversity

Knowledge of the diversity within a species is important not only for use as genetic resources but also for evolution analysis. Wild type plants have a lot of potential genes that are valuable for agriculture, moreover diverse genes could help the species survive in various environments, including extreme climatic events or serious diseases. It is necessary to analyze gene diversity and to collect germplasm simultaneously for the future.

3.4 Molecular Linkage Maps: Strategies, Resources and Achievements

3.4.1 Brief History of Mapping Efforts

Although forage crops are very important for human life as are major crops, the genomic studies of forage crops has lagged behind other crops. Now many researchers who work on genomic studies of forage crops are trying to adopt information and materials of other crops studies for time, cost and labor and achieving remarkably rapid progress in genomic studies such as linkage map construction. For instance, the use of common markers from other crops make the genomic comparison among them possible, and this information will be useful for future breeding. On the other hand, the development of genomic resources, e.g., EST and genomic libraries for Italian ryegrass are advancing as is new marker information from linkage maps.

3.4.2 Molecular Linkage Maps

As described before, Hayward et al. (1998) constructed a linkage map using a perennial ryegrass × Italian ryegrass hybrid and assumed that most of the linkage groups obtained were derived from polymorphism in the IRG parent.

The first linkage map based entirely on Italian ryegrass was published by Vandewalle et al. (2003) in order to develop and evaluate a strategy for the application of DNA-marker data to breeding. They constructed a consensus map that was derived from four populations using mainly AFLPs and few microsatellites, sequence-tagged site (STS) markers. The mapping was carried out with the software JoinMap 3.0 (Stam 1993; Van Ooijen and Voorrips 2001) under the cross-pollination (CP) model and Kosambi map function, using F_1 populations (110 genotypes each) which were obtained by pair-cross. The main 10 linkage groups (LGs) were included from 283 to 349 marker ranges and from 793 cM to 1055 cM map length ranges. Most LGs of the maps could be aligned with The International Lolium Genome Initiative (ILGI) reference map (Jones et al. 2002), using for some group alignments an intermediate map built for perennial ryegrass (Muylle 2003). They presumed that in order to get a better coverage of the chromosome, more data points were needed.

Two other linkage maps were constructed using an annual x perennial ryegrass population (Warnke et al. 2004) and a two-way pseudo-testcross F_1 population of Italian ryegrass (Inoue et al. 2004b). Warnke et al. (2004) produced 91 progenies from a three-generation population of the perennial ryegrass cultivar Manhattan and annual ryegrass cultivar Floregon. A total of 235 AFLP markers, 81 RAPD markers, 16 comparative grass RFLPs, 106 SSR markers, two isozyme loci and two morphological characteristics were used to construct the maps and were analyzed by JoinMap 3.0 under the CP model and Kosambi map function. The female and male maps each had seven LGs, and the map length was 712 cM and 537 cM, respectively. Inoue et al. (2004b) constructed the linkage map using 82 individuals of a two-way pseudo-testcross F_1 population derived from a pair cross between single individuals of two Italian ryegrass cultivars. The 385 (mostly RFLP) markers were selected from the 1226 original markers (867 AFLPs, 274 RFLPs, and 85 telomeric repeat associated sequence (TAS)) were grouped into seven LGs using JoinMap 3.0 under the CP model and Haldane map function. In this study, the Italian ryegrass genomic library (*Pst*I digested) and the Poaceae anchor probes (Van Deynze et al. 1998) from cDNA of rice, barley and oat were used in the polymorphism survey. The results of the Poaceae anchor markers mapping indicated the possibility of a synteny study in the future.

A fourth linkage map was developed by extensive use of SSR markers derived from a genomic library enriched for (CA)n-containing SSR repeats (Hirata et al. 2006). A two-way pseudo-testcross F_1 population consisting of 60 individuals derived from a single cross between two Italian ryegrass individuals was used. The total of 378 SSR, 48 RFLP and 121 AFLP markers were mapped on the two parental linkage maps, seven LGs each, which cover a genome size of 1684 cM, using JoinMap 3.0 under Kosambi map function. Several SSR clusters were found on LG2, LG5, LG6 and LG7 for both parental maps. They concluded that the SSR clusters still need to confirm by using large mapping population.

The latest linkage map that was enhanced using expressed sequence tag (EST)-cleaved amplified polymorphic sequence (CAPS) markers was constructed using the same population of Hirata et al. (2006) as shown in Fig. 3-2 (Miura et al. 2007). Seventy four EST-CAPS markers were produced using 12 endonucleases and 69 EST-CAPS markers were mapped on the previous map finally. Miura et al. (2008) also mapped total 10 disease resistance gene analog (RGA) segments using STS markers that were designed for efficient generation of the RGA fragments on the previous map.

Recently, the transferability of cereal EST-SSR markers to ryegrass was verified using the annual x perennial ryegrass population mentioned above (Sim et al. 2009), also the transferability of tall fescue EST-SSR markers were evaluated and the amplification in 12 grass species were revealed (Mian et al. 2005). Diversity array Technology (DArT) array for genome profiling has demonstrated the utility of genetic mapping within the *Festuca-Lolium* complex (Kopecky et al. 2009), and vice versa. STS markers that were converted from RFLP markers of Italian ryegrass were amplified in closely related species, perennial ryegrass (*L. perenne* L.), meadow fescue (*F. pratensis*) and tall fescue (*F. arundinacea*) (Inoue and Cai 2004).

3.5 Molecular Mapping of Complex Traits

There are some studies that tried to reveal "crown rust resistance" gene position on the Italian ryegrass chromosomes (Fujimori et al. 2004; Sim et al. 2007; Studer et al. 2007). Crown rust is caused by the fungus *Puccinia coronate* f. sp. *Lolii* and is a serious foliar disease in Italian ryegrass as well as perennial ryegrass. Italian ryegrass infected by crown rust has severe damages in terms of yield and quality as forage and appearance as turfgrass. Therefore, crown rust resistance genes are a major target for mapping. Studer et al. (2007) used a two-way pseudo-testcross population consisting

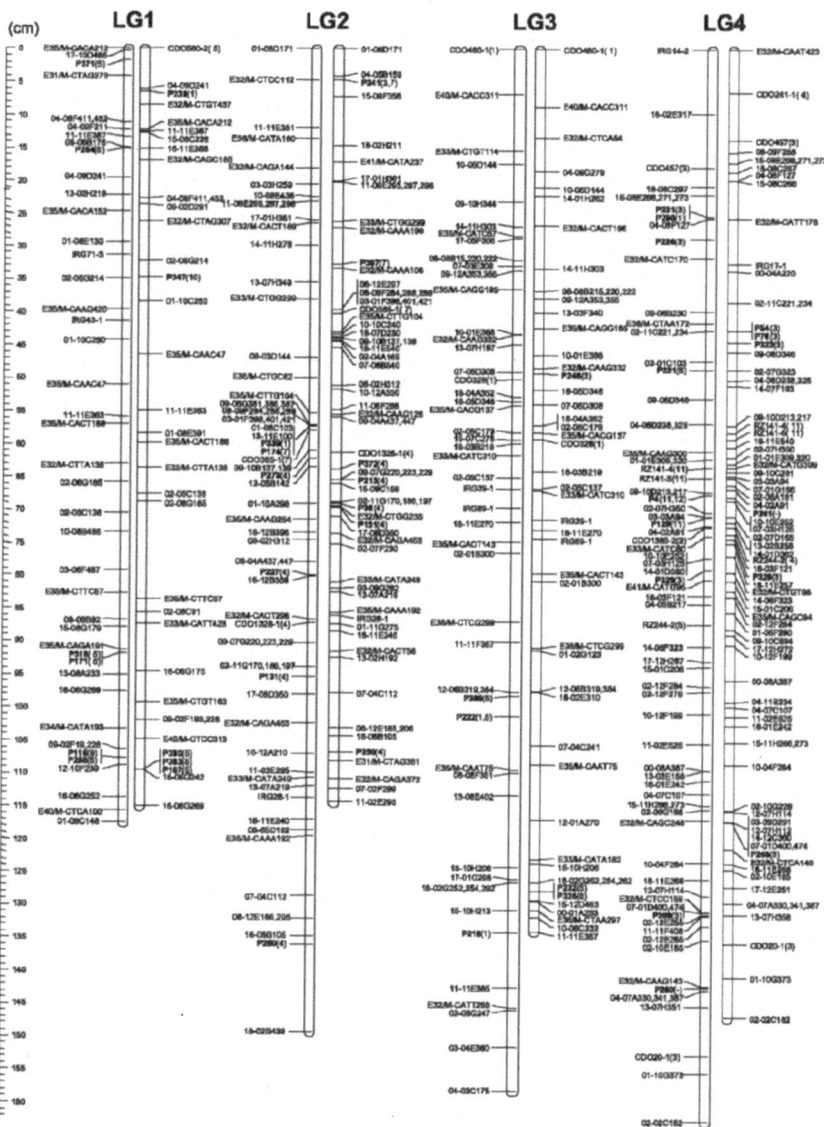

Figure 3-2 contd....

Figure 3-2 contd.

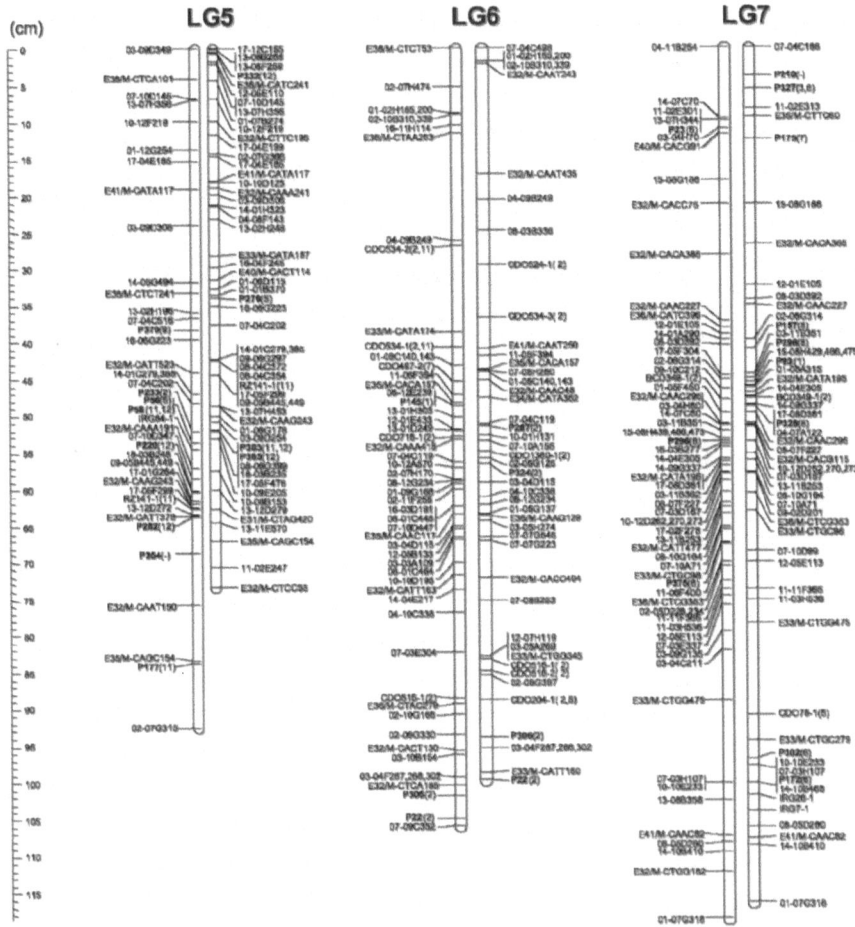

Figure 3-2 High-density molecular linkage map of Italian ryegrass (from Miura et al. 2007). Distribution of EST-CAPS markers on SSR, AFLP, and RFLP linkage maps of Italian ryegrass (Fujimori et al. 2000; Hirata et al. 2006). The maps for the female (11S2) and male (11F3) parents are shown on the left and right, respectively, in each of seven linkage groups (LG1-7). EST-CAPS markers begin with P (shown in bold). AFLP markers are indicated by a combination of *EcoRI/MseI* primer pair (e.g., E35/MCACA) and amplified product size in base pairs. SSR markers are indicated by a combination of clone name (e.g., 17–10D) and amplified product size (last three digits show the size in base pairs). Italian ryegrass genomic RFLP, barley, oat and rice anchor markers are indicated as IRGx, BCDx, CDOx and RZx, respectively. The rice chromosome locations of EST-CAPS and anchor markers are indicated in parentheses following the marker names. The locations of the anchor markers come from Van Deynze et al. (1998).

of 306 Italian ryegrass individuals and performed multisite field evaluation and artificial inoculation in controlled environments to identify QTLs controlling resistance to crown rust. Two major QTLs were consistently detected on linkage group (LG) 1 and LG 2, explaining up to 56% of total phenotypic variance based on a linkage map consisting of 368 AFLP and SSR markers. QTLs of crown rust resistance were also located on LG1 and LG2 in perennial ryegrass. This comparative mapping could enhance the reliability of results and be a clue for future studies. They also suggested that the possibility of the existence of specific local pathogen population from the minor QTLs that were detected. Sim et al. (2007) evaluated crown rust resistance under natural conditions at two locations (WI and IL in USA) over two years. They used a three-generation Italian x perennial ryegrass interspecific population including 156 progeny which was developed from a cross between two Italian x perennial ryegrass hybrids, MFA and MFB. Two QTLs on LG2 and LG7 in MFA map (resistant parent) were detected consistently regardless of year and location. The others were located on LG3 and LG6 in MFB map (susceptible parent). The QTL on LG2 was likely to correspond to those reported mentioned above, the QTL on LG7 suggested the possibility of ortholoci for resistance genes between ryegrass and oat.

Gray leaf spot (GLS) resistance QTLs were detected using the same population that Sim et al. used as mentioned above (Curley et al. 2005). Two kinds of isolates were analyzed, one ryegrass GLS and a rice-infection laboratory strain, the QTLs were detected on four linkage groups. Of three potential QTLs detected using the ryegrass isolate, the one with the strongest effect for resistance was located on linkage group 3 of the MFB parent, explaining between 20 and 37% of the phenotypic variance. Another QTL was detected on linkage group 6 of the MFA parent, explaining between 5 and 10% of the phenotypic variance. Genomic information such as ortholoci about intra-family (inter-genus) plants may be useful for prediction or could be a clue in molecular mapping studies, mainly comparative mapping studies in outcrossing and high-diversity plants. Generally, agronomically important genes, for example yield, were conserved among plants and inherited stably. The results of the differences in QTL locations among studies were not surprising because of the possibility of different gene expressions, but concerned with same target traits, depending on environment, genetic background, pathogenic race and study method. Another disease, bacterial wilt is also one of the major diseases of ryegrasses and fescues. Seedling screening in glasshouse and field trial was performed to investigate genetic control of resistance to bacterial wilt using a two-way pseudo-testcross population consisting of 306 F_1 individuals (Studer et al. 2006). Analysis of QTL based on high-density genetic linkage map consisting of 368 AFLP and SSR markers revealed a single major QTL on LG4 explaining 67% of the total phenotypic variance in glasshouse screening. In addition, a minor QTL was observed on LG5. Field experiments confirmed the major QTL on LG4 to explain 43% (in 2004) to 84% (in 2005) of phenotypic variance and revealed additional minor QTLs on LG1, LG4 and LG6.

QTLs of lodging resistance and related traits were detected in a two-way pseudo-testcross F_1 population consisting of 220 individuals (Inoue et al. 2004a). Evaluation of lodging resistance at the cultivar level was performed in a previous study and seven traits (plant height, culm weight, culm diameter, culm strength, tiller number, pushing resistance, and rate of lodging resistance) showed significant correlation with lodging resistance and were selected in this chapter. A total of 17 QTLs for all traits except culm weight and lodging score were detected on six of seven LGs and 18 QTLs for lodging scores evaluated at nine different times were detected on all LGs.

3.6 Molecular Breeding

The final goal of most studies is breeding new cultivars using molecular techniques. The characteristics of Italian ryegrass as forage and turf have been improved by many traditional breeding programs so far and new agricultural plant varieties that are accepted by the National Authorities are used widely. The summary of value for cultivation and use (VCU) tests and assessments for National List are 1) yield; total dry matter yield and seasonal dry matter yield, 2) quality; digestibility, 3) resistance to disease; resistance to infection by mildew, crown rust, Drechslera, stem rust, Rhychosporium, RMV, brown rust, BYDV, and bacterial wilt, 4) reaction to environment; ground cover and resistance to winter damage.

3.6.1 Marker-assisted Gene Introgression

Tall fescue (*F. arundinacea*) and meadow fescue (*F. pratensis*) are two closely related species of Italian and perennial ryegrass and are known to carrying many ideal traits such as resistance to crown rust, drought resistance, winter hardiness and frost tolerance (Roderick et al. 2003; Humphreys et al. 2005; Armstead et al. 2006; Kosmala et al. 2007). In perennial ryegrass, introgression mapping using molecular markers has been performed already (King et al. 1998; Armstead et al. 2001), the recombination between *Festuca* and *Lolium* were identified and utilized for breeding (Yamada et al. 2005).

The genes for drought resistance were transferred from *F. glaucescens* (*Fg*) into *L. multiflorum* (*Lm*) (Humphreys et al. 2005). Nine AFLP markers that gave discriminatory *Fg*-specific bands and a STS marker that cosegregated with *Fg* introgression and the Fg-specific AFLP markers were used for tracing *Fg* segments including drought resistance genes. The procedures of introgression of the genes into *Lm* were, the first crossing was *Lm* (4x) x *Fg* (4x), the second crossing was *Lm* (2x) x F_1 hybrid (4x), the third crossing was *Lm* (2x) x BC_1F_1 (3x). BC_2F_1 populations were evaluated for drought resistance for two years, BC_3F_1 mapping family was derived from BC_2F_1 (2x, drought resistance line) x *Lm* (2x, drought sensitive) and third drought resistance test was performed in BC_3F_1 mapping family. Forty-three percent of the BC_3F_1 population had all 10 of the *Fg*-derived markers. From a total of 122 plants tested, the 10 BC_3F_1 plants that survived and recovered from drought and high temperature stress each contained the entire set of 10 *Fg* markers.

Crown rust resistance (CRres) had been introgressed from meadow fescue (*F. pratensis*; *Fp*) into an Italian ryegrass (*L. multiflorum*; *Lm*) background (Armstead et al. 2006). Eventually, they produced BC_3 introgression mapping population segregating for CRres using the following procedures. (i) CRres from diploid *Fp* was transferred into CR-susceptible tetraploid *Lm* background, (ii) the subsequent backcrossing was transferring into CR-susceptible diploid *Lm* background. CRres from *Fp* was mapped with AFLP and STS markers to a terminal segment of chromosome 5. Based on comparative mapping work, this region of the *Lolium/Festuca* genome shares a degree of conserved synteny with regions of the rice genome on Chromasome11 and C12. Sequences from rice Chromosome12 were used as templates for identifying further STS markers that co-segregated with CRres.

In a study by Roderick et al. (2003), genomic *in situ* hybridization (GISH) was used to identify the introgressed fescue chromosome segment. Using GISH the resistance locus could be physically mapped to the midpoint of a short arm. The resistance plants in two BC_3 lines all carried an introgressed segment on a single chromosome, which in one of the lines was confined to the short arm of the chromosome. Genes for winter hardiness and frost tolerance were introgressed from tall fescue into winter-sensitive Italian ryegrass using backcross (Kosmala et al. 2007). GISH was performed on six winter-hardy plants of BC_2 and was shown to be significantly more frost tolerant than the Italian ryegrass control. Five plants carried 13 intact Italian ryegrass chromosomes and one Italian ryegrass chromosome with a single introgressed tall fescue terminal chromosome segment.

The stay-green gene was transferred to Italian ryegrass by using marker-assisted selection (Moore et al. 2005). Synthetic autotetraploid Italian ryegrass ($2n = 4x = 28$) plant carrying four doses of the wild type gene (YYYY) of yellowing in senescing leaves was pollinated with *F. pratensis* homozygous for the recessive gene (stay-green, yy). The progenies (YYy) were crossed with Italian ryegrass (YY), the cross produced YY or Yy genotypes. Italian ryegrass (yy) x BC_1 (YY or Yy) were test-crossed for GISH analysis. The frequency recombinants were revealed from GISH analysis, the one that carried a single *F. pratensis* chromosome segment, the genotype carrying the smallest introgression was selected for AFLP analysis. The genotype of BC_1 was crossed with Italian ryegrass (YY), and the BC_2 population was used for mapping with AFLP analysis. The stay-green gene and 28 *F. pratensis*-specific markers were mapped and 12 of them were converted STS and the STS markers could distinguish between the *Lolium* and *Festuca* genotype. Those markers were tested for the stay-green locus in marker-assisted selection programs.

Although the marker-assisted study is inadequate in Italian ryegrass, reports indicate that reliable marker-assisted molecular breeding in ryegrass is possible. As some comparative mapping studies, e.g., compared with cereal crops maps, supported the reliability for limited research of Italian ryegrass (Sim et al. 2005; Jo et al. 2008).

3.6.2 Gene Pyramiding and Limitations and Prospects of MAS

How efficiently we accumulate and reliably fix target loci are important points in the progress of breeding (Hospital et al. 1997; Servin et al. 2004). They computed the prospective genotypic heritability in each generation, population size for needs and duration of gene pyramiding focusing on QTL. The following steps were suggested , first reliable DNA markers (i.e., closely linked target QTL and flanking marker is desirable) were determined to trace the ideal genotype through generations. Select each breeding parent with ideal genotype after checking the existence of markers determined above, keep mating individual until all ideal genes are collected into one genotype (by which time, the ideal gene alleles may be heterozygous). In the fixing genotype step, i.e., making homozygous, the best way is making DH line, if not, try to self regarded the plant habit. Getting homozygous loci on ideal gene alleles is most important to eventually produce the ideal genotype. The ideal gene alleles are needed to chase securely in all generations with markers. Obtaining the alleles information about which parent the alleles are derived from is a matter of course. For outcrossing plants, the fixation step takes time because of its heterozygous habit. The possibility of recombination between markers and target genes should be kept in mind.

3.6.3 Transgenic Breeding

There are some reports about transgenic Italian ryegrass for breeding of biotic and abiotic stress tolerance. Chitinase hydrolyzes a structural component of the fungal cell wall, thereby weakening the fungi. Overexpression of the chitinase gene in transgenic plants would lead to enhanced resistance to a wide range of fungal pathogens, which have been alluded to by many researchers. Takahashi et al. (2005) introduced the rice chitinase (*RCC2*) gene into calli of Italian ryegrass by particle bombardment. Total 39 *RCC2* transgenic plants were obtained, all of them showed higher chitinase activity than non-transformed plants. Ten days after inoculation with crown rust (*Puccinia coronate*) spores, smaller lesions and a smaller total lesion area were observed in most transgenic detached leaves.

In an abiotic stress study, Takahashi et al. (2006) investigated the expression levels of the replacement histone H3 gene under several stress conditions such as cold, heat and highsalinity (NaCl) using transgenic Italian ryegrass. In plants, comprehensive (like replacement histone H3) gene expression analysis reveals that the expression of many genes is induced in response to environmental stress (Seki et al. 2002; Oono et al. 2003). The replacement histone H3 gene was isolated from the EST database of Italian ryegrass (Ikeda et al. 2004). The replacement histone H3 gene is expressed in the root, stem, leaf, peduncle and spikelet, but very low expression level was observed in leaves. However, the gene expression in leaves was significantly increased by abiotic stimuli mentioned above.

These results indicated the possibility of using transgenic Italian ryegrass for breeding as an alternative way of gene introgression.

3.7 Structural Genomics

Ikeda et al. (2004) reported the generation of EST sequences from seven cDNA libraries of Italian ryegrass. These cDNA libraries were constructed from various tissues (leaf, stem, inflorescence and callus) and leaves under abiotic and biotic stresses such as heat-shock, cold-shock and powdery-mildew infection. Of the 5922 comparable sequences, 3244 clones (54.8%) showed sequence similarity with known genes from plants and other organisms. Other unique sequences showed the specificity depending on cDNA resources. For example, of 863 leaf ESTs, 28.1% showed sequence similarity to photosynthesis-related genes, and 47.8% were considered as metabolism-related genes. From the analysis of 966 stem ESTs, they identified a subset of genes as putative genes for cell-wall related proteins. Some ESTs of each abiotic and biotic stress resource showed sequence similarities of abiotic and biotic stress-related genes and proteins.

The majority of plant disease resistance genes encode a predicted nucleotide-binding site (NBS) and leucine-rich repeat (LRR) (Takken and Joosten 2000), several conserved amino acid motifs in NBS-LRR type disease resistance genes have been identified (Meyers et al. 1999). Large-scale sequencing of disease resistance gene analogs (RGAs) in Italian ryegrass was carried out using degenerate primers and a nested PCR strategy (Ikeda 2005). A total of 9344 clones showed significant levels of homology to either known resistance genes or RGA derived from other plant species using primer sets based on conserved amino acid motifs mentioned above.

According to PlantGDB sequence data, to date 5968 ESTs, 109 STSs, 39 proteins, and 67 GSS were counted and are available for use (2010 April).

Information of structural genomics is important for future studies such as functional and transcriptom analysis.

3.8 Functional Genomics: Transcriptomics

Function and transcriptom analyses are significant for understanding gene activity. There is one expression profiling study aimed at identifying the candidate gene of bacterial wilt caused by *Xanthomonas translucens* pv. *graminis* (*Xtg*) in Italian ryegrass (Rechsteiner et al. 2006). They sampled cDNA at six time-points after infection in a partially resistant genotype of Italian ryegrass infected with *Xtg*. cDNA-amplified fragment length polymorphism (AFLP) was used to analyze and 173 differentially expressed transcript-derived fragments (TDF) were detected. Out of which 37 TDFs were isolated and four of the TDFs were revealed to be homologous to various other genes expressed in response to bacterial and fungal infection. RT-PCR and quantitative RT-PCR (qRT-PCR) confirmed the expression patterns found by cDNA-AFLP. The results indicated the development of molecular tools to improve resistance breeding in Italian ryegrass.

3.9 Role of Bioinformatics as a Tool

The comparative analysis of plant genomes is a significant research method since the remarkable similarity in genomic regions were discovered among closely related genomes. Of course each DNA content may vary, however, information of other plant genomes is useful. Recently many types of genome database were constructed and enhanced by increasing new data ?day by day. The following database list may be useful in forage grass molecular study.

3.9.1 Gene and Genome Databases

The PlantGDB (http://www.plantgdb.org/) database provides the genomic information including EST, STS, etc. The EST sequence data are also available in Gramene (http://www.gramene.org/). In the FoggDB (http://ukcrop.net/perl/ace/search/FoggDB) of UK CropNet, the sequence and phenotypic data of *L. perenne*, *L. multiflorum* and *F. pratensis* are available.

3.9.2 Comparative Genome Databases

In the near future comparative mapping data will be included as they become available in the FoggDB.

3.9.3 Gene Expression Databases

The transcript data are available in chromDB (http://www.chromdb. org/).

3.9.4 Molecular Marker and Genetic map Databases

The data of around 200 loci mapped within the named species and 20 genetic linkage maps including three Italian maps are in the FoggDB.

3.9.5 Protein or Metabolome Databases

The protein data are available in Gramene (http://www.gramene.org/).

3.10 Future Prospects

3.10.1 Potential for Expansion of Productivity

Molecular breeding for biotic and abiotic stress in plants shows promise to increase biomass such as yield and dry weight in the changing climate and environment. For example, disease and pest resistant plants can survive disease epidemics, pests, drought, freezing. Heat tolerant plants can expand their range of use in more severe environments. For this purpose, to conserve the genetic diversity as gene materials and to collect genome information not only for forage grasses but also for whole plants and sustainable projects incorporating conventional breeding strategies should continue vigorously. Finally, stable cultivars system of low cost and high yield will be essential for farmers.

3.10.2 Potential for Expansion into other Geographical Areas

If the potential of plant subsistence area is expanded by molecular studies, some special areas such as high altitude and high latitude will be available for use in the future.

3.10.3 New Uses

Now, forage grass industry, as a part of agriculture, faces a new phase. Forage grasses are not only for feedstuff and soil stability enhancer any longer, they are significant parts of human life and bioremediation. The application of the forage grass to biofuel is a great opportunity for new projects, which give high value to forage grass. The main criteria for any biofuel crops are high biomass yields achieved with low input costs in an environmentally friendly manner (Bouton 2009). These criteria will be one of the goals of forage grass breeding as biofuel feedstock in the near future. Also the next objectives of breeding with MAS are enhancing animal nutrition and improving livestock product quality, reducing pollution risks and health problems, and potential for alternative use of grassland such as for amenity purposes (Humphreys 2005a). Breeders and researchers should keep in mind the state of forage grass of the future including bioremediation which is the most important for next generation breeding.

References

Armstead IP, Bollard A, Morgan WG, Harper JA, King IP, Jones RN, Forster JW, Hayward MD, Thomas HM (2001) Genetic and physical analysis of a single *Festuca pratensis* chromosome segment substitution in *Lolium perenne*. Chromosoma 110: 52–57.

Armstead IP, Harper JA, Turner LB, Skot L, King IP, Humphreys MO, Morgan WG, Thomas HM, Roderick HW (2006) Introgression of crown rust (*Puccinia coronata*) resistance from meadow fescue (*Festuca pratensis*) into Italian ryegrass (*Lolium multiflorum*): genetic mapping and identification of associated molecular markers. Plant Pathol 55: 62–67.

Boller B, Peter-Schmid MKI, Tresch E, Tanner P, Schubiger FX (2009) Ecotypes of Italian ryegrass from Swiss permanent grassland outperform current recommended cultivars. Euphytica 170: 53–65.

Bouton JH (2009) Molecular breeding to improve forages for use in animal and biofuel production systems. Molecular Breeding of Forage and Turf 1–13.

Cresswell A, Sackville Hamilton NR, Roy AK, Viegas BM (2001) Use of amplified fragment length polymorphism markers to assess genetic diversity of *Lolium* species from Portugal. Mol Ecol 1: 229–241.

Curley J, Sim SC, Warnke S, Leong S, Barker R, Jung G (2005) QTL mapping of resistance to gray leaf spot in ryegrass. Theor Appl Genet 111: 1107–1117.

De Haan H (1955) Origin of westerwolths ryegrass (*Lolium multiflorum westerwoldicum*). Euphytica 4: 206–210.

Fujimori M, Hayashi K, Hirata M, Ikeda S, Takahashi Y, Mano Y, Sato H, Takamizo T, Mizuno K, Fujiwara T, Sugita S (2004) Molecular breeding and functional genomics for tolerance to biotic stress. Proceedings of the 3rd International Symposium, Molecular Breeding of Forage and Turf, dallas, Texas, and Ardmore, Oklahoma, USA, May, 18–22, 2003. 21–35.

Hayward MD, Forster JW, Jones JG, Dolstra O, Evans C, McAdam NJ, Hossain KG, Stammers M, Will J, Humphreys MO, Evans GM (1998) Genetic analysis of *Lolium*. I. Identification of linkage groups and the establishment of a genetic map. Plant Breed 117: 451–455.

Hayward MD, Mcadam NJ, Jones JG, Evans C, Evans GM, Forster JW, Ustin A, Hossain KG, Quader B, Stammers M, Will JK (1994) Genetic-markers and the selection of quantitative traits in forage grasses. Euphytica 77: 269–275.

Hirata M, Cai H, Inoue M, Yuyama N, Miura Y, Komatsu T, Takamizo T, Fujimori M (2006) Development of simple sequence repeat (SSR) markers and construction of an SSR-based linkage map in Italian ryegrass (*Lolium multiflorum* Lam.). Theor Appl Genet 113: 270–279.

Hospital F, Moreau L, Lacoudre F, Charcosset A, Gallais A (1997) More on the efficiency of marker-assisted selection. Theor Appl Genet 95: 1181–1189.

Humphreys J, Harper JA, Armstead IP, Humphreys MW (2005) Introgression-mapping of genes for drought resistance transferred from *Festuca arundinacea* var. *glaucescens* into *Lolium multiflorum*. Theor Appl Genet 110: 579–587.

Humphreys MO (2005a) Genetic improvement of forage crop—past, present and future. J Agri Sci 143: 441–448.

Humphreys MO (2005b) Genetic improvement of forage crops—past, present and future. J Agri Sci 143: 441–448.

Hutchinson J, Rees H, Seal AG (1979) An assey of the activity of supplementary DNA in *Lolium*. Heredity 43: 411–421.

Ikeda S (2005) Isolation of disease resistance gene analogs from Italian ryegrass (*Lolium multiflorum* Lam.). Grassland Sci 51: 63–70.

Ikeda S, Takahashi W, Oishi H (2004) Generation of expressed sequence tags from cDNA libraries of Italian ryegrass (*Lolium multiflorum* Lam.). Grassland Sci 49: 593–598.

Inoue M, Cai HW (2004) Sequence analysis and conversion of genomic RFLP markers to STS and SSR markers in Italian Ryegrass (*Lolium multiflorum* Lam.). Breed Sci 54: 245–251.

Inoue M, Gao Z, Cai H (2004a) QTL analysis of lodging resistance and related traits in Italian ryegrass (*Lolium multiflorum* Lam.). Theor Appl Genet 109: 1576–1585.

Inoue M, Gao Z, Hirata M, Fujimori M, Cai H (2004b) Construction of a high-density linkage map of Italian ryegrass (*Lolium multiflorum* Lam.) using restriction fragment length polymorphism, amplified fragment length polymorphism, and telomeric repeat associated sequence markers. Genome 47: 57–65.

Jo YK, Barker R, Pfender W, Warnke S, Sim SC, Jung GH (2008) Comparative analysis of multiple disease resistance in ryegrass and cereal crops. Theor Appl Genet 117: 531–543.

Jones ES, Mahoney NL, Hayward MD, Armstead IP, Jones JG, Humphreys MO, King IP, Kishida T, Yamada T, Balfourier F, Charmet G, Forster JW (2002) An enhanced molecular marker based genetic map of perennial ryegrass (*Lolium perenne*) reveals comparative relationships with other *Poaceae* genomes. Genome 45: 282–295.

King IP, Morgan WG, Armstead IP, Harper JA, Hayward MD, Bollard A, Nash JV, Forster JW, Thomas HM (1998) Introgression mapping in the grasses. I. Introgression of *Festuca pratensis* chromosomes and chromosome segments into *Lolium perenne*. Heredity 81: 462–467.

Kopecky D, Bartos J, Lukaszewski AJ, Baird JH, Cernoch V, Kolliker R, Rognli OA, Blois H, Caig V, Lubberstedt T, Studer B, Shaw P, Dolezel J, Kilian A (2009) Development and mapping of DArT markers within the *Festuca-Lolium* complex. BMC Genomics 10: 473.

Kosmala A, Zwierzykowski Z, Zwierzykowska E, Luczak M, Rapacz M, Gasior D, Humphreys M (2007) Introgression mapping of genes for winter hardiness and frost tolerance transferred from *Festuca arundinacea* into *Lolium multiflorum*. J Hered 98: 311–316.

Lacefield G, Collins M, Henning J, Phillips T, Rasnake M, Spitaleri R, Grigson D, Turner K (2003) Annual Ryegrass. Cooperative Extension Service, University of Kentucky, College of Agriculture, ARG 179 (www.ca.uky.edu).

Lopes V, Reis A, Barata A, Nunes E (2009) Morphological characterization of Portuguese Italian ryegrass. Cent Eur Agri 10: 89–100.

Lübbersted T, Andreasen B, Holm PB (2003) Development of ryegrass allele-specific (GRASP) markers for sustainable grassland improvement—a new EU framework V project. Czech J Genet Plant Breed 39: 125–128.

McGrath S, Hodkinson TR, Barth S (2007) Extremely high cytoplasmic diversity in natural and breeging populations of *Lolium* (Poaceae). Heredity 99(5): 531–544.

Meyers BC, Dickerman AW, Michelmore RW, Sivaramakrishnan S, Sobral BW, Young ND (1999) Plant disease resistance genes encode members of an ancient and diverse protein family within the nucleotide-binding superfamily. Plant J 20: 317–332.

Mian MAR, Saha MC, Hopkins AA, Wang ZY (2005) Use of tall fescue EST-SSR markers in phylogenetic analysis of cool-season forage grasses. Genome 48: 637–647.

Miura Y, Ding C, Hirata M, Takahashi W (2008) Genetic mapping of disease resistance gene analogs from the Italian ryegrass (*Lolium multiflorum* Lam.) genome. Breed Sci 58: 469–473.

Miura Y, Hirata M, Fujimori M (2007) Mapping of EST-derived CAPS markers in Italian ryegrass (*Lolium multiflorum* Lam.). Plant Breed 126: 353–360.

Moore BJ, Donnison IS, Harper JA, Armstead IP, King J, Thomas H, Jones RN, Jones TH, Thomas HM, Morgan WG, Thomas A, Ougham HJ, Huang L, Fentem T, Roberts LA, King IP (2005) Molecular tagging of a senescence gene by introgression mapping of a stay-green mutation from *Festuca pratensis*. New Phytol 165: 801–806.

Muylle H (2003) Genetic analysis of crown rust resistance in ryegrass (*Lolium* spp.) using molecular markers. en Toegepaste Biol Wet. Universiteit Gent, Belgium.

Oertel C, Matzk F (1999) Introgression of crown rust resistance from *Festuca* spp. into *Lolium multiflorum*. Plant Breed 118: 491–496.

Oono Y, Seki M, Nanjo T, Narusaka M, Fujita M, Satoh R, Satou M, Sakurai T, Ishida J, Akiyama K, Iida K, Maruyama K, Satoh S, Yamaguchi-Shinozaki K, Shinozaki K (2003) Monitoring expression profiles of Arabidopsis gene expression during rehydration process after dehydration using ca 7000 full-length cDNA microarray. Plant J 34: 868–887.

Peter-Schmid MKI, Boller B, Kolliker R (2008a) Habitat and management affect genetic structure of *Festuca pratensis* but not *Lolium multiflorum* ecotype populations. Plant Breed 127: 510–517.

Peter-Schmid MKI, Kolliker R, Boller B (2008b) Value of permanent grassland habitats as reservoirs of *Festuca pratensis* Huds. and *Lolium multiflorum* Lam. populations for breeding and conservation. Euphytica 164: 239–253.

Rechsteiner MP, Widmer F, Kolliker R (2006) Expression profiling of Italian ryegrass (*Lolium multiflorum* Lam.) during infection with the bacterial wilt inducing pathogen Xanthomonas translucens pv. graminis. Plant Breed 125: 43–51.

Roderick HW, Morgan WG, Harper JA, Thomas HM (2003) Introgression of crown rust (*Puccinia coronata*) resistance from meadow fescue (*Festuca pratensis*) into Italian ryegrass (*Lolium multiflorum*) and physical mapping of the locus. Heredity 91: 396–400.

Seki M, Narusaka M, Ishida J, Nanjo T, Fujita M, Oono Y, Kamiya A, Nakajima M, Enju A, Sakurai T, Satou M, Akiyama K, Taji T, Yamaguchi-Shinozaki K, Carninci P, Kawai J, Hayashizaki Y, Shinozaki K (2002) Monitoring the expression profiles of 7000 Arabidopsis genes under drought, cold and high-salinity stresses using a full-length cDNA microarray. Plant J 31: 279–292.

Servin B, Martin OC, Mezard M, Hospital F (2004) Toward a theory of marker-assisted gene pyramiding. Genetics 168: 513–523.

Sim S, Chang T, Curley J, Warnke SE, Barker RE, Jung G (2005) Chromosomal rearrangements differentiating the ryegrass genome from the Triticeae, oat, and rice genomes using common heterologous RFLP probes. Theor Appl Genet 110: 1011–1019.

Sim S, Diesburg K, Casler M, Jung G (2007) Mapping and comparative analysis of QTL for crown rust resistance in an Italian x perennial ryegrass Population. Phytopathology 97: 767–776.

Sim SC, Yu JK, Jo YK, Sorrells ME, Jung G (2009) Transferability of cereal EST-SSR markers to ryegrass. Genome 52: 431–437.

Stam P (1993) Construction of Integrated Genetic-Linkage Maps by Means of a New Computer Package—Joinmap. Plant J 3: 739–744.

Stebler FG, Volkart A (1913) Die besten Futterpflanzen. Erster Band, 4. Auflage. K. J. Wyss.

Studer B, Boller B, Bauer E, Posselt UK, Widmer F, Kolliker R (2007) Consistent detection of QTLs for crown rust resistance in Italian ryegrass (*Lolium multiflorum* Lam.) across environments and phenotyping methods. Theor Appl Genet 115: 9–17.

Studer B, Boller B, Herrmann D, Bauer E, Posselt UK, Widmer F, Kolliker R (2006) Genetic mapping reveals a single major QTL for bacterial wilt resistance in Italian ryegrass (*Lolium multiflorum* Lam.). Theor Appl Genet 113: 661–671.

Takahashi W, Fujimori M, Miura Y, Komatsu T, Nishizawa Y, Hibi T, Takamizo T (2005) Increased resistance to crown rust disease in transgenic Italian ryegrass (*Lolium multiflorum* Lam.) expressing the rice chitinase gene. Plant Cell Rep 23: 811–818.

Takahashi W, Oishi H, Ikeda S, Takamizo T, Komatsu T (2006) Molecular cloning and expression analysis of the replacement histone H3 gene of Italian ryegrass (*Lolium multiflorum*). J Plant Physiol 163: 58–68.

Takken FLW, Joosten MHAJ (2000) Plant resistance genes: their structure, function and evolution. Eur J Plant Pathol 106: 699–713.

Van Ooijen JW, Voorrips R (2001) Joinmap version 3.0, software for the calculation of genetic linkage maps. Wageningen.

Vandewalle M, Calsyn E, Van Bockstaele E, Baert J, De Riek J (2003) DNA-markers for yield and quality traits in Italian ryegrass. Czech J Genet Plant Breed 39: 140–146.

Van Deynze AE, Sorrells ME, Park WD, Ayres NM, Fu H, Cartinhour SW, Paul E, McCouch SR (1998) Anchor probes for comparative mapping of grass genera. Theor Appl Genet 97: 356–369.

Warnke SE, Barker RE, Jung G, Sim SC, Mian MAR, Saha MC, Brilman LA, Dupal MP, Forster JW (2004) Genetic linkage mapping of an annual x perennial ryegrass population. Theor Appl Genet 109: 294–304.

Wilman D, Walters RJK, Baker DH, Williams SP (1992) Comparison of 2 Varieties of Italian Ryegrass (*Lolium-Multiflorum*) for Milk-Production, when fed as silage and when grazed. J Agri Sci 118: 37–46.

Yamada T, Forster JW, Humphreys MW, Takamizo T (2005) Genetics and molecular breeding in *Lolium/Festuca* grass species complex. Grassland Sci 51: 89–106.

Tall Fescue

Malay C. Saha

ABSTRACT

Tall fescue [*Festuca arundinacea* Schreb. syn. *Lolium arundinaceum* (Schreb.) Darbysh.] is a cool-season perennial grass with C_3 photosynthesis and commonly grown throughout the temperate regions of the world. It belongs to the family Poaceae, sub family Pooideae and tribe Poeae. Tall fescue has three distinct morphotypes, e.g. Continental, Mediterranean and rhizomatous. The Continental fescue remains in active growth throughout the year and constitutes the vast majority of acreage in North America. The Mediterranean fescue displays summer-dormancy and is less winter hardy. Successful crosses between the two morphotypes often times show irregular meiotic pairing resulting from extensive multivalent formation. Tall fescue can be cultivated in a wide range of soil conditions including acid soils with low pH, low available phosphorus, shallow and eroded soils. Traditionally tall fescue possesses wild type endophyte which is beneficial to the plant but toxic to the grazing animal. Interest on tall fescue cultivation decreased due to toxicosis problems. Breeding efforts were shifted to develop endophyte free cultivars. However, the discovery of novel endophytes reintroduced endophytes in the tall fescue breeding objectives.

Tall fescue is an allohexaploid ($2n = 6x = 42$) with genome constitution of $PPG_1G_1G_2G_2$ and genome size between $5.27–5.83 \times 10^6$ kb. The chloroplast genome is 136,048 bp. As a consequence of the high level of self-incompatibility, each genotype is genetically unique. Within population variation is much wider than among population variation. Due to the complex genetics and low heritability, improvement of quantitative

Forage Improvement Division, The Samuel Roberts Noble Foundation, Ardmore, OK 73401, USA.
E-mail: *mcsaha@noble.org*

traits is difficult and time consuming. The additive genetic variation is captured through the recurrent selection to develop improved cultivars. Genetic linkage maps and quantitative trait loci associated with few important traits have been reported. Using recent advancements in genotyping technologies, phenotyping instrumentation, advanced populations, computer softwares and analysis tools; improving both of quantitative and qualitative traits in tall fescue is possible through marker-assisted breeding. Transgenic tall fescue plants with improved forage and turf qualities and resistant to both biotic and abiotic factors have been developed. Without substantial changes in the regulatory process and transgenic development methodologies, it is expected that transgenic tall fescue cultivars will not reach the marketplace in the near future.

Key words: Tall fescue, Endophyte, Morphotype, Breeding, Phenotype, Genotype, Marker-assisted selection, Transgenics

4.1 Introduction

Festuca is the largest genus in the Poaceae family which contains more than 500 species of temperate grasses (Inda et al. 2008). The vast majority of the species are allopolyploids and the ploidy levels vary from diploid ($2n = 2x = 14$) to dodecaploid ($2n = 12x = 84$) (Loureiro et al. 2007). Tall fescue (*Festuca arundinacea* Schreb. syn. *Lolium arundinaceum* (Schreb.) Darbysh.) is one of the most agriculturally important and widely distributed species within the *Festuca* genus. It is a cool-season perennial with C_3 photosynthesis, usually grown throughout the temperate regions of the world. Tall fescue is widely adapted in North America for forage and landscaping purposes. It has been cultivated in 18 M ha in the United States (Buckner et al. 1979).

4.2 Taxonomy, Botany, Adaptation

4.2.1 Taxonomy, Classification, and Botany

Traditionally tall fescue has been classified under the family – Poaceae, sub family – Pooideae, tribe – Poeae, genus – *Festuca*, and species – *arundinacea*. However, there has been a debate among plant systematists about the taxonomic classification of tall fescue. Agrostologists in the past decade have stated that the genus *Festuca* is comprised of several often marginally related lineages (Craven et al. 2009). Thus, it has been classified under the genus *Lolium* (Darbyshire 1993). In the Catalogue of New World Grasses, tall fescue is classified in the genus *Schedonorus*. The National Plant Data Center recognized tall fescue as *Schedonorus phoenix* within PLANTS, but was declared as *Schedonorus arundinaceus* in their updates (Henson and

Safley 2009). Tall fescue is still popularly known as *Festuca arundinacea*. Fescue germplasm pool is classified into five botanical varieties: tetraploid *glaucescens* Boiss; hexaploid *genuina* Schreb.; octaploid *atlantigena* (St.-Yves) Auquier; and decaploid *cirtensis* (St-Yves) J. Gamisans and *letourneuxiana* (St.-Yves) Torrecilla & Catal.

Tall fescue can better be described as a species complex consisting of three distinct morphotypes, e.g., Continental, Mediterranean and rhizomatous (Fig. 4-1). Each of these morphotypes has distinct agronomical, morphological and physiological attributes. Continental germplasm originates from central and northern Europe eastward through a portion of Asia. Continental germplasms are relatively winter hardy and often have wide leaves. These remain in active growth throughout the year and thus, constitute the vast majority of tall fescue acreage in North America. Most of published tall fescue studies are based on Continental germplasms. Mediterranean germplasm is traced to southern Europe, the Middle East, and North Africa. Mediterranean ecotypes generally do not show winter dormancy and grow well at moderately low temperatures and/or short nights. However, Mediterranean types are less winter hardy (Burner et al. 1988) and often have narrower leaves than Continentals. The Mediterranean fescue displays summer dormancy. Rhizomatous tall fescue originates from Northwest Spain and Portugal (Borrill et al. 1971). Continental

Figure 4-1 Shoot and root growth of tall fescue morphotypes; Continental (left), rhizomatous (middle), and Mediterranean (right).

Color image of this figure appears in the color plate section at the end of the book.

germplasms can also form rhizomes (Jernstedt and Bouton 1985). However, rhizomes in rhizomatous germplasm are longer and more prevalent than the Continentals. Rhizomatous germplasm does not appear to be as winter hardy as the Continental (Diesburg and Carlson 1983). Rhizomatous fescue has superior spreading ability thus suitable for use as turf.

Cross incompatibility and/or infertile hybrids indicated independent origins of the morphotypes from different progenitors. Successful crosses between Continental and Mediterranean morphotypes often show irregular meiotic pairing resulting from extensive multivalent formation (Hunt and Sleper 1981). Hybrids between rhizomatous and Mediterranean morphotypes are usually sterile and form numerous univalents at meiosis (Jauhar 1991). Phylogenetic analysis based on low-copy nuclear and chloroplast genes suggests independent evolution of the Mediterranean and Continental morphotypes (Hand et al. 2010). Though *F. pratensis* and *F. arundinacea* var. *glaucescens* were identified as probable progenitors to Continental and rhizomatous morphotype, no obvious candidate progenitors of Mediterranean tall fescues were identified. The internal transcribed spacer (ITS) sequence based dendrogram positioned Continental samples in the "European" and rhizomatous in "Maghrebian" subclades, which suggests a distinct separation of the two morphotypes by a large evolutionary distance (Hand et al. 2012).

There are two distinct functional groups, i.e., forage and turf types. Forage-types can be either Continental or Mediterranean types with large dark green leaves. These are characterized by coarse leaves, upright growth habit and tall plants. Turf-types have finer dark green leaves, short growth stature and dense tillers. Turf types are typically Continental germplasm that can tolerate frequent and low mowing. Rhizomatous germplasms can spread fast thus have potential for use as turf. Tall fescue inherently possess better shade tolerance than many other grass species, this widens its popularity as a turf species.

Tall fescue plants have dense tillers and tall growth habit (Fig. 4-2). Plants produce short rhizomes but have a bunch-type growth habit. It spreads primarily by erect tillers. It has broad, dark green basal leaves. Leaf blades are glossy on the underside and slightly serrated on the margins. The leaf sheath is smooth and the ligule is a short membrane. Individual tillers or stems terminate in an inflorescence which reaches 90 to 120 cm in height. The inflorescence is a compact panicle with lanceolate spikelets. Each spikelet is about 0.75 to 1.25 cm long. The plants are vernalized in winter and flowers in spring. Seeds become mature in early summer. Seeds are elliptic in shape and 4 to 7 mm long.

Figure 4-2 A tall fescue plant in the breeding nursery of Noble Foundation, Ardmore, Oklahoma. Seed heads were collected for harvesting seed.

Color image of this figure appears in the color plate section at the end of the book.

4.2.2 Distribution

Though originally from Europe and North Africa, tall fescue has been in cultivation all over the world. It was adapted throughout continental Europe between northern Italy in the South, Scandinavia in the North, Hungary to the East, and United Kingdom in the West. It is well adapted to the cool and moist climate of the UK and less frequent in Turkey. In the Caucasus in the southern part of European USSR, tall fescue has been found in the wild state. Due to its inherent ability to adapt in wide geographic conditions tall fescue cultivation has been extended to a wide area throughout the Africas. The Atlas Mountain regions in Morocco and Tunisia became an area of adaptation. However, many of the tall fescue populations collected in Morocco are identified as octaploid. Tall fescue is a productive winter grass in South Africa. Japan, China and Korea are the leading Asian countries in tall fescue cultivation. It is widely adapted throughout Japan. Among the Middle-East countries, Iran has distinct germplasm resources and the country has an active tall fescue breeding program. Tall fescue is adapted to extensive areas of Australia and New Zealand.

It was introduced in USA in the early 1800's. The transition zone; includes eastern Nebraska, Kansas, Oklahoma, Missouri and Arkansas; southern Iowa, Illinois, Indiana, Ohio; Kentucky, Tennessee, Virginia, West Virginia, North and South Carolina, Mississippi, Alabama and Georgia became the primary areas of growing forage tall fescue. However, use as turf extended from the east coast to the west coast of the U.S. Tall fescue cultivation was further extended to Canada. The coastal provinces of British Columbia and Nova Scotia became the primary areas of adaptation. Brandon, Manitoba (49° 52′ N 99° 59′ W) was considered the northern limit of survival (Burns and Chamblee 1979). Tall fescue has been successfully cultivated at the higher elevations of tropical and subtropical regions of South America including Colombia (Crowder et al. 1959). In the pampas of Argentina, tall fescue became an important forage species. Its cultivation is widespread in Chile as a part of grass and legume mixtures.

4.2.3 Adaptation

Tall fescue is a cool-season perennial grass thus, the best adaptation is found in the temperate regions of the world. However, below freezing temperatures, short days and low light intensities during winter months limit winter production of tall fescue in a temperate climate. In such a climate, tall fescue mainly grows in late spring, summer and early autumn (Burns and Chamblee 1979). The mean monthly maximum temperature for optimum growth is 36°C. Top growth continues when soils are as cold as 4°C. The Continental tall fescue has better winter hardiness than the Mediterranean morphotype thus more adaptive in the temperate climates. However, the Mediterranean morphotype can persist better in hot and dry summers through induction of summer dormancy.

Tall fescue is a deep rooted perennial grass thus adapted to a wide range of soil conditions, but performs best on well drained clay soils. Heavy to medium textured, deep and moist soils with high organic matter are the best for high biomass production. It can be found growing in low, damp pastures and wet meadowlands throughout Europe, North Africa and North America. Tall fescue can grow fairly well on soils low in fertility, but is better adapted to fertile conditions. It is tolerant to acidity, alkalinity and salinity with the optimum pH range lies between 5.5 and 8.5. Tall fescue is suitable to grow in excessively or poorly drained soils. It can tolerate long periods of flooding (24 to 35 d).

Soil moisture has a big influence on tall fescue growth and persistence especially in warmer climates where persistence is determined mainly by soil moisture availability. It can grow well with a minimum precipitation of 375 to 450 mm. In regions where soil moisture is limiting and temperature is high, tall fescue will not survive long. The major factors limiting the adaptation are abiotic factors (high and low temperatures and rainfall) and geographic (mainly altitude), with less influence from biotic factors (Burns and Chamblee 1979).

4.3 Agronomy

4.3.1 Cultivation

Tall fescue cultivation has manyfold advantages. It can provide better land utilization, reduced soil erosion, and enhanced economics of livestock production. It can be cultivated in a wide range of soil conditions including acid soils with low pH, low available phosphorus, shallow and eroded soils. The plants are capable of making physiological and morphological changes to adjust or tolerate chemical and physical stresses. A vast inventory of cultivars has been developed through decades of breeding efforts. The cultivar best adapted to the area of cultivation needs to be selected. High-quality seed with high germination percentage and free from weed seed is a key factor in cultivation. Like most of the perennial crops, tall fescue seedlings are slow in establishment. Thus planting time is very important especially for fall plantings. In temperate climates, the seed should be planted early enough in the season so that the crop can be established well before freezing. The seed should be planted shallow, usually between 0.5 to 1.5 cm, depending on soil texture and moisture (Fribourg and Milne 2009). Seeding rate for pasture and hay establishment varies from 13 to 22 kg ha^{-1}. During early establishment, weeds especially the annual species are serious threats for better plant stand. Herbicides are commonly used for effective weed controls. Specifics on tall fescue cultivation can be obtained from "Tall Fescue for the Twenty-first Century" (Fribourg et al. 2009).

4.3.2 Pest and Diseases

Stem rust (*Puccinia graminis* Pers.:Pers. subsp. *graminicola* Z. Urban), Gray leaf spot (*Magnaporthe grisea*), Brome mosaic virus (BMV) are some of the major diseases of tall fescue. In recent years stem and leaf rust of tall fescue appears as a major disease of tall fescue (Fig. 4-3). Rust is particularly important for seed crops. Seed yield losses of up to 40% were reported in Oregon, USA (Barker et al. 2003). Stem rust outbreak was first observed in Oregon seed fields in 1987 (Welty and Mellbye 1989). The disease was widespread throughout the seed production area in the Oregon valley with disease severity rated from moderate to severe in 22 of 29 cultivars. Most of the common tall fescue cultivars were found to be susceptible to the disease. Applications of fungicides are common methods of disease control (Pscheidt 1996). Several applications are usually necessary for effective control of

Figure 4-3 Severe rust infected tall fescue plant grown in field experiment at Ardmore Oklahoma, USA. Photo taken on July 16, 2012.

Color image of this figure appears in the color plate section at the end of the book.

the disease depending on the earliness and rate of epidemic development (Pscheidt and Ocamb 2001). The cost of application of fungicides to control the disease was estimated as US$3.9 million a year. The return on rust control treatments for each dollar spent ranged from US$7.5 ha^{-1} to US$13.4 ha^{-1} (Gingrich and Mellbye 2004).

BMV can cause serious economic damage to tall fescue. A BMV strain, F-BMV, severely restrict the growth, development and yield of tall fescue plants. In greenhouse studies, 40% fewer tillers and 42% less dry matter production were recorded in susceptible plants than the resistant plants. Lesser extent of damage, i.e., 25% less tillers and 36% less dry matter production, were observed in field compared to greenhouse (Mian et al. 2005b). Gray leaf spot is especially important for turf-type tall fescue. Several isolates of the pathogen are able to develop typical leaf spot symptoms. "KY31" was found susceptible to the disease whereas "Coronado" and "Coyote" were recognized resistant to the disease. The resistant genotypes of the two cultivar exhibited longer incubation and latent periods, reduced rates of disease progress and lesion expansion, lower disease incidence, mean lesion length, area under lesion expansion curve (Tredway et al. 2003). Details on pest and diseases of tall fescue are available in respective chapters in "Tall Fescue for the Twenty-first Century".

Endophyte infected (E+) tall fescue showed better resistance to pest and diseases compared to the endophyte free (E–) plants. Endophytes provide a natural control to the plants thus the E+ plants have very little or no insect problem. The toxic alkaloids produced by the endophyte have insecticidal properties. Studies on chemical compounds produced by the endophytes and their role in controlling pest and diseases in tall fescue and related species are important and interesting research areas. However, it is often found difficult to get conclusive results due to the presence of many measurable variables in the trials. In addition to several abiotic factors, at least three biotic components, e.g., endophyte, plant and pest, and their interactions play a role in such studies.

4.4 Germplasm Resources

The genetic diversity in tall fescue and related species complex is extensive. Each seed is genetically distinct thus individual plants in a population are highly heterozygous. The National Plant Germplasm System (NPGS) of the United States Department of Agriculture (USDA) maintains germplasm collections of different crop in their Germplasm Resources Information Network (GRIN). A total of 1,049 *Festuca* accessions have been listed in the GRIN database of which 937 are available for distribution but for the rest 112 are not currently available (http://www.ars-grin.gov/cgi-bin/npgs/html/taxon.pl?412634 verified Sept. 10, 2012). Data on collection site, date the samples submitted, improvement status, source history, etc. on all these accessions can be obtained from the online services. Information on germplasm maintenance can also be obtained from the website. In addition, small scale collections have been maintained by institutions and organizations (e.g., Noble Foundation, Ardmore, Okla., USA; Aberystwyth University, Ceredigion, Wales, UK; AgResearch, New Zealand, etc.) all over the world.

4.5 Conventional Breeding

Tall fescue is an outcrossing polyploid with a high level of self-incompatibility. Wide genetic variabilities exist both within and among populations. Both morphological and molecular marker studies suggested wider within population variation than among population variation. This is a common feature in outcrossing polyploid species including perennial ryegrass (*Lolium perenne*; Huff 1997), orchard grass (*Dactylis glomerata*; Ubi et al. 2003), and harding grass (*Phalaris aquatica*; Mian et al. 2005a). Wider variability for many important traits exists in tall fescue populations. The phenotype of a plant is determined by its genetic composition, the growing environment, and the interaction of genotype and environment. Classical breeding of any crop species is relied on phenotypic selection. Plants with desirable traits are selected from natural populations and then used as parents in breeding programs.

One important consideration of a breeding program is how a trait is inherited from generation to generation. Heritability is a method of estimating the possible genetic gain of a trait in subsequent generations. In heritability analyses, the relative contribution of differences in genetic and non-genetic factors to the total phenotypic variance in a population is estimated. Traits with high heritability can be improved without difficulty compared to traits with low heritability. Narrow sense heritability estimate is more important than the broad sense heritability as the former measures the ratio of additive genetic variation to phenotypic. Traits controlled by one or a few genes, qualitative traits, usually have high heritability estimates. Traits controlled by many genes each having a minor effect, quantitative traits, have low heritability. Stem rust in tall fescue is considered as a qualitative trait as it is controlled by a major gene (Barker et al. 2003). Yield and persistence are the quantitative traits. Heritability estimates of forage yield were between 0.36 to 0.77 (Reeder et al. 1986).

4.5.1 Breeding Procedures

As each genotype in a population is genetically distinct, selection of superior genotypes can frequently generate new populations. Traditional breeding systems include: i) diverse germplasm collection, ii) evaluation at multiple environments, iii) selection of superior genotypes, iv) performance evaluation, v) seed multiplication and vi) release as a new cultivar. Most of the early released cultivars, including the most popular cultivar KY31 were developed in this manner. Alta and KY31 are the two oldest tall fescue varieties still in cultivation. Alta was selected from a tall fescue stand in Oregon in 1923 and KY31 is an increase from tall fescue found in 1931 on a Kentucky farm where it was established for 50 yr.

The additive genetic variation is generally the most important genetic component to tall fescue breeders. Additive genetic variation is captured through the breeding procedures such as recurrent selection to develop improved cultivars. Recurrent selection is a common breeding procedure followed in the improvement of cross-pollinated species. Plants with superior agronomic qualities are selected, intermated in isolation and the resulting populations are evaluated and advanced in each cycle of selection. Substantial improvements can be achieved even for traits with low heritability. Most of the tall fescue cultivars both forage and turf types, released after 1980, were developed following the recurrent selection procedure from advanced breeding populations (Hopkins et al. 2009). Cultivars that are produced through random mating of selected parental plants are commonly known as synthetic cultivars. Unlike cultivars of self-pollinated species, the synthetic cultivars are not genetically pure. However, a certain level of phenotypic homogeneity needs to be attained to pass variety designation. At the same time, a high level of heterozygosity should be maintained to maximize adaptability and agronomic performance. A typical schematic diagram of tall fescue breeding procedure followed in several breeding programs is presented in Fig. 4-4.

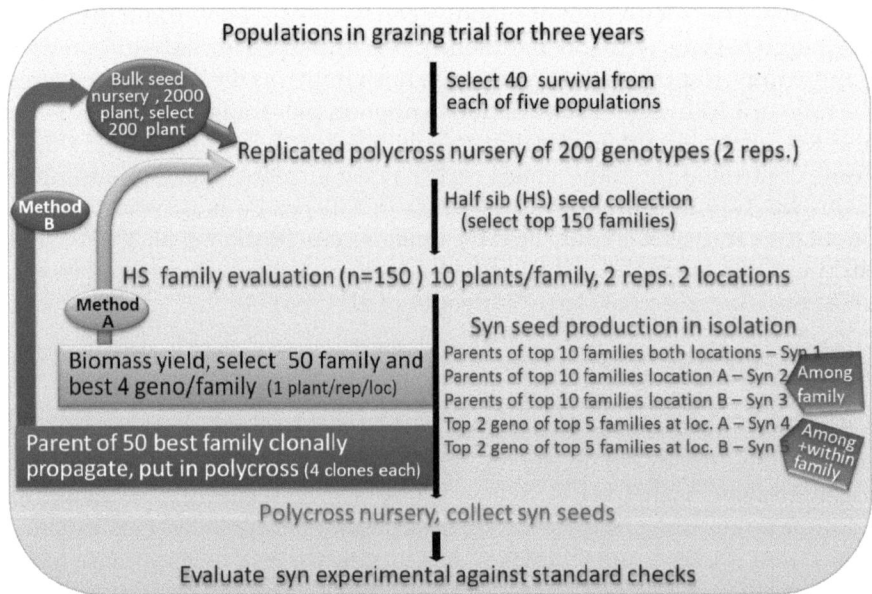

Figure 4-4 A half-sib family evaluation and synthetic cultivar development scheme for improving persistence in tall fescue.

4.5.2 *Traits of Interest*

Yield is always a key trait in any breeding program. Yield is a complex trait contributed by many phenotypes. Improved tall fescue yield can be achieved by decreased dark respiration rate (Sleper 1985), increased weight per tiller (Zarrough et al. 1983), increased leaf area expansion rate (Nelson and Sleper 1983). As tall fescue is a perennial crop, persistence has been a major goal in most breeding programs. The persistence of perennial grasses largely depends on their ability to tolerate drought stress. Tall fescue cultivars do not generally persist for more than 2 to 4 yr in the south central US due to drought and heat stress (Hopkins 2005). In the Continental tall fescue improved persistence can be achieved by introducing drought and grazing tolerance especially in areas where summer is very hot and dry. Drought tolerance is an important trait in tall fescue breeding. Drought tolerance can be achieved via indirect selection of traits, such as, increased relative water content (Elmi and West 1995), decreased carbon[13] isotope discrimination (Johnson and Yangyang 1999), better root architecture (Torbert et al. 1990; Bonos et al. 2004), increased osmotic potential (Elmi and West 1995), etc. Mediterranean tall fescue remains dormant during summer which is an effective avoidance mechanism to extreme summer conditions thus provides better persistence. In turf-type, close mowing can improve persistence. Heavy and/or continuous grazing can substantially reduce tall fescue persistence. Persistence of populations under heavy grazing pressure can be evaluated in either space planted or seeded sward plots (Hopkins 2005).

Increased digestibility is an important breeding objective. Improved digestibility was found to have a major positive impact on animal performance (Casler and Vogel 1999). Attempts have been undertaken at the Noble Foundation to improve forage digestibility of tall fescue using both traditional and molecular breeding approaches. Increased palatability is another goal of forage tall breeding (Buckner and Burrus 1968). Leaf softness is an indication of palatability. Over the past decades, cultivars with soft leaf like "Barcel", "Adora", "Quantum", "Advance", "BarOptima", etc. have been developed. Maturity of plants has greater impact on tall fescue quality. Late maturity improves palatability of tall fescue (van Santen 1992). Delayed maturity can be achieved through appropriate breeding methods (Watson and McLean 1991). Several diseases, e.g., stem and leaf rust, gray leaf spot, etc. largely decrease yield and persistence of tall fescue. Two cycles of recurrent selection improved resistance to stem rust (Barker et al. 2003). Seed yield is another important trait for tall fescue breeding.

4.5.3 Cultivars

Tall fescue breeding prior to 1981 did not pay much importance on endophyte infection due to lack of knowledge. During the period from 1981 to 1995, substantial knowledge was gained on the role of fungal endophyte in tall fescue production and utilization. Breeding efforts were shifted to develop endophyte free cultivars (Pedersen and Sleper 1988). During the late 90s novel endophytes revolutionized the tall fescue breeding objectives. According to the rules of the Association of Official Seed Certifying Agencies (AOSCA) and the Organization for Economic Cooperation and Development (OECD) a large number of tall fescue cultivars were eligible for certifications. A detailed list of these cultivars can be viewed in Hopkins et al. (2009). Though the list covers both forage and turf-type cultivars but it is largely dominated by the turf-types. "Jesup MaxQ" and "TexomaMaxQ II" are the tall fescue cultivars with novel endophytes and available for commercial cultivation. Seed cost for cultivars with novel endophytes can be up to four times the cost for seed with toxic endophyte and/or endophyte free cultivars. This is mainly due to the cost of developing and marketing this new technology. Novel endophytes provide mutualistic benefits to the plants and at the same time increased animal performance. The cost of stand establishment can be recovered within three to seven years (Gunter and Beck 2004). Cultivars with novel endophytes can bring a breakthrough in tall fescue cultivation and utilization.

4.6 Molecular Resources and Breeding

It is expected that tall fescue breeding will enter a new era with the application of biotechnologies such as genomics and marker assisted selection in near future. Molecular technologies, such as molecular markers, genomics and transgenics are yet to be developed and adapted in tall fescue breeding programs. Strict and expensive regulatory processes pose a prohibitive barrier for testing and releasing transgenic tall fescue cultivars.

4.6.1 Nuclear Genome

Tall fescue is an allohexaploid with basic chromosome number $2n = 6x = 42$ and the genome constitution of PPG1G1G2G2. The P genome is derived from *F. pratensis* ($2n = 2x = 14$) and the G1G2 genomes from *F. arundinacea* var *glaucescens* (Sleper 1985; Humphreys et al. 1995). Tall fescue nuclear genome is estimated between $5.27–5.83 \times 10^6$ kb (Seal 1983), which is more than 12 times larger than the genome of rice (*Oryza sativa* L.). Unlike the genomes of legume species, e.g., soybean and lotus, the tall fescue genome is rich in GC rather than AT content (Brown-Guedira et al. 2000). Several ploidy levels have been reported in tall fescue which ranges from diploid to dodecaploid. The Mediterranean tall fescue is also a hexaploid but octaploids are also reported in Moroccan collections.

4.6.2 Chloroplast Genome

Chloroplasts are key organelles primarily responsible for photosynthesis in plant cells. Chloroplast genomes are highly conserved among taxonomically related species. The chloroplast genome is small and circular also known as plastome. A group of students at the Middle Tennessee State University decoded the chloroplast genome of tall fescue (Cahoon et al. 2010). Chloroplasts from KY31 plants were isolated following the protocol of Jansen et al. (2006). Large (3–5 kb) amplicons and shotgun libraries were constructed and sequenced using a CEQ 8000 genetic analyzer (Beckman-Coulter). After assembly of sequence data, small gaps of 2–200 bp were spanned and sequenced using fescue-specific primers and completed the circular sequence.

The tall fescue chloroplast genome is 136,048 bp in size and has typical quadripartite structure (GeneBank ID FJ466687). A small (11,300 bp) and a large (80,560 bp) single copy regions and an inverted repeat region (22,600 bp) constitute the fescue chloroplast genome. A total of 75 protein-coding genes, 29 tRNAs, 4 rRNAs, and a hypothetical coding region were identified which comprised about 56% of the chloroplast genome. Intergenic and intron spaces made up rest of the sequences. Seven larger than 30 bp dispersed repeats were identified. Several genes (e.g., *rpoB, atpF, ccsA, ndhD, ndhA, rps14, and ycf4*) have translation or sequence discrepancies with other grass species. Premature stop codons and missing start and stop codons have been identified in RNA editing sites. It was evident that the fescue plastome lacks functional *rps14* and *ycf4* genes (Cahoon et al. 2010).

Tall fescue chloroplast genome was compared with all sequenced grass plastomes. The gene order is highly congruent with other grasses within the Bambusoideae-Oryzoideae-Pooideae (BOP) clade. The chloroplast genomes of maize, sorghum and sugarcane were fairly larger (≥140,384 bp) than that of tall fescue and other members of the BOP clade. Several large deletions (primarily seven) in intergenic regions are responsible for the majority of size discrepancy. *Lolium perenne* plastome is the closest relative of tall fescue. However, a number of changes have been occurred especially in the intergenic regions since these species diverged. Phylogenetic analysis of 11 grass species based on 61 translated coding regions identified a strongly associated clade with members of tribes, Aveneae, Poeae, Triticeae, and Brachypodium.

4.6.3 Marker Development

Molecular markers are indispensible tool for genomic research. Molecular markers are highly heritable, well distributed throughout a genome, highly polymorphic between genotypes, and can be easily detected with advanced technologies. Application of molecular markers for the genetic improvement of tall fescue was initiated in the early 90s. Randomly Amplified Polymorphic DNA (RAPD), Restriction Fragment Length Polymorphism (RFLP), and Amplified Fragment Length Polymorphism (AFLP) markers were developed. These markers were used for the analysis of genetic diversity (Xu et al. 1994; Mian et al. 2002; Mian et al. 2005c), phylogenetic studies (Xu and Sleper 1994; Charmet et al. 1997), characterization of *Festuca × Lolium* hybrids (Chen et al. 1995; Wang et al. 2003), and cultivar identification (Eizenga et al. 1998). The markers were also used for the construction of genetic linkage maps (Xu et al. 1995; Saha et al. 2005) and comparative mapping (Chen et al. 1998).

Microsatellite or simple sequence repeat (SSR) markers were eventually developed (Saha et al. 2004, 2006) and used for the construction of genetic linkage maps (Saha et al. 2005) and evaluating their cross species applicability (Saha et al. 2004; Mian et al. 2005c; Zeid et al. 2010). Trinucleotide motifs were the most abundant type of SSRs (70%) in tall fescue ESTs. CCG/GGC and GA/CT was the most abundant tri and di-nucleotide repeats, respectively (Saha et al. 2004). Recently, single nucleotide polymorphism (SNP) markers have been developed and used for the morphotype classification and phylogenetic analysis of *Festuca* spp. (Hand et al. 2010, 2012). However, the progress made in the development and utilization of molecular markers in tall fescue remains far behind than other major crop species.

DArT genotyping service was developed to provide cost-effective genotyping facilities across the agricultural sector for the efficient use of genetic resources for crop improvement (http://www.diversityarrays.com/molecularprincip.html). DArT technologies consists of complexity reduction of DNA, construction of libraries, microarraying species specific libraries onto glass slides, hybridization of DNA from species of interest to respective slides, acquisition of hybridization signals through scanners, and data analysis. DArT technologies were used for comparative genomics, genome-wide linkage disequilibrium and association analyses, whole-genome molecular analysis, marker-based linkage analysis, QTL mapping, and diversity analysis (http://www.diversityarrays.com/conferenceabstracts.html). In addition to major crop species, DArT array was also developed for five species of *Festuca-Lolium* complex including tall fescue. The *Festuca-Lolium* DArT array (DArTFest) contains 7,680 probes derived from methyl-filtered genomic representations. Until now, a total of 80,354,304 assays were implemented in 2,783 genotypes of *Festuca* and *Lolium* species. DArT markers were assigned to each of the seven chromosomes of *F. pratensis* (Kopecký et al. 2009). Tall fescue mapping populations developed for deciphering drought tolerance and summer-dormancy were assayed with DArT markers. A set of 93 entries of turf type tall fescue was also evaluated with DArT markers for assessing genetic diversity (Baird et al. 2012).

4.6.4 Development of Libraries

A total of 141,309 ESTs have been developed from *Festuca* species and published in dbEST of which 63,808 from *F. arundinacea*, 74,111 from *F. pratensis* and rest from other spp. (http://www.ncbi.nlm.nih.gov/sites/entrez?db=nucest&term=festuca&pmfilter_MDatLimit=10+Years; verified September 12, 2012). At Noble Foundation, 41,516 ESTs were generated from nine tall fescue cDNA libraries. The libraries were developed from tissues of different plant organs at developmental growth stages and abiotic stress conditions. About 5,000 ESTs from each of the eight cDNA libraries were developed and the ninth library constructed from field stressed plants was represented by 1,125 ESTs. In addition, 2,495 ESTs were generated from heat responsive gene transcripts between heat-tolerant and heat-sensitive fescue genotypes of which 656 were singlets and the remaining 1,839 were grouped into 434 clusters (Zhang et al. 2005). These ESTs were used for the development of SSR markers and can be potentially used for the development of SNP markers.

A $(GA/CT)_n$ enriched genomic library was developed following the protocol suggested by Hamilton et al. (1999). Sequencing of clones generated 5,320 genomic sequences (can be viewed online in NCBI). Genomic DNA from a pool of 31 KY31 plants was used for the library construction and a biotinylated $(GA)_n$ oligonucleotide was used to capture and enrich the library. The sequences were used for the development of SSR markers. EST-SSRs are associated with expressed genes which are concentrated in the gene rich regions and usually less polymorphic. On the other hand, the genomic SSRs are well distributed throughout the genome and usually more polymorphic.

4.6.5 Genetic Linkage Map

Genetic linkage map is an important molecular tool which arranges molecular markers in linear order within the genome. Markers are placed in maps based on their pair-wise recombination frequencies. Bi-parental mapping populations are developed by crossing two distinct parents that are used for the calculation of marker recombination frequencies. In self-pollinated species; backcross, F_2, and recombinant inbred lines (RIL) are the most commonly used mapping populations. Pseudo F_1 populations are used to construct genetic linkage maps in self-incompatible out-crossing species. Genetic linkage maps have been developed for most of the agriculturally important plant species.

Genetic linkage mapping efforts in tall fescue are impaired by the genome complexity, heterozygosity within clones and a high level of self-incompatibility (Xu et al. 1991). Tall fescue is a hexaploid with three genomes, thus, greater number of possible genotype combinations are possible. Homologous chromosomes pairing and double reduction ratios are largely unknown. Thus, accurate estimation of recombination frequencies is difficult. High level of genetic variability in populations ensures marker polymorphisms. Unlike hexaploid wheat, which displays a very low level of intraspecific polymorphism (Röder et al. 1998), a large number of polymorphic markers can be developed in tall fescue with less effort (Saha et al. 2005). However, like other polyploid species, construction of linkage maps in tall fescue is complicated because one or both parents may be heterozygous at a certain locus, markers may be dominant or co-dominant and the linkage phase of marker alleles is usually unknown (Maliepaard et al. 1997). The two-way pseudo-testcross procedure is an efficient way of constructing genetic linkage maps in out-crossing species (Ritter et al. 1990; Hemmat et al. 1994). Genetic linkage maps developed in tall fescue followed the pseudo-testcross procedure.

The first mapping population in tall fescue was developed by crossing two distinct genotypes, HD28-56 and a selection from KY31. The F_2 mapping population was genotyped with 108 RFLP markers (Xu et al. 1995). Marker segregation data clearly proposed a disomic inheritance in tall fescue which was also suggested earlier by cytogenetic (Sleper 1985) and isozyme studies (Lewis et al. 1980). Thus, tall fescue is an allohexaploid. A total of 95 markers were arranged in 19 linkage groups (LGs) and covered 1,274 centimorgan (cM) of the tall fescue genome (Xu et al. 1995). On average, each marker was identified in 17.9 cM. Five of the seven homoeologous LGs were identified using the markers that segregated at more than one locus. Genome-specific probe set was used to assign the LGs in the three genomes (PG1G2).

A PCR marker based genetic linkage map of tall fescue was constructed using a pseudo-F_1 mapping population developed by crossing HD28-56 × R43-64 genotypes. The population was genotyped with 773 AFLP and 343 SSR markers (Saha et al. 2005). Markers segregated in 1:1 and 3:1 Mendelian segregation ratios were used to construct the linkage groups. Markers present in both parents and segregated in 3:1 ratio were useful for identifying homologous groups (Maliepaard et al. 1998). The HD28-56 map included 558 loci placed in 22 LGs and covered 2,013 cM of the genome. The R43-64 map comprised 579 loci grouped in 22 LGs with a total map length of 1,722 cM. The marker density varied from 3.61 cM (HD28-56) to 2.97 (R43-64) cM per marker. Homologous LGs of the parental maps were combined to form the integrated LGs. The integrated map covers 1,841 cM on 17 LGs, with an average of 54 loci per LG, and has an average marker density of 2.0 cM per marker. The parental LGs that carried common markers were allocated to the same homoeologous groups. Six of the seven predicted homeologous groups were identified following this strategy (Saha et al. 2005). Variation in recombination frequencies were observed between the two parents. The male parent showed reduced recombination compared to the female parent (1.6 and 1.8 chiasmata per LG, respectively).

Recombination events were not evenly distributed throughout the genome. Clustering of markers was evident in some regions and gaps of 10–21 cM were noticed in other regions. Like the tall fescue RFLP map (Xu et al. 1995), high levels of segregation distortion (23% of total markers) were observed in this mapping population which was also reported in other outcrossing polyploid species (Warnke et al. 2004; Brummer et al. 1993; Wang et al. 1994). Gametic selection and/or faulty chromosome pairing, an association between heterozygosity and plant vigor, and the selection of one parental type are the key factors responsible for segregation distortion (Xu et al. 1995). Distorted markers were distributed in both the parental maps, thus indicating that both the male and female gametophytes and/or sporophytes are involved in segregation distortion. Markers with segregation distortion were mainly distributed in four LGs of the consensus map.

The bi-parental map discussed above was dominated by AFLP markers. Even an AFLP marker associated with a trait of interest is very difficult to reproduce. Hence, a functional map of tall fescue was essential to construct. The mapping population was genotyped with additional SSR and STS markers developed at the Noble Foundation. Parental linkage maps were constructed following the same strategy discussed earlier. The female (HD28-56) parent map consists of 21 LGs and covered a total length of 1,580 cM with a marker density of 5.7 cM/marker. The male (R43-64) parent map covered 1,364 cM with a marker density of 5.8 cM/marker (Saha et al. 2009). This map was used for the detection of quantitative trait loci (QTL) and for the identification of molecular markers associated with traits of interest. A couple of other bi-parental mapping populations, developed for the detection of loci associated with summer dormancy and drought tolerance, were genotyped with SSR, STS, and DArT markers. Parental maps were constructed and QTL detection is in progress (unpubl. data). Genotyping by sequencing of a tall fescue population developed for deciphering the drought tolerant is in progress (unpubl. data).

4.6.6 QTL Identification

Understanding the genetic basis of quantitative traits is a major goal in any molecular breeding program. Detection of QTL associated with traits of interest is an important step towards marker-assisted breeding. Literature on QTL studies in different plant species has grown enormously over the past few decades (Bernardo 2008). Significant improvement in marker technologies, low-cost genotyping, advanced phenotyping, statistical analyses and computer software have enabled detection of both major and minor QTL in populations. Association between molecular marker and phenotypic variation in a population is the basis of QTL detection. Traditionally biparental mapping populations are used for the detection of QTL. Development of genome-wide association studies (GWAS) provides the opportunity of detecting QTL in unstructured populations, i.e., germplasm collections. The same QTL for heading date were detected in both bi-parental and association mapping populations in barley. However, larger population size (>384 individuals) is critical in GWAS studies (Wang et al. 2012). Recently nested association mapping population has been used for the detection of minor QTL for flowering time in maize (Buckler et al. 2009).

Most of the earlier QTL studies were based on RFLP and SSR based genetic linkage maps which are low in density and thus were not able to precisely detect all QTL associated with a trait and their genomic locations. Ultra high density SNP map constructed from next generation sequencing data detected more QTL associated with different traits and precisely positioned those in the genome (Yu et al. 2011). The high-density maps are very helpful in understanding the genetic basis of quantitative traits and for fine mapping and cloning of QTL.

Interval mapping, that uses the regression or maximum likelihood approach, is a popular method for detection of QTL (Lynch and Walsh 1997). However, if there are multiple QTL present within the marker bracket of target QTL, estimates will be confounding. Composite interval mapping (CIM) has been proposed as an alternative approach which accounts for the effects of other QTL if present within the marker bracket of target QTL (Jansen 1993). This procedure is similar to interval mapping using multiple regressions with the other markers added as cofactors in the model. Buckler et al. (2009) used joint stepwise regression and joint inclusive composite interval mapping (JICIM) for detecting QTL in the maize NAM population. JICIM were found to be effective and added 20 more minor QTL for each trait.

Mapping QTL in tall fescue is still in its infancy. Until now, only a few QTL studies have been reported. The HD28-56 × R43-64 population was used for detection QTL associated with forage digestibility. Thirteen LGs of the parental maps contributed significant QTL. However, QTL on LGs 1, 4, and 19 of HD28-56 map and LGs 15 and 18 of R43-64 map were consistently identified across years (Saha et al. 2009). Individual QTL explained up to 4.4% effect of *in vitro* dry matter digestibility. Two putative QTL associated with brown patch disease were identified in a tall fescue population developed by crossing a resistant and susceptible genotype (Bokmeyer 2009). The QTL on LG21 explained 25.6 and 32.5% of the phenotypic variation in two years of evaluation. The QTL on LG18 accounted for 23.8% of the phenotypic variation in combined analysis. A tall fescue population developed by crossing a Mediterranean (103-2) and a Continental (R43-64) genotype was used for identification of QTL associated with cold tolerance (Dierking 2010). Location and year specific as well as across environment QTL for fall growth and winter survival were identified on homologous LGs in the parental maps.

4.6.7 Marker Assisted Breeding (MAB)

Genetic improvement of quantitative traits is the target of most breeding programs. However, due to the complex genetics and low heritability, improvement of quantitative traits is difficult and time consuming especially in perennial crops. Recent advances in molecular marker technologies and phenotyping have enabled the dissection of complex traits. Desired alleles can be reliably and efficiently identified in mapping populations and transferred in target populations using MAB. Prerequisites for MAB are a robust set of informative markers and accurate phenotyping of the target traits in an effective population. Using recent advancement in genotyping technologies, e.g., SNP genotyping, genotyping by sequencing, etc., thousands of molecular markers can be generated within a short time in an inexpensive way. Breakthroughs in phenotyping instrumentation, e.g., PhenoFab from KeyGene, Li-Cor LI-6400XT portable photosynthesis system, etc., make it possible to accurately measure many phenotypes. Advances in plant populations, e.g., association mapping and nested association populations, computer software and analysis tools, make it possible to identify key alleles controlling key traits. All these advancements open up the opportunities for improving both quantitative and qualitative traits in target species through MAB.

Marker-assisted selection (MAS) can be used to improve economically important traits in tall fescue (Xu et al. 1995). However, only a few known efforts have been initiated for the genetic improvement of tall fescue through MAS. Alleles associated with high and low IVDMD were used as positive and negative markers selections, respectively (Saha et al. 2009). Plants selected with MAS were evaluated for two years at two Ardmore, Oklahoma locations. The positive marker-selected plants showed 2–3% higher and negative marker selected plants showed 1–3% lower IVDMD than the population from which they were selected. Breeding populations were developed by random matting of the plants selected with desired markers. Evaluation of these populations against the standard check cultivars are in progress at Georgia and Oklahoma locations. Molecular markers associated with stem rust resistance in tall fescue have been identified (Saha, unpubl. data). Effectiveness of these markers was evaluated in a segregating population. Co-segregation of few markers with resistance and susceptibility was observed. Plants selected with markers have been planted in field experiments to confirm their effectiveness. It is expected that utilizing the recent breakthroughs in genotyping and phenotyping technologies, more efforts will be initiated for the genetic improvement of tall fescue using MAS.

4.7 Comparative Mapping and Cross-species Amplification of Molecular Markers

Grass genomes are highly conserved (Gale and Devos 1998). Marker colinearity and synteny within the members of the *Poaceae* family have been well documented (e.g., Ahn and Tanksley 1993; Van Deynze et al. 1995; Jones et al. 2002; Alm et al. 2003; Kuleung et al. 2004). Identification of putative orthologous loci controlling agronomic traits is possible through comparative mapping (Paterson et al. 1995; Devos and Gale 1997). Information developed in model species can be used for the genetic improvement of other species. Comparative analyses recommended that ryegrass and meadow fescue were highly orthologous and co-linear to rice, oat, maize and sorghum (Jones et al. 2002; Alm et al. 2003). Three chromosome-I specific maize clones were detected on the homoeologous LG3 (3a, 3b, and 3c) in tall fescue maps (Xu et al. 1995). Evolutionary relationship between meadow and tall fescue has been demonstrated through comparative RFLP mapping (Chen et al. 1998). Eight markers that were mapped in LG1 of meadow fescue were also present in LG1 of tall fescue. Chen et al. (1998) suggested that the P genome of diploid meadow fescue diverged substantially in hexaploid tall fescue during evolution.

High levels of cross-species amplification and transferability of markers derived from coding sequences demonstrated that markers developed in one species can be effectively used in related grass species (Zeid et al. 2010). A set of 919 EST-based primers developed from seven grass species were tested in 16 cereal, turf and forage grass species. A total of 340 primers generated PCR amplicons in at least half of the species in the panel (Zeid et al. 2010). At least 43% and a maximum of 65% of tall fescue EST-SSR markers amplified across 12 grass species (Mian et al. 2005c; Saha et al. 2004). About 48% of tall fescue genomic SSR primers were amplified in rice and 34% of these markers were polymorphic in a rice mapping population (Saha et al. 2006).

Tall fescue EST-SSR markers were successfully mapped in ryegrass LGs (Wranke et al. 2004). Sequence-based comparative maps developed for rice-wheat (Sorrells et al. 2003) and sorghum-rice (Klein et al. 2003) enhanced map resolution considerably. Sequence databases are not well developed in forage grass species that includes tall fescue. Thus comparative genomic studies between tall fescue and other well studied grass species can be very useful for the genetic improvement of this important species.

4.8 Endophyte

Most tall fescue plants in wild and naturalized stands are infected with a fungus called endophyte. Fungal endophytes belong to the sexual *Epichloë* and asexual *Neotyphodium* species. Endophytes mainly localized in above ground plant parts especially in pseudo-stems. Although localized in the aboveground portion, it seems to influence the root function. Endophyte benefits the plant through increased resistance to insects, diseases, drought and other stresses (Popay and Jensen 2005). Responses to abiotic stress depend on specific host-endophyte associations (Bacon 1993). Endophyte infected plants showed improved resistance to root-knot nematodes (*Meloidogyne marylandi*). In nutrient limited soil conditions E+ plants showed better growth than the E- plants. E+ plants also showed a greater N economy (Lyons et al. 1990). Tall fescue plants infected with endophytes have many morphological and yield advantages. Shoot mass and tiller numbers often are greater in E+ than in E– plants (West et al. 1993). E– plants usually showed reduced root mass compared to E+ plants (Elmi et al. 2000). Field trials conducted in different geographic conditions showed that E+ populations yielded more and persisted longer than E- stands in Texas (Read and Camp 1986), Louisiana (Coombs et al. 1999), Arkansas (West et al. 1993) and Georgia (Bouton et al. 1993). Mechanisms of host-endophyte interactions include a range of cellular metabolism and whole-plant responses. Osmotic adjustment in the tiller growth zone was much higher in E+ than E- plants (Elmi et al. 2000). E+ plants showed greater stomatal conductance and less leaf rolling than E-plants in field evaluations (Elbersen and West 1996).

Endophyte in tall fescue is disseminated only by seed. Planting infected seeds insuring a high level of endophyte infection in the new stands. Endophytes colonize in developing seeds after anthesis. It moves from germinating seed into seedling, colonizes mainly leaf sheaths, meristems, and internodes of elongating stems. Though endophyte can persist well in plants, but can be easily lost during seed storage. Endophyte viability is lost if the seeds are stored at high temperature and high humidity (Welty et al. 1987). Seed physical properties and storage conditions are critical for endophyte viability. Seed stored on-farm at ambient conditions for a year or more can kill the endophyte inside it (Williams et al. 1984). Planting of these seeds is the reason for the development of E- pastures. Seed production from endophyte infected plants, appropriate processing, packaging and storage techniques and proper storage conditions can ensure delivering seed with viable endophyte to the end user.

Endophytes though mutualistic to the tall fescue plants have negative impacts on livestock performance and metabolism. Endophytes produce a range of bioactive alkaloids including the ergot alkaloids, peramine, lolines, lolitrems and terpendoles (Bush et al. 1997; Schardl et al. 2011). Peramine and loline have insecticidal properties thus provide defenses to their host. Livestocks grazed on endophyte infected tall fescue suffer from low weight gain, poor milk production, difficult births and gangrenous hooves, etc. Some of the alkaloids, e.g., ergot alkaloids and lolitrem B, are responsible for the harmful effects to grazing animals. The ergot alkaloids are responsible for fescue toxicosis and the lolitrem B causes ryegrass staggers (Schardl et al. 2011). Symptoms of toxicosis in livestock include reduced forage intake, reproductive inefficiencies, rough hair coat, formation of hard fat in mature cows, and loss of hooves or the tip of ears and tail. Affected livestock usually graze during cooler parts of the day and stand in water or shade during hot days. Alkaloid production by endophytes is a function of nutrient inputs (Gentry et al. 1969). Nitrogen has large effects on alkaloid production. Higher nitrogen fertilization increases alkaloids production especially perloline. Application of P and K lowers the perloline concentrations.

The two tall fescue morphotypes appear to harbor distinctly different symbiotic fungal endophytes of the *Epichloë* type (Clement et al. 2001). The morphotypes have distinct *Neotyphodium* endophyte profiles. *N. coenophialum* is commonly found in Continental tall fescue but *Neotyphodium* sp. FaTG-2 and *Neotyphodium* sp. FaTG-3 are found only in Mediterranean tall fescue (Takach et al. 2012). The *N. coenophialum* has distinct morphotype, chemotype, alkaloid production, isozyme and microsatellite markers profiles which are very different from the other two *Neotyphodium* sp. (Moon et al. 1999). Endophyte colonization pattern varies in tall fescue morphotypes. In Mediterranean tall fescue, endophyte has been detected in leaf blade but it was not readily observed in the blade tissue of the Continental type (Takach et al. 2012).

The widespread cultivation of KY31 is partly related to endophyte infection. In the past, tall fescue breeding activities took place without the knowledge of endophytes. The toxicosis problem and the role of endophytes in toxicosis largely shifted the fescue breeding objectives. In the 1980s, emphasis was paid on developing endophyte free cultivars. Planting E-tall fescue cultivars were becoming popular. However, under stress situation high risk of stand losses was observed in E-tall fescue (Christensen et al. 1997). Endophyte-free tall fescue is only a viable option where plant stress is minimized (Asay et al. 2001; Brummer and Moore 2000; Hopkins and Alison 2006). An alternative strategy is to reinfect the E-germplasms with non-toxic endophyte strains commonly known as "Novel endophyte" (Bacon and Siegel 1988). Novel endophytes do not produce toxic alkaloids but retain agronomic benefits to the grass. Techniques were developed to reinfect E-tall fescue seedlings with novel endophytes in the laboratory (Latch et al. 1985). Superior germplasms can be infected with different novel endophyte strains and superior combination can be selected. "Jesup MaxQ®" is a tall fescue cultivar with a novel endophyte (Bouton et al. 2002). It is commercially available in the US and marketed as Jesup MaxP® in Australia. "TexomaMaxQ II", a Continental-type tall fescue with novel endophyte was recently released for commercial cultivation in the USA.

Tall fescue cultivars infected with a novel endophyte persist like wild-type endophyte infected plants (Bouton et al. 2002) and at the same time leads to excellent animal weight gain (Watson et al. 2004; Hopkins and Alison 2006). Details on the use of endophytes in tall fescue cultivar development can be obtained in Bouton and Easton (2005), and Bouton and Hopkins (2003).

4.9 Transgenics

Genetic transformation has become an important and useful tool for understanding basic biological processes and development of transgenic cultivars. Through the transgenic process, transgenes with desirable characteristics are introduced into a target species so that the species can express the new character. It has been successfully implemented for the improvement of major crop species, especially corn and soybean. Herbicide and insect resistant transgenic corn has been widely cultivated in the USA. Genetic transformation has also been used for the improvement of forage and turf grass species (reviewed in Wang and Ge 2006). Genetic transformation creates unique genetic variation that could not be otherwise achieved by traditional breeding methods. Thus, genetic transformation can accelerate and/or complement conventional breeding efforts. Two methods, microprojectile bombardment or simply biolistics and *Agrobacterium*-mediated transformation, are usually practiced for the development of transgenic plants.

Genetic transformation in tall fescue was initiated in the early 1990's. The first transgenic tall fescue plant was developed with a transformed protoplast (Wang et al. 1992). Several other studies on direct transfer of genes to protoplasts have also been reported (e.g., Dalton et al. 1995; Kuai et al. 1999). Development of transgenic tall fescue using biolistic transformation of embryogenic cultures was reported by Spangenberg et al. (1995). Several other laboratories also used the protocol for the development of transgenic tall fescue plants (e.g., Cho et al. 2000; Wang et al. 2001; Chen et al. 2003). *Agrobacterium* mediated transformation protocol became popular due to its manyfold advantages over the other methods. Transgenic tall fescue plants developed through *Agrobacterium* mediated transformation was first reported in 2005 (Dong and Qu 2005; Wang and Ge 2005). Eventually, tall fescue became one of the most intensively studied monocot forage and turf grass species for transgenic research.

Transgenic tall fescue plants with improved forage and turf qualities and resistant to both biotic and abiotic factors have been developed. Protein quality of tall fescue plant was altered with the introduction of sunflower seed albumin-8 gene (Wang et al. 2001). Transgenic plants with modified expression of cinnamyl alcohol dehydrogenase (CAD) and caffeic acid O-methyltransferase (COMT) genes, involved in lignin biosynthesis, showed reduced lignin concentration and altered composition thus increased forage digestibility (Chen et al. 2002). Introduction of *ipt* gene into turf type tall fescue significantly enhanced tillering ability, increasing chlorophyll a and b content, and improving cold tolerance (Hu et al. 2005). Transgenic tall fescue with rice *Pi9* gene conferred resistance against gray leaf spot (Dong et al. 2007). Overexpression of *AtNHX1* gene improves salt tolerance in transgenic tall fescue (Tian et al. 2009). Heat and methyl viologen stress tolerance of tall fescue can be improved through overexpression of the Arabidopsis *2-Cys Prx* (Kim et al. 2010). Enhanced tolerance to abiotic stresses was observed in tall fescue plants transformed with *MsHsp23* gene (Lee et al. 2012). Transgenic tall fescue cultivars with value added traits are expected to have major benefits to the growers all over the world.

Performance of transgenic tall fescue plants were evaluated in field experiments under strict regulatory requirements (Wang et al. 2004). Pollen viability and dispersal have been evaluated. Despite the beneficial agronomic and economic benefits, regulatory issues are the major limitations for deployment of transgenics in commercial cultivation. Though transgenics in major crops (e.g., corn, soybean, cotton, and canola) have gained much popularity, deregulation and commercialization of other species seem extremely difficult. Besides, sky rocketing expenses of regulatory process make it almost impossible for commercialization of transgenic crops with low economic potential. Tall fescue is an outcrossing polyploid. It is cross compatible to many species including meadow fescue, tetraploid fescue and perennial rye grass. Pollen mediated transgene flow will be a major concern for the deregulation of tall fescue. Without substantial changes in the regulatory process and transgenic development methodologies, it is expected that transgenic tall fescue cultivars will not reach the marketplace within the next several years.

4.10 Conclusion

Tall fescue is one of the most important cool-season perennial grasses in the temperate regions of the world. However, interest on tall fescue for use in pasture cultivations went down due to the toxicosis problems associated with fungal endophytes. In the past few decades, significant improvements have been reported in endophyte genetics and genomics. Tall fescue cultivars with novel endophytes have been developed. It is expected that the glory of tall fescue will return soon. Improvement of tall fescue was impaired by the genetic complexities of important traits and long selection cycles. Recent breakthroughs in molecular biology, marker technologies and phenotyping protocols show the prospect of effective improvement of this outcrossing, polyploid perennial crop. Incorporation of all these advancements through collaborative approaches is expected to redesign tall fescue for the future world.

References

Alm V, Fang C, Busso CS, Devos KM, Vollan K, Grieg Z et al. (2003) A linkage map of meadow fescue (*Festuca pratensis* Huds.) and comparative mapping with other Poaceae species. Theor Appl Genet 108: 25–40.

Ahn SN, Tanksley SD (1993) Comparative linkage maps of the rice and maize genomes. Proc Natl Acad Sci USA 90: 7980–7984.

Asay KH, Jensen KB, Waldron BL (2001) Responses of tall fescue cultivars to an irrigation gradient. Crop Sci 41: 350–357.

Bacon CW (1993) Abiotic stress tolerance (moisture, nutrients) and photosynthesis in endophyte infected tall fescue. Agric Ecosystems Enf 44: 123–141.

Bacon CW, Siegel MR (1988) Endophyte parasitism of tall fescue. J Production Agric 1: 45–55.

Baird JH, Kopecký-Lukaszewski AJ, Green RL, Bartoš J, Dolezel J (2012) Genetic diversity of turf-type tall fescue using diversity array technology. Crop Sci 52: 408–412.

Barker RE, Pfender WF, Welty RE (2003) Selection for stem rust resistance in tall fescue and its correlated response with seed yield. Crop Sci 43: 75–79.

Bernardo R (2008) Molecular markers and selection for complex traits in plants: Learning from the last 20 years. Crop Sci 48: 1649–1664.

Bokmeyer J (2009) Inheritance Characteristics of Brown Patch Resistance in Tall Fescue. PhD Dissertation, Rutgers University, New Brunswick, NJ.

Bonos SA, Rush D, Hignight K, Meyer WA (2004) Selection for deep root production in tall fescue and perennial ryegrass. Crop Sci 44: 1770–1775.

Borrill M, Tyler BF, Lloyd-Jones M (1971) Studies in *Festuca* 1 A chromosome atlas. Bovinae and Scariosae Cytologia 36: 1–14.

Bouton JH, Easton S (2005) Endophytes in forage cultivars. In: Roberts CA et al. (eds) *Netyphodium* in Cool-season Grasses, Current Research and Applications. Black-Well Publ, Ames, IA, pp 327–340.

Bouton JH, Gates RN, Belesky DP (1993) Yield and persistence of tall fescue in the southeastern coastal plain after removal of its endophyte. Agron J 85: 52–55.

Bouton JH, Hopkins AA (2003) Commercial applications of endophytic fungi. In: White JF et al. (eds) Clavicipitalean Fungi: Evolutionary Biology, Chemistry, Biocontrol, and Cultural Impacts. Marcel Dekker, NY, pp 495–516.

Bouton JH, Latch GCM, Hill NS et al. (2002) Reinfection of tall fescue cultivars with non-ergot alkaloid producing endophytes. Agron J 94: 567–574.

Brown-Guedira GL, Thomson JA, Nelson RL, Warburton ML (2000) Evaluation of genetic diversity of soybean introductions and North American ancestors using RAPD and SSR markers. Crop Sci 40: 815–823.

Brummer EC, Bouton JH, Kochert G (1993) Development of an RFLP map in diploid alfalfa. Theor Appl Genet 86: 329–332.

Brummer EC, Moore KJ (2000) Persistence of perennial cool-season grass and legume cultivars under continuous grazing by beef cattle. Agron J 92: 466–471.

Buckler ES, Holland JB et al. (2009) The genetic architecture of Maize flowering time. Science 325: 714–718.

Buckner RC, Burrus PB (1968) Registration of Kenwell tall fescue. Crop Sci 8: 398.

Buckner RC, Powell JB, Frakes RV (1979) Historical development. In: Bush LP, Buckner RC (eds) Tall fescue Agronomy Monograph. ASA, CSSA, SSSA, Madison, pp 1–8.

Burner DM, Balasko JA, O'Brien PM (1988) Attributes of tall fescue germplasm of diverse geographic origin. Crop Sci 28: 459–462.

Burns JC, Chamblee DS (1979) Adaptation. In: Buckner RC, Bush LP (eds) Tall Fescue. ASA, CSSA, SSSA, Madison, pp 9–30.

Bush LP, Wilkinson HH, Schardl CL (1997) Bioprotective alkaloids of grass-fungal endophyte symbioses. Plant Physiol 114: 1–7.

Cahoon AB, Sharpe RM, Mysayphonh C, Thompson EJ, Ward AD, Lin A (2010) The complete chloroplast genome of tall fescue (*Lolium arundinaceum*; Poaceae) and comparison of whole plastomes from the family Poaceae. American J Botany 97: 49–58.

Casler MD, Vogel KP (1999) Accomplishments and impact from breeding for increased forage nutritional value. Crop Sci 39: 12–20.

Charmet G, Ravel C, Balfourier F (1997) Phylogenetic analysis in the *Festuca-Lolium* complex using molecular markers and ITS rDNA. Theor Appl Genet 94: 1038–1046.

Chen C, Sleper DA, Johal GS (1998) Comparative RFLP mapping of meadow and tall fescue. Theor Appl Genet 97: 255–260.

Chen C, Sleper DA, West CP (1995) RFLP and cytogenetic analysis of hybrids between *Festuca mairei* and *Lolium perenne*. Crop Sci 35: 720–725.

Chen L, Auh C, Chen F, Cheng X, Jlkoe H, Dixon RA, Wang Z-Y (2002) Lignin deposition and associated changes in anatomy, enzyme activity, gene expression and ruminal degradability in stems of tall fescue at different developmental stages. J Agric Food Chem 50: 5558–5568.

Chen L, Auh C, Dowling P, Bell J, Chen F, Hopkins A, Dixon RA, Wang Z-Y (2003) Improved forage digestibility of tall fescue (*Festuca arundinacea*) by transgenic down-regulation of cinnamyl alcohol dehydrogenase. Plant Biotechnol J 1: 437–449.

Cho YG, Ishii T, Temnykh S, Chen X, Lopovich L, McCouch SR, Park WD, Ayres N, Cartinhour S (2000) Diversity of microsatellites derived from genomic libraries and GenBank sequences in rice (*Oryza sativa* L). Theor Appl Genet 100: 713–722.

Christensen MJ, Ball OJ-P, Bennett RJ et al. (1997) Fungal and host genotype effects on compatibility and vascular colonization by *Epichloe festucae*. Mycological Res 101: 493–501.

Clement SL, Elberson LR, Youssef NN, Davitt CM, Doss RP (2001) Incidence and diversity of *Neotyphodium* fungal endophytes in tall fescue from Morocco, Tunisia, and Sardinia. Crop Sci 41: 570–576.

Cooombs DF, Joost RE, Loyacano AF, Pitman WD (1999) Endophyte-free tall fescue potential for pastures on fertile clay loam soils in central Louisiana. Professional Ani Scientist 15: 230–237.

Craven KD, Clay K, Schardl CL (2009) Systematics and morphology. In: Fribourg HA et al. (eds) Tall Fescue for the Twenty-first Century. Agron Monogr 53, ASA, CSSA, SSSA, Madison, WI, pp 11–30.

Crowder LV, Vanegas JA, Lotero JC, Michelin A (1959) The adaptation and production of species and selections of grasses and clover in Colombia. J Range Manag 12: 225–230.

Dalton SJ, Bettany AJE, Timms E, Morris P (1995) The effect of selection pressure on transformation frequency and copy number in transgenic plants of tall fescue (*Festuca arundinacea* Schreb.). Plant Sci 108: 63–70.

Darbyshire S (1993) Realignment of *Festuca* subgenus *Schedonorus* with the genus *Lolium* (Poaceae). Novon 3: 239–243.

Devos KM, Gale MD (1997) Comparative genetics in the grasses. Plant Mol Biol 35: 3–15.

Dierking RM (2010) Physiological Responses of Continental and Mediterranean Tall Fescue to Cold Stress and Identification of Underling QTL for Fall Growth and Winter Survival. PhD Dissertation, University of Missouri, MO.

Diesburg KL, Carlson IT (1983) Rhizomatous Spreading Ability and Seed Set in Wide Crosses of Tall Fescue (*Festuca arundinacea* Schreb.). Agron Abst, ASA, Madison, WI.

Dong S, Qu R (2005) High efficiency transformation of tall fescue with *Agrobacterium tumefaciens*. Plant Sci 168: 1453–1458.

Dong S, Tredway LP, Shew HD, Wang G, Sivamani E, Qu R (2007) Resistance of transgenic tall fescue to two major fungal diseases. Plant Sci 173: 501–509.

Eizenga GC, Schardl CL, Phillips TD, Sleper DA (1998) Differentiation of tall fescue monosomic lines using RFLP markers and double monosomic analysis. Crop Sci 38: 221–225.

Elbersen HW, West CP (1996) Growth and water relations of field-grown tall fescue as influenced by drought and endophyte. Grass and Forage Sci 51: 333–342.

Elmi AA, West CP (1995) Endophyte infection effects on stomatal conductance, osmotic on adjustment and drought recovery of tall fescue. New Phytol 31: 61–67.

Elmi AA, West CP, Robbins RT et al. (2000) Endophyte effects on reproduction of a root-knot nematode (*Meloidogyne marylandi*) and osmotic adjustment in tall fescue. Grass and Forage Sci 55: 166–172.

Fribourg HA, Hannaway DB, West CP (2009) Tall Fescue for the Twenty-first Century. Agron Monogr 53, ASA, CSSA, SSSA, Madison, WI.

Fribourg HA, Milne G (2009) Establishment and renovation of old sods for forage. In: Fribourg HA et al. (eds) Tall Fescue for the Twenty-first Century. Agron Monogr 53, ASA, CSSA, SSSA, Madison, WI, pp 67–84.

Gale MD, Devos KM (1998) Comparative genetics in the grasses. Proc Natl Acad Sci USA 95: 1971–1974.

Gentry CE, Chapman RA, Henson L et al. (1969) Factors affecting the alkaloid content of tall fescue (*Festuca arundinacea*) in thin layer chromatography and paper chromatography. Agron J 61: 313–316.

Gingrich GA, Mellbye ME (2004) The effect of fungicides on seed yield and economic returns in perennial ryegrass In: William C, Young III (eds) Seed Pdn Res at Oregon State Univ, USDA-ARS cooperating http: //cropandsoiloregonstateedu/seed-ext/pub/2004.

Gunter SA, Beck PA (2004) Novel endophyte-infected tall fescue for growing beef cattle. J Anim Sci 82 (E Suppl): E75–E82.

Hamilton MB, Pincus EL, DiR-Fiore A, Fleischer RC (1999) A universal linker and ligation procedures for construction of genomic DNA libraries enriched for microsatellites. Biotechniques 27: 500–507.

Hand ML, Cogan NOI, Forster JW (2012) Molecular characterization and interpretation of genetic diversity within globally distributed germplasm collections of tall fescue (*Festuca arundinacea* Schreb) and meadow fescue (*F. pratensis* Huds.) Theor Appl Genet 124: 1127–1137.

Hand ML, Cogan NO, Stewart AV, Forster JW (2010) Evolutionary history of tall fescue morphotypes inferred from molecular phylogenetics of the *Lolium-Festuca* species complex. BMC Evol Biol 10: 303 doi: 101186/1471-2148-10-303.

Hemmat M, Weeden NF, Manganaris AG, Lawson DM (1994) A molecular marker linkage map for apple. J Heredity 85: 4–11.

Henson J, Safley M (2009) NRCS plant information and conservation practice standards. In: Fribourg HA et al. (eds) Tall Fescue for the Twenty-first Century. Agron Monogr 53, ASA-CSSA-SSSA, Madison, WI.

Hopkins AA (2005) Grazing tolerance of cool-season grasses planted as seeded sward plots and spaced plants. Crop Sci 45: 155–1564.

Hopkins AA, Alison MW (2006) Stand persistence and animal performance for tall fescue endophyte combinations in the south central USA. Agron J 98: 1221–1226.

Hopkins AA, Saha MC, Wang Z-Y (2009) Breeding, genetics, and cultivars. In: Fribourg HA et al. (eds) Tall Fescue for the Twenty-first Century. Agron Monogr 53, ASA, CSSA, SSSA, Madison, WI, pp 339–366.

Hu Y, Jia W, Wang J, Zhang Y, Yang L, Lin Z (2005) Transgenic tall fescue containing the *Agrobacterium tumefaciens ipt* gene shows enhanced cold tolerance. Plant Cell Rep 23: 705–709.

Huff DR (1997) RAPD Characterization of heterogenous perennial ryegrass cultivars. Crop Sci 37: 557–564.

Humphreys MW, Thomas HM, Morgan WG, Meredith MR, Harper JA, Thomas H, Zwierzykowski Z, Ghesquiere M (1995) Discriminating the ancestral progenitors of hexaploid *Festuca arundinacea* using genomic in situ hybridization. Heredity 75: 171–174.

Hunt KL, Sleper DA (1981) Fertility of hybrids between two geographic races of tall fescue. Crop Sci 21: 400–404.

Inda LA, Segarra-Moragues JG, Müller J, Peterson PM, Catalán P (2008) Dated historical biogeography of the temperate Loliinae (Poaceae, Pooideae) grasses in the northern and southern hemispheres. Mol Phylogenetics and Evol 46: 932–957.

Jansen RC (1993) Interval mapping of multiple quantitative trait loci. Genetics 135: 205–211.

Jansen RK, Kaittanis C, Saski C, Lee SB, Tomkins J, Alverson AJ, Daniell H (2006) Phylogenetic analyses of *Vitis* (Vitaceae) based on complete chloroplast genome sequences: Effects of taxon sampling and phylogenetic methods on resolving relationships among rosids. BMC Evolutionary Biology 6: 32.

Jauhar PP (1991) Recent cytogenetics of the *Festuca-Lolium* complex. In: Tsuchiya T, Gupta PK (eds) Chromosome Engineering in Plants: Genetics, Breeding, Evolution, vol 2B, Elsevier Science Publishers, Amsterdam, pp 325–362.

Jernstedt JA, Bouton JH (1985) Anatomy, morphology, and growth of tall fescue rhizomes. Crop Sci 25: 539–542.

Johnson RC, Yangyang L (1999) Water relations, forage production, and photosynthesis in tall fescue divergently selected for carbon isotope discrimination. Crop Sci 39: 1663–1670.

Jones ES, Mahoney NL, Hayword MD, Armstead IP, Jones JG, Humphreys MO, King IP, Kishida T, Yamada T, Balfourier F, Charmet G, Forster JW (2002) An enhanced molecular marker based genetic map of perennial ryegrass (*Lolium perenne*) reveals comparative relationships with other Poaceae genomes. Genome 45: 282–295.

Kim K-H, Alam I, Lee K-W, Sharmin SA, Kwak S-S, Lee SY, Lee B-H (2010) Enhanced tolerance of transgenic tall fescue plants overexpressing 2-Cys perosiredoxin against methyl viologen and heat stress. Biotechnology Letters 32: 571–576.

Klein PE, Klein RR, Vrebalov J, Mullet JE (2003) Sequence-based alignment of sorghum chromosome 3 and rice chromosome 1 reveals extensive conservation of gene order and one major chromosomal rearrangement. Plant J 34: 605–621.

Kopecký D, Bartoš J, Lukaszewski AJ, Baird JH, Černoch V, Kölliker R, Rognli OA, Blois H, Caig V, Lübberstedt T, Studer B, Shaw P, Dolezel J, Kilian A (2009) Development and mapping of DArT markers within the Festuca—Lolium complex. BMC Genomics 10: 473.

Kuai B, Dalton SJ, Bethany AJE, Morris P (1999) Regeneration of fertile transgenic tall fescue plants with a stable highly expressed foreign gene. Plant Cell Tissue Organ Ult 58: 149–154.

Kuleung C, Baenziger PS, Dweikat I (2004) Transferability of SSR markers among wheat, rye, and triticale. Theor Appl Genet 108: 1147–1150.

Latcj GCM, Christensen MJ, Gaynor DL (1985) Aphid detection of endophyte infection in tall fescue (*Festuca arundinacea*). NZ J Agroc Res 28: 129–132.

Lee K-W, Choi GJ, Kim K-Y, Ji HJ, Park HS, Kim Y-G, Lee BH, Lee S-H (2012) Transgenic expression of *MSHsp23* confers enhanced tolerance to abiotic stresses in tall fescue. Asian-Aust J Anim Sci 25: 818–823.

Lewis EJ, Humphreys MW, Caton MP (1980) Disomic inheritance in *Festuca arundinacea* Schreb. Z Pflanzenzucht 84: 335–341.

Loureiro J, Kopecký D, Castro S, Santos C, Silveira P (2007) Flow cytometric and cytogenetic analyses of Iberian Peninsula *Festuca* spp. Plant Systematics and Evol 269: 89–105.

Lynch M, Walsh B (1997) Genetics and Analysis of Quantitative Traits. Sinauer Associates Inc., Sunderland, USA.

Lyons PC, Evans JJ, Bacon CW (1990) Effects of the fungal endophyte *Acremonium coenophialum* on nitrogen accumulation and metabolism in tall fescue. Plant Phys 92: 726–732.

Maliepaard C, Jansen J, van Ooijen JW (1997) Linkage analysis in a full-sib family of an outbreeding plant species: overview and consequences for applications. Genet Res 70: 237–250.

Mian MAR, Hopkins AA, Zwonitzer JC (2002) Determination of genetic diversity in tall fescue with AFLP markers. Crop Sci 42: 944–950.

Mian MAR, Saha MC, Hopkins AA, Wang Z (2005c) Use of tall fescue EST-SSR markers in phylogenetic analysis of cool-season forage grasses. Genome 48: 637–647.

Mian MAR, Zwonitzer JC, Chen Y, Saha MC, Hopkins AA (2005a) AFLP diversity within and among hardinggrass populations. Crop Sci 45: 2591–2597.

Mian MAR, Zwonitzer JC, Hopkins AA, Ding XS, Nelson RS (2005b) Response of tall fescue genotypes to a new strain of Brome mosaic virus. Plant Dis 89: 224–227.

Moon CD, Tapper BA, Scott B (1999) Identification of *Epichloë* endophytesin planta by a microsatellite-based PCR fingerprinting assay with automated analysis. Appl Environ Microbiol 65: 1268–1279.

Nelson CJ, Sleper DA (1983) Using leaf-area expansion rate to improve yield of tall fescue. In: Proc XIV Intern Grassld Cong, Lexington, KY, pp 413– 416.

Pedersen JF, Sleper DA (1988) Considerations in breeding endophyte-free tall fescue forage cultivars. J Prod Agric 1: 127–132.

Paterson AH, Lin YR, Li S, Schertz KF, Doebley JF, Pinson SRM, Liu SC, Stansel JW, Irvine JE (1995) Convergent domestication of cereal crops by independent mutations at corresponding genetic loci. Sci 269: 1714–1717.

Popay AJ, Jensen JG (2005) Soil biota associated with endophyte-infected tall fescue in the field. NZ Plant Protection 58: 117–121.

Pscheidt JW (1996) Pacific Northwest Plant Disease Control Handbook. Oregon State Univ Ext Serv, Corvallis, OR.

Pscheidt JW, Ocamb CM (2001) Pacific Northwest Plant Disease Control Handbook Oregon. State Univ Ext Serv, Corvallis, OR.

Read JC, Camp BJ (1986) The effect of the fungal endophyte *Acromonium coenophialum* in tall fescue *Festuca arundinacea* on animal performance toxicity and stand maintenance. Agron J 78: 848–850.

Reeder LR Jr, Nguyen HT, Sleper DA, Brown JR (1986) Genetic variability of mineral concentration in tall fescue grown under controlled conditions. Crop Sci 26: 514–518.

Ritter E, Gebhardt C, Salamini F (1990) Estimation of recombination frequencies and construction of RFLP linkage maps in plants from crosses between heterozygous parents. Genetics 135: 645–654.

Röder MS, Korzun V, Wendehake K, Plaschke J, Tixier M-H, Leroy P, Ganal MW (1998) A microsatellite map of wheat. Genetics 149: 2007–2023.

Saha MC, Cooper JD, Mian MAR, Chekhovskiy K, May GD (2006) Tall fescue genomic SSR markers: development and transferability across multiple grass species. Theor Appl Genet 113: 1449–1458.

Saha MC, Kirigwi F, Chekhovskiy K, Black J, Hopkins A (2009) Molecular mapping of QTLs associated with important forage traits in tall fescue. In: Yamada T, Spangenberg G (eds). Molecular Breeding of Forage and Turf. Springer Science+Business Media, LLC, NY, pp 251–257.

Saha MC, Mian MAR, Eujayl I, Zwonitzer JC,Wang L, May GD (2004) Tall fescue EST-SSR markers with transferability across several grass species. Theor Appl Genet 109: 783–792.

Saha MC, Mian MAR, Zwonitzer JC, Chekhovskiy K, Hopkins AA (2005) An SSR- and AFLP-based genetic linkage map of tall fescue (*Festuca arundinacea* Schreb.). Theor Appl Genet 110: 323–336.

Seal AG (1983) DNA variation in Festuca. Heredity 50: 225–236.

Schardl CL, Young CA, Faulkne,r JR, Florea S, Pan J (2011) Chemotypic diversity of epichloae fungal symbionts of grasses. Fungal Ecol 5: 331–344.

Sleper DA (1985) Breeding tall fescue. J Plant Breed Rev 3: 313–342.

Sorrells ME, La Rota M, Bermudez-Kandianis CE et al. (2003) Comparative DNA sequence analysis of wheat and rice genomes. Genome Res 13: 1818–27.

Spangenberg G, Wang Z-Y, Wu XL, Nagel J, Iglesias VA, Potrykus I (1995) Transgenic tall fescue (*Festuca arundinacea*) and red fescue (*F. rubra*) plants from microprojectile bombardment of embryogenic suspension cells. J Plant Physiol 145: 693–701.

Takach JE, Mittal S, Swoboda GA et al. (2012) Genotypic and chemotypic diversity of neotyphodium endophytes in tall fescue from Greece. App Env Microbiol 78: 5501–5510.

Tian L, Huang C, Yu R, Liang R, Li Z, Zhang L, Wang Y, Zhang X, Wu Z (2009) Overexpression of *AtNHX1* confers salt-tolerance of transgenic tall fescue. African J Biotech 5: 1041–1044.

Torbert HA, Edwards JH, Pedersen JF (1990) Fescues with large roots are drought tolerant. Appl Agric Res 5: 181–187.

Tredway LP, Stevenson KL, Burpee LL (2003) Components of resistance to *Magnaporthe grisea* in 'Coyote' and 'Coronado' tall fescue. Plant Dis 87: 906–912.

Ubi BE, Kolliker R, Fujimori M, Komatsu T (2003) Genetic diversity in diploid cultivars of rhodesgrass determined on the basis of amplified fragment length polymorphism markers. Crop Sci 43: 1516–1522.

Van Deynze AE, Ducovsky J, Gill KS, Nelson JC, Sorrells ME, Dvorak J, Gill BS, Lagudah ES, McCouch SR, Apples R (1995) Molecular-genetic maps for chromosome 1 in *Triticeae* species and their relation to chromosomes in rice and oats. Genome 38: 45–59.

van Santen E (1992) Animal preference of tall fescue during reproductive growth in the spring. Agron J 84: 979–982.

Wang GL, Mackill DJ, Bonman JM, McCouch SR, Champoux MC, Nelson RJ (1994) RFLP mapping of genes conferring complete and partial resistance to blast in a durably resistant rice cultivar. Genetics 136: 1421–1434.

Wang H, Smith KP, Combs E, Blake T, Horsley RD, Muehlbauer GJ (2012) Effect of population size and unbalanced data sets on QTL detection using genome-wide association mapping in barley breeding germplasm. Theor Appl Genet 124: 111–124.

Wang JP, Bughrara SS, Sleper DA (2003) Genome introgression of *Festuca mairei* into *Lolium perenne* detected by SSR and RAPD Markers. Crop Sci 43: 2154–2161.

Wang Z-Y, Ge Y (2005) Agrobacterium-mediated high efficiency transformation of tall fescue (*Festuca arundinacea* Schreb). J Plant Physiol 162: 103–113.

Wang Z-Y, Ge Y (2006) Recent advances in genetic transformation of forage and turf grasses. In Vitro Cell Dev Biol Plant 42: 1–18.

Wang Z-Y, Ge YX, Scott M, Spangenberg G (2004) Viability and longevity of pollen from transgenic and non-transgenic tall fescue (*Festuca arundinacea*) (Poaceae) plants. Am J Bot 91: 523–530.

Wang Z-Y, Takamizo T, Iglesias VA, Osusky M et al. (1992) Transgenic plants of tall fescue (*Festuca arundinacea* Schreb) obtained by direct gene transfer to protoplasts. BioTechnology 10: 691–696.

Wang Z-Y, Ye XD, Nagel J, Potrykus I, Spengenberg G (2001) Expression of a sulphur-rich sunflower albumin gene in transgenic tall fescue (*Festuca arundinacea* Schreb) plants. Plant Cell Rep 20: 213–219.

Warnke SE, Barker RE, Jung G, Sim S-C, Mian MAR, Saha MC, Brilman LA, Dupal MP, Forster JW (2004) Genetic linkage mapping of an annual-perennial ryegrass population. Theor Appl Genet 109: 294–304.

Watson CE Jr, McLean SD (1991) Response to divergent selection for anthesis date in tall fescue. Crop Sci 31: 422–424.

Watson RH, McCann MA, Parish JA et al. (2004) Productivity of cow-calf pairs grazing tall fescue pastures infected with either the wild-type endophyte or a nonergot alkaloid producing endophyte strain, AR542. J Animal Sci 82: 3388–3393.

West CP, Izekor E, Turner KE et al. (1993) Endophyte effects on growth and persistence of tall fescue along a water-supply gradient. Agron J 85: 264–270.

Welty RE, Azevedo MD, Cooper TM (1987) Influence of moisture content temperature and length of storage on seed germination and survival of endophytic fungi in seeds of tall fescue and perennial ryegrass. Phytopath 77: 893–900.

Welty RE, Mellbye ME (1989) *Puccinia graminis* subsp *graminicola* indentified on tall fescue in Oregon. Plant Dis 73: 775.

Williams MJ, Backman PA, Clark EM et al. (1984) Seed treatments for control of the tall fescue endophyte *Acremonium coenophialum*. Plant Dis 68: 49–52.

Xu WW, Sleper DA, Hoisington DA (1991) A survey of restriction fragment length polymorphisms in tall fescue and its relatives. Genome 34: 686–692.

Xu WW, Sleper DA (1994) Phylogeny of tall fescue and related species using RFLPs. Thero Appl Genet 88: 685–690.

Xu WW, Sleper DA, Krause GF (1994) Genetic diversity of tall fescue germplasm based on RFLPs. Crop Sci 34: 246–252.

Xu WW, Sleper DA, Chao S (1995) Genome mapping of tall fescue *(Festuca arundinacea* Schreb.) with RFLP markers. Theor Appl Genet 91: 947–955.

Yu H, Xie W, Wang J, Xing Y, Xu C et al. (2011) Gains in QTL detection using an ultra-high density SNP map based on population sequencing relative to traditional RFLP/SSR markers. PLosONE 6: e17595 doi: 101371/journalpone0017595.

Zarrough KM, Nelson CJ, Coutts JH (1983) Relationship between tillering and forage yield of tall fescue. I Yield. Crop Sci 23: 333–337.

Zeid M, Yu JK, Goldowitz I, Denton ME, Costich DE, Jayasuriya CT, Saha MC, Elshire R, Benscher D, Breseghello F, Munkvold J, Varshney RK, Belay G, Sorrells ME (2010) Cross-amplification of EST-derived markers among 16 grass species. Field Crops Res 118: 28–35.

Zhang Y, Mian MAR, Chekhovskiy K, So S, Kupfer D, Lai H, Roe BA (2005) Differential gene expression in *Festuca* under heat stress conditions. J Expt Bot 56: 897–907.

5

Meadow Fescue

Kovi Mallikarjuna Rao and Odd Arne Rognli*

ABSTRACT

Meadow fescue (*F. pratensis* Huds. [=syn *Lolium pratense* (Huds.) Darbysh.]) is a diploid ($2n =14$) outbreeding species that belongs to the genus *Festuca*, together with *Lolium* the most important genera of forage grasses in temperate regions. Meadow fescue is a forage grass species with high quality dry matter yields, good winter survival and persistency, and is suitable both for frequent-cutting conservation regimes and grazing. Genetic variation within meadow fescue seems to be rather low which might be due to extensive gene flow, restricted diversity due to bottle-necks with subsequent radiation from a few refugia, gene flow from a few sown broad-based cultivars to natural populations, or most probably a combination of these factors. The *Lolium-Festuca* species complex is unique since it is possible to combine the genomes in different Festulolium hybrid combinations. Fescues in general have evolved superior adaptations to abiotic stresses, e.g., winter survival in *F. pratensis*. *Lolium* species are known for superior nutritive quality, rapid establishment and growth, but lack persistency under harsh environmental conditions. Complementation of traits in Festulolium hybrids is thus a very interesting strategy for developing novel germplasm and cultivars with improved quality and persistency, which can contribute in making fodder production more economically and environmentally sustainable. Relatively modest genomic resources

Department of Plant and Environmental Sciences/CIGENE, Norwegian University of Life Sciences, NO-1432 Ås, Norway.
*Corresponding author: *odd-arne.rognli@umb.no*

have been developed for meadow fescue compared with other grasses, e.g., *L. perenne*. However, this will likely be improved in the near future with application of the new sequencing technologies, combined with an efficient utilization of the close relationship with *Lolium* spp. through comparative genomics approaches.

Key words: Meadow fescue, *Festuca pratensis* Huds., *Lolium-Festuca* complex, Abiotic and Biotic stresses

5.1 Basic Information on *Festuca pratensis* Huds.

5.1.1 Brief History of F. pratensis Huds.

Natural and managed (semi-natural) grasslands are of great ecological and agricultural importance worldwide. In humid temperate regions, permanent pastures and meadows are multi-species plant communities, often dominated by graminaceous plant species. These grasslands evolved over thousands of years as a result of anthropogenic influence and would, for the most part, revert to forest without management. In many parts of western Europe, semi-natural grassland has greatly declined during the last part of the 20th century due to intensification of agriculture with subsequent conversion into arable land or forest after abandonment of grazing and mowing (Rognli et al. 2013). Meadow fescue (*Festuca pratensis* Huds.) is a significant component of species-rich permanent pastures and meadows in temperate regions with cool climates, ensuring high forage yield under harsh climatic conditions where other productive forage grass species are unable to grow.

Meadow fescue is considered native to Europe and Eurasia (Hultén and Fries 1986). It is distributed throughout the climatic regions of oceanic North West Europe and transitional oceanic/continental zone of central Europe (Borrill et al. 1976). It constitutes a significant proportion of the species composition of rich permanent pastures and hay fields in eastern Europe and in the alpine regions of Europe. Studies of the species composition of grasslands in Switzerland in the late 20th century found proportions of *F. pratensis* between 9 and 36% in fertilized meadows and pastures (Stebler and Schröter 1887; ref in Kölliker 1998). Under mild climatic conditions, other high quality grasses, e.g., *Lolium* spp. replace meadow fescue but these species are not adapted to harsher climates with lower temperatures and long-lasting snow cover. Meadow fescue was probably introduced to Scandinavia from Europe and West Asia, and has since become naturalized. The species was also introduced to North America, Australia, New Zealand and Japan.

5.1.2 Academic and Economic Importance

Meadow fescue (*F. pratensis* Huds.) is a forage grass species with high quality dry matter yields, good winter survival and is suitable both for frequent-cutting conservation regimes and grazing (Casler and van Santen 2001). The productivity, persistency and nutritive values of meadow fescue appear to be closely related with grazing frequency and intensity. A study by Brink et al. (2009) examined the effects of frequency and intensity by implementing two frequencies of grazing at two residual sward heights. A residual sward height is a measurement that determines the amount of vegetative matter which remains above ground after grazing. Compared to tall fescue (*F. arundinacea* Schreb.) and orchard grass (*Dactylis glomerata* L.) meadow fescue has lower annual yields in terms of overall biomass productivity. Persistency is an important attribute in grasses that are grown in regions subjected to adverse growing conditions such as cold winters and short, rather warm summers. Although meadow fescue is less productive than tall fescue and orchard grass, its persistency is equal or superior to orchard grass meaning that it has the potential to replace orchard grass in North American pastures (Brink et al. 2009). Crude protein did not vary among cultivars of meadow fescue, tall fescue and orchard grass, but meadow fescue had superior digestibility of Neutral Detergent Fiber (NDFD) (Brink et al. 2010). NDFD is positively correlated with dry matter intake and milk production in dairy cattle.

The cultivars of meadow fescue that have resulted from recent breeding programs have promising potential to be included in temperate grassland production systems. In northern Europe meadow fescue is rarely sown in pure stand; most often multi-species seed mixtures containing meadow fescue, timothy (*Phleum pratense* L.) and red clover (*Trifolium pratense* L.) as the most important constituents are sown to establish short-term leys. The greatest and arguably most important attribute that meadow fescue provides is its superior nutritive value that is positively associated with animal performance, and its superior winter survival. These attributes make meadow fescue a good alternative to typical grasses that are currently utilized in temperate regions such as in North America (Casler et al. 1998). Even in Japan, which has a severe winter climate, meadow fescue is promising as a grazing species and breeders recently developed the winter-hardy and high-yielding cultivar "Makibasa kae" (Yamada 2011).

5.1.3 Botanical Description

F. pratensis Huds. [=syn *Lolium pratense* (Huds.) Darbysh.] (Darbyshire 1993) is a diploid (2*n* =14) out-breeding species that belongs to the genus *Festuca*, together with *Lolium* the most important genera of forage grasses in temperate regions. *F. pratensis* is closely related to *Lolium* spp. (Jauhar 1993) and it was suggested that all outbred *Lolium* species and *F. pratensis* diverged from a common diploid ancestor (Charmet et al. 1997). However, *Festuca* and *Lolium* plants are clearly differentiated by the shape of the inflorescence (*Festuca* forms a panicle; *Lolium* forms an ear). The average estimated DNA content of the unreplicated haploid genome of *F. pratensis* is 3.55 ± 1.33 pg (Smarda et al. 2008; Bennett and Leitch 2012), corresponding to a genome size of about 3.47 Gb. Tetraploid *F. pratensis* plants can be produced by means of colchicine treatments and Kostoff (1949) found that tetraploid plants showed poor fertility. Simonsen (1975) found that tetraploid plants from a population open pollinated for two generations had equal or higher seed set that diploids from the source population. Induced tetraploidy is routinely used to develop *Festulolium* introgression lines by crossing *F. pratensis* ($4x$) x *Lolium perenne* ($2x$) and backcrossing the triploid hybrid to *L. perenne* (Ghesquière et al. 2010). A tetraploid cytotype, *F. pratensis* var. *apennina* (De Not.) ($2n = 4x = 28$), is found at higher altitudes (above 1300 m a.s.l.) in the Alps and is well-adapted to the harsh conditions (Jauhar 1993).

The variation present in cpDNA of meadow fescue is very restricted, only three haplotypes demonstrating distinct geographical structuring were found among meadow fescue genotypes representing the present distribution area of the species (Fjellheim et al. 2006). The putative origin of these haplotypes points toward migration of meadow fescue from alpine regions of Iberia, the Alps and the Caucasus after the last glaciation. Lower cpDNA variation in meadow fescue than in polyploidy fescues, e.g., tall fescue that derives from it, indicates that meadow fescue went through a bottleneck during or after the last glaciation. Today, *F. pratensis* is widespread in the open agricultural landscape but otherwise appears confined to naturally open habitats such as river banks and wetlands, and its populations may have been decimated when dense forests dominated in the previous interglacial (Fjellheim et al. 2006). cpDNA variability in *L. perenne* seems to be extensive (Balfourier et al. 2000; McGrath et al. 2007), and surprising considering the close relationship between meadow fescue and perennial ryegrass. This indicates that *L. perenne* is a younger species which has not been affected by bottlenecks. Balfourier et al. (2000) concluded that the geographic structuring of cpDNA variation in *L. perenne* pointed towards migration of from the Fertile Crescent after the last glaciation, similar to the major cereal species.

5.2 Diversity Analysis

Ecological factors and agricultural practices have created a vast biodiversity that can only be conserved by protecting the habitats and using management methods close to those that created the diversity (Peeters 2004). The genus

Festuca comprises about 600 species and is the most species rich and broadly distributed genus of grasses adapted to a range of habitats both natural and cultivated (Smarda et al. 2008). Various *Festuca* species are main components of natural (permanent), semi-natural, naturalized and improved grasslands. Genetic diversity can be conserved either *ex situ* or *in situ*. *Ex situ* conservation store germplasm collections in genebanks which are intended to represent the genetic diversity. It is effective for protecting the species or a relatively small number of threatened cultivars or ecotypes. In grass breeding stations, superior genotypes are often maintained as clones by vegetative propagation. However, it is very expensive to maintain the vast genetic diversity often characteristic of a species, especially when diversity has to be maintained by vegetative propagation (Peeters 2004). *In situ* conservation maintains ecotypes in their natural habitats, e.g., by traditional management of permanent grasslands which allow evolutionary adaptation (Frankel et al. 1995; Maxted et al. 1997). Biodiversity in grasslands is threatened by intensification of forage production due to the introduction of fertilizers and resowing with improved cultivars which happened in the last part of the 20th century (Brown 1992; Tscharntke et al. 2005). It is estimated that as much as 90% of the species rich and diverse semi-natural grasslands has been lost in Europe (Hansson and Fogelfors 2000), especially in lowland areas. Although a large number of accessions of cultivars, wild- and semi-wild populations of forage grass species (for Festuca see Rognli et al. 2010) are stored in several genebanks around the world, there are major challenges as regards securing long-term storage and rejuvenation without creating genetic drift. Taken together this has led to the establishment of several *in situ* projects for active conservation of selected semi-natural pastures and meadows, and on-farm projects for conservation of cultivars through active use. An example of on-farm conservation is a project in Norway aiming at conservation and development of new cultivars of timothy, meadow fescue and red clover (Daugstad 2012).

Large variation for traits of adaptive significance such as heading time growth habit, or resistance to various diseases are seen in permanent grasslands which harbor forage grass ecotype populations that are adapted to individual macro- and micro-habitats (Wilkins 1991; Fjellheim et al. 2007). This important variation has been used as the main source of genetic variation for forage crop breeding since its beginning in Europe in the early 20th century (Humphreys et al. 2005). With the growing interest in the use of meadow fescue, detailed genetic and morphological characterization of potentially useful germplasm is essential. Several investigations of molecular and phenotypic diversity of meadow fescue ecotypes and genebank accessions have been reported (Kölliker et al. 1998, 1999; Casler and van Santen 2000; Fjellheim and Rognli 2005a,b; Fjellheim et al. 2007; Peter-Schmid et al. 2008a,b; Hand et al. 2012). In general the studies show that most of the molecular variation exists within populations (80–90%), thus there is very little differentiation between populations. This is expected for obligate outbreeding species like meadow fescue with strong gametophytic self-incompatibility. Rather restricted genetic variation among ecotypes of meadow fescue might be caused by extensive gene flow, restricted diversity due to bottle-necks with subsequent radiation from a few refugia (Fjellheim et al. 2006), gene flow from a few sown broad-based cultivars (naturalized grasslands), or most probably a combination of these factors. Kölliker et al. (1998) found that fertilization and frequent defoliation led to a reduction in genetic variability within natural meadow fescue populations in Switzerland, and molecular and phenotypic diversity were considerably lower within cultivars of meadow fescue compared to cultivars of *L. perenne* and *D. glomerata* (Kölliker et al. 1999). Molecular diversity analysis by amplified fragment length polymorphism (AFLP) of local populations from Norway show that they are structured into three groups, western, southern, and inland, probably reflecting different routes of introduction of the species into Norway (Fjellheim and Rognli 2005a). Similar geographic structuring was also found for 12 Swiss ecotype populations using simple sequence repeat (SSR) markers (Peter-Schmid et al. 2008b). The Norwegian inland populations are closely related to the cultivars and have most probably been established as a result of migration from sown meadows (Fjellheim and Rognli 2005a). The analyses of Nordic meadow fescue local populations and cultivars found little variation between local populations and cultivars, and the level of variation within cultivars was even higher than within local populations (Fjellheim and Rognli 2005b). Fjellheim et al. (2007) found larger phenotypic diversity for a number of traits within local meadow fescue populations than within cultivars. In a comparison of diverse Swiss ecotypes with cultivars several ecotype populations were superior to cultivars in important phenotypic traits (Peter-Schmid et al. 2008b). This demonstrates the value of natural habitats as reservoirs of genetic resources for breeding. Such grassland genetic resources might need to be conserved using an *in situ* conservation strategy.

5.3 Molecular Linkage Maps: Strategies, Resources and Achievements

Developing efficient molecular marker resources are crucial for building linkage maps and further utilizing them in marker assisted breeding (MAB). Molecular DNA markers based on Southern hybridization such as RFLP (restriction fragment length polymorphism) as well as PCR-based markers such as RAPD (random amplified polymorphic DNA), AFLP, and SSRs have been developed for grass species. Yamada and Kishida (2003) applied rice cDNA-RFLP probes to forage grasses in order to investigate genetic variation within and between varieties of grasses and to identify variety-specific RFLP markers for use in breeding programs exploiting intergeneric hybridization of *Lolium* and *Festuca*. RFLP analysis is a highly labor-intensive methodology compared to the PCR-based methods. At the Noble Foundation in USA 1,800 EST-SSR primers from tall fescue were developed and used for genetic diversity analysis, construction of genetic linkage maps, quantitative trait locus (QTL) mapping, marker-assisted breeding, and tested for cross-amplication in several grass species (Saha et al. 2004). Momotaz et al. (2004) analyzed the genetic polymorphism of multiple genotypes derived from taxa of the *Lolium/Festuca* complex using these distinct sets of SSR markers and applied these data to investigate introgression and genetic relatedness in *Festulolium* accessions. Recently, Tamura et al. (2009) developed the intron-flanking EST markers for the genetic analysis and molecular breeding of *Lolium*, *Festuca*, and their intergeneric hybrid, Festulolium. Primer sets designed from *Lolium/Festuca* ESTs showed high similarity to unique rice genes and were used to amplify insertion-deletion (indel) type markers and cleaved amplified polymorphic sequence (CAPS) markers that could distinguish between *Lolium perenne* and *Festuca pratensis*. Thus, comparative genomic approaches can be used to develop markers in the related species.

A diversity arrays technology (DArT) array for five grass species, *F. pratensis*, *F. arundinacea*, *F. glaucescens*, *L. perenne*, and *L. multiflorum*, has been developed by Kopecky et al. (2009). The DArTFest array contains 7,680 probes derived from methyl-filtered genomic representations of the five species. In a first marker discovery experiment using about 40 genotypes of each species, 3,884 polymorphic markers were detected, varying from 821 to 1,852 for each single genotype. The DArTFest array could facilitate the development of genetic maps in *Festuca* and *Lolium*, analyses of genetic diversity, and monitoring of the genomic constitution of *Festuca* × *Lolium* hybrids. To test the usefulness of DArTFest array for physical mapping, DArT markers have been physically mapped to each of the seven chromosomes of *F. pratensis* using monosomic and disomic chromosome substitution lines of *F. pratensis* into *L. multiflorum* (Kopecký et al. 2009), and to chromosome bins on these chromosomes. Kopecký et al. (2011) also showed how hybrid genome constitution could be studied at a resolution which has never been achieved before with other markers or *in situ* hybridization techniques, i.e., GISH and FISH. They detected a minimum number of polymorphic DArT markers which discriminated between the parental *Lolium* and *Festuca* genome contribution to five Festulolium cultivars. Bartos et al. (2011) mapped 149 DArT markers on the meadow fescue genetic map of which 20 markers were also placed on a *L. multiflorum* map. Comparative analysis was performed between

L. multiflorum, rice and *Brachypodium* using the sequenced DArT markers. They also identified 96 DArT markers associated with freezing tolerance in a Festulolium population (FuRs0357), and found genomic loci which co-localized with chromosome segments and QTLs previously shown to be associated with freezing tolerance. This work clearly demonstrates the potential of annotated DArTFest array genomic resources in genetic studies of the *Festuca-Lolium* complex and comparative genomics with model grass species.

In meadow fescue, RFLP markers were developed from a genomic DNA library and used in combination with AFLP and a few isozyme markers to develop the *F. pratensis* linkage map (Alm et al. 2003). The genetic linkage map of meadow fescue was constructed by using a full-sib family of a cross between a genotype from a Norwegian population (HF2) and a genotype from a Yugoslavian cultivar (B14) (Alm et al. 2003). The two-way pseudo-testcross procedure was used to develop separate maps for each parent, as well as a combined map. The combined map consisted of 466 markers, RFLPs and AFLPs, and a few isozymes and SSRs, with a total length of 658.8 cM with an average marker density of 1.4 cM/marker. A high degree of orthology and colinearity was observed between meadow fescue and the Triticeae genome(s) for all linkage groups, and the authors proposed to designate the individual linkage groups 1F–7F in accordance with the orthologous Triticeae chromosomes. As expected, the meadow fescue linkage groups were highly orthologous and co-linear with *Lolium*. Studies of chromosomal rearrangements relative to Triticeae and rice showed that the meadow fescue genome has a more ancestral configuration than any of the Triticeae genomes. This is especially evident in chromosome 4F which is completely orthologous to rice chromosome 3 in contrast to the Triticeae where this rice chromosome is distributed over homoeologous groups 4 and 5 chromosomes.

Currently SSR markers maybe the best marker system due to their abundance, ubiquitous distribution in plant genomes, high level of reproducibility, ease of PCR-based analysis, and detection of co-dominant multiallelic loci; this makes them especially attractive in making combined maps in outbreeding species where two-way pseudo testcross mapping populations are commonly employed. However, due to the rapid development of cost-effective sequencing technologies, the focus today is on developing and application of single nucleotide polymorphism (SNP) markers which can be developed in several ways and genotyped using a range of platforms. Sequencing (Roche 454 GS-Flex) of cDNA from cold acclimated crown tissues of the two *F. pratensis* mapping parents (B14/16 and HF2/7) has identified about 15,000 SNPs which is an important marker resource. Alignment with EST sequences of *L. perenne* makes it possible to develop high-density species specific markers for high-resolution mapping of the genome constitution of Festulolium hybrids (Sandve et al. unpubl. results).

5.4 Molecular Mapping of Complex Traits

Heading time, freezing tolerance, winter survival and drought tolerance are some of the complex traits of meadow fescue that contribute to seasonal dry matter yield production and forage quality. QTLs for frost and drought tolerance, and for winter survival in the field, have been mapped in meadow fescue using the "B14/16 × HF2/7" mapping family (Alm et al. 2011). Major QTLs for drought tolerance traits mapped on linkage groups 1F, 3F, 4F, and 5F and for frost tolerance/winter survival on linkage groups 1F, 2F, 5F, and 6F. QTLs for several of the stress tolerance traits mapped to the same regions on *Festuca* chromosomes 1F, 4F, and 5F. Comparative mapping with Triticeae indicates that two frost tolerance/winter survival QTLs on linkage group 5F correspond to the frost tolerance loci *Fr-A1* and *Fr-A2* on wheat homoeologous group 5A chromosomes (Fig. 5-1). *Festuca* orthologs of genes known be associated with abiotic stress tolerance in Triticeae, e.g., *FpCBF6*, *FpIRI1*, *FpVRN1* and dehydrin genes, co-locate with the QTLs and indicate conservation across grass species. Fang (2003) mapped 34 chromosomal regions containing QTLs for seed yield and component traits in meadow fescue. QTLs for a number of related traits clustered in a few chromosomal segments, most evident on linkage groups 1F, 4F, and 5F indicating that there must be one or a few major gene(s) in these regions affect reproductive development with pleiotropic effects on many traits. Co-located QTLs for panicle fertility and seed yield were detected on chromosomes 1F, 2F, 4F, and 6F, and comparative mapping with cereal species identified a number of putatively orthologous QTLs. Ergon et al. (2006) mapped QTLs controlling vernalization requirement, heading time and number of panicles in meadow fescue. Genes with strong effect on vernalization requirement are present along most of linkage group 4F; in one of the regions a meadow fescue ortholog (*FpVRN1*) of the wheat VRN1 gene is located. QTLs involved in the induction of flowering are found on linkage groups 1F, 4F, 5F, 6F and 7F in the "B14/16 × HF2/7" mapping family; several of them co-locate with QTLs for seed yield components and are found in syntenic positions on chromosomes of other Poaceae species. Further analyses of gene expression of *FpVRN1* and other MADS-box genes involved in regulation of flowering time using *F. pratensis* populations segregating for vernalization requirement were published by Ergon et al. (2012).

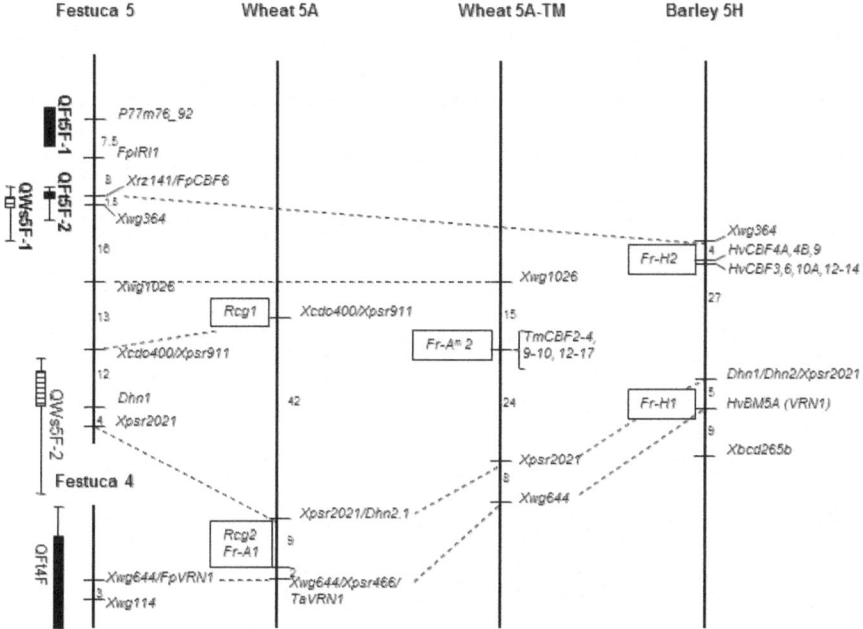

Figure 5-1 A comparative QTL-map of *Festuca* chromosomes 4 and 5 containing frost tolerance (QFt-) and winter survival (QWs-) QTLs with the homoeologous Triticeae group 5 chromosomes. The Triticeae group 5 maps and positions of QTLs and regulatory genes have been compiled from the following sources: Sutka and Snape (1989) Euphytica 42: 41–44; Pan et al. (1994) Theor Appl Genet 89: 900–910; Sutka et al. (1999) Theor Appl Genet 99: 199–202; Vágújfalvi et al. (2000) Mol General Genet 263: 194–200; Sarma et al. (2000) Genome 43: 191–198; Choi et al. (2002) Theor Appl Genet 98: 1234–1247; Vágújfalvi et al. (2003) Mol Genet Genom 269: 60–67; Francia et al. (2004) Theor Appl Genet 108: 670–680; Skinner et al. (2006) Theor Appl Genet 112:832-842; and Miller et al. (2006) Mol Genet Genom 275: 193–203 (from Alm et al. 2011; Theor Appl Genet 123: 369–382, with permission).

5.5 Proteomics

Only one comprehensive proteome study has been carried out in *F. pratensis*. Kosmala et al. (2009) studied leaf protein accumulation before and after 2, 8, 26 hr, and 3, 5, 7, 14 and 21 d of cold acclimation (CA) of genotypes with high (HFT) and low (LFT) freezing tolerance. Differential changes in the proteome during CA were studied using high-throughput two-dimensional electrophoresis (2-DE) in combination with electrospray ionization mass spectrometry. A total number of 41 (5.1%) proteins showed a minimum of 1.5-fold difference in abundance during CA, and the largest differences in protein abundance (28.1%) appeared most often already on day 2 of CA. At the time of maximum freezing tolerance differences (on day 21 of CA), 10 out of the 41 proteins (24%) had >1.5 fold differences in abundance between HFT and LFT genotypes. The majority of differentially accumulated *F. pratensis* proteins were proteins which are directly involved in photosynthesis, stressing the importance of the link between the function of the photosynthetic apparatus under cold stress and the degree of freezing tolerance. Several of the *F. pratensis* proteins identified, i.e., chloroplast-localized Ptr ToxA binding protein 1, globulin 2, 50S ribosomal protein L10 from chloroplasts, 30S ribosomal protein S10 from chloroplasts, ADP (adenosine diphosphate)-glucose pyrophosphatase, and ADP-ribosylation factor 1, had not been reported to be expressed during CA and development of freezing tolerance, even in model species.

5.6 Breeding of Meadow Fescue

Meadow fescue has many positive traits like tolerance towards abiotic and biotic stresses, good persistency and adaptation to grazing and frequent cutting and high nutritive quality. The majority of the European meadow fescue cultivars originate from eastern/central and northern Europe. Cultivars of meadow fescue are fairly long-lived and breeding is characterized by a rather slow but continuous improvement; e.g., for traits like winter survival and persistency. Casler and Van Santen (2000) bred meadow fescue for crown rust resistance, and crown rust resistance has been introgressed from meadow fescue into Italian ryegrass (Armstead et al. 2006). The common European catalogue of cultivars of meadow fescue lists a total number of 84 cultivars (http://ec.europa.eu/food/plant/propagation/catalogues/agri2009/15.html). Casler and can Santen (2000) mentioned that out of 233 *F. pratensis* accessions listed by the Germplasm Resources Information Network (http://www.ars-grin.gov/npgs/), 79 either had a cultivar name or otherwise appear to be derived from breeding programs. They also state that there were four cultivars on the market in the USA in 1994, two of them developed in Canada. Meadow fescue is cultivated in 88 regions and republics of former USSR countries, and 38 cultivars have the state permission for utilization (Dzyubenko and Dzyubenko 2009) (www.agroatlas.ru/en/content/cultural/Festuca_pratensis_K/). In Japan a number of meadow fescue cultivars have been developed, and they have been bred with plant materials introduced from Europe. Takai et al. (2001) described a new cultivar "Harusakae", a synthetic composed of eight clones originating from the Scandinavian cultivars "Boris", "Leto, "Tammisto", and " Salten". "Harusakae" has excellent winter hardiness including freezing tolerance and resistance to speckled snow-mold (*Typhula ishikariensis*). It also contains endophytes of the non-toxic loline alkaloid types. In Sapporo, "Harusakae" showed 12% higher dry matter yield than the older cultivar "Tomosakae" in frequent-cutting trials (six cuttings per season). Induced tetraploidy does not seem to improve agronomic characteristics the same way it does in, e.g., *Lolium* species. Autotetraploid meadow fescue has low tillering capacity and Simonsen (1975) found 7% lower dry matter yields compared to diploids.

The *Lolium-Festuca* species complex is unique since it is possible to combine the genomes in different Festulolium hybrid combinations which resulted in many promising commercial cultivars (see Ghesquière et al. 2010). Fescues in general have evolved superior adaptations to abiotic stresses, e.g., drought tolerance in *F. arundinacea* and winter survival in *F. pratensis*. *Lolium* species are known for superior nutritive quality, rapid establishment and growth, but lack persistency under harsh environmental conditions. Complementation of traits in Festulolium hybrids is thus a very interesting strategy for developing novel germplasm and cultivars with improved quality and persistency, which can contribute in making fodder production more economically and environmentally sustainable.

5.6.1 Breeding Meadow Fescue for Disease Resistance

A major disease in meadow fescue, which may reduce yield and quality considerably, is net blotch (*Drechslera dictyoides*). Leaf diseases usually cause injury in later cuts and especially in seed fields. Net blotch is a very serious disease in seed production fields and the prevalence of this disease has increased in recent years in Norway (Havstad 2009). Meadow fescue is generally quite resistant against the low-temperature fungi snow mold (*Microdochium nivale*), gray snow mold (*Typhyla incarnata*), speckled snow mold (*Typhula ishikariensis*), and sclerotinia snow mold (*Sclerotinia borealis*). Bacterial wilt (*Xanthomonas campestris pv. graminis*) is a serious disease and considered the main reason for shortened persistency of meadow fescue and ryegrasses in Switzerland (Michel 2001). Resistance to *Xanthomonas* is tested using artificial inoculation, and an improvement by recurrent selection was reported, although progress leveled off after four to six cycles of selection (Boller et al. 2001). Based on greenhouse screening of resistance to bacterial wilt in Italian ryegrass (*Lolium multiflorum* Lam.), a single major QTL which explained 67% of the phenotypic variation was mapped on linkage group 4 (LG4) (Studer et al. 2006). This locus also explained a major part of the variation under field conditions which is very encouraging.

5.6.2 Breeding Techniques

Recently with advancement in breeding techniques, fescue cultivars are much improved populations developed by either phenotypic mass selection of ecotypes or a breeding population, or synthetics. Synthetic populations are constructed from parental clones or remnant seed following progeny testing using half-sib or full-sib families and this is sometimes combined with among- and within-family selection (Casler and Brummer 2008).

Ecotype selection was the earliest method used to develop fescue cultivars and is still considered an important breeding method (Fjellheim and Rognli 2005a). "Norild", one of the most recent Norwegian cultivars, resulted from intercrossing surviving plants after several years of natural selection within a local population. Cultivars from ecotypes are either developed directly from multiplication of ecotypes or more commonly by phenotypic mass selection. Most of the earliest fescue cultivars were developed in this manner (Hopkins et al. 2007). The oldest Nordic cultivar, "Svalöfs Sena" from Sweden, dates back to 1917 and some of the newer cultivars are still developed from ecotypes (Fjellheim and Rognli 2005b). Ecotype selection can generate new populations in short periods of time with the high levels of genetic variation present in the germplasm (Valay and Van Santen 1999).

Clonal selection is another effective technique for selecting traits with high heritability, like heading date and disease susceptibility, but is unlikely to be effective for traits demonstrating genotype × environment interactions like herbage yield and persistency. Single clone selection has improved the selection for disease resistance in several turfgrass species including fine fescues. These plants are typically maintained for 1–2 yr at a height of 5 cm before superior clones are selected for further improvement based on turf quality (i.e., fine leaf texture, high shoot density, dark green color, clean mowing quality) and the absence of disease (Bonos et al. 2006).

Recurrent phenotypic selection is another popular breeding technique used to develop improved cultivars. It can give considerably uniform cultivars from diverse germplasm. Superior genotypes with desirable traits are selected from a population or diverse germplasm, the selected plants are intercrossed in isolation, and this cycle is repeated for multiple generations. Recurrent selection appears to be a powerful means of accumulating favorable alleles in a population and is often the method of choice for improving traits with low heritability (Rognli et al. 2010).

5.6.3 Transgenic Breeding

Genetic improvement of forage grasses such as fescues by conventional plant breeding is slow since many species are predominantly, if not completely, allogamous wind-pollinated grasses. Genetic engineering and the production of transgenic plants offer the opportunity to generate unique genetic variation. Application of genetic engineering to forage plant improvement has been focused on the development of transformation events with unique genetic variation and in studies on the molecular dissection of plant biosynthetic pathways and developmental processes of high relevance for forage production (Spangenberg et al. 2001). Methods for transforming many of the most important forage grass species have been developed. However, although meadow fescue was one of the first forage grasses to be regenerated from *in vitro* cultures (Wang et al. 1993), development of efficient transformation protocols have met with little success. Spangenberg et al. (1995) described the production of stably transformed meadow fescue plantlets from protoplasts. The plants grew *in vitro* but did not survive in the greenhouse. Recently the first successful protocol for *Agrobacterium tumefaciens* mediated transformation of embryogenic calli derived from mature embryos of meadow fescue was described (Gao et al. 2009). The transformation efficiency was 2% with 82% of the resultant lines expressing the transgene. *Agrobacterium*-mediated transformation is considered superior to particle bombardment. The establishment of an efficient transformation protocol could pave the way for manipulation of traits like disease resistance and feed quality characteristics in meadow fescue. However, the current biosafety and regulatory regimes make transgenic breeding less of an option. Meadow fescue is one of the few forage grass species where gene flow has been studied in field experiments and used to model gene flow as part of a risk assessment of releasing transgenic wind-pollinated grasses (Nurminiemi et al. 1998; Rognli et al. 2000).

5.7 Genome Sequencing Initiatives

A bacterial artificial chromosome (BAC) library of *F. pratensis* was developed at IBERS, Aberystwyth (Donnison et al. 2005). It has been used in comparative genomic strategies to clone candidate orthologous sequences to the CONSTANS-like rice *Hd1* (*Se1*) gene in *Lolium perenne* and *Festuca pratensis* (Armstead et al. 2005), and to isolate orthologous sequences for frost tolerance and lignin biosynthesis candidate genes in *F. pratensis* (Rudi et al. 2011). About 74,130 meadow fescue expressed sequence tags (ESTs) are available in public databases (http://www.plantgdb.org/prj/ESTCluster/progress.php) in addition to 372 ESTs detected by Rudi et al. (2011). The EST sequences can be downloaded from PlantGDB resource tool (http://www.plantgdb.org/download/download.php?dir=/Sequence/ESTcontig/Festuca_pratensis/current_version). At IBERS, UK, a set of monosomic substitution lines, each carrying one *F. pratensis* chromosome and 13 *L. perenne* chromosomes, have been developed (King et al. 1998; Harper et al. 2011), and backcrossed to *L. perenne* to create a series of recombinants with introgressed segments of *F. pratensis* chromosomes (King et al. 2007). This is a very useful genomic resource that is being utilized to develop physical maps by "introgression mapping" (King et al. 2007; King et al. 2013). Similarly individual *Festuca* chromosomes have been introgressed into autotetraploid *L. multiflorum* (Kopecky et al. 2008). They found high homoeologous chromosome pairing for each chromosome indicating few chromosomal rearrangements. However, since it is easy to distinguish between *Festuca* and *Lolium* genom segments using GISH the DNA repeats must have diverged substantially during evolution while sequences involved in pairing have been conserved.

For several years European research groups have tried to establish funding for genome sequencing of *L. perenne* and *F. pratensis*. An ongoing *L. perenne* physical mapping project at the University of Aberystwyth (IBERS), Wales, UK and a *de novo* sequencing project at the University of Aarhus, Denmark are now being integrated and will, pending sufficient funding, deliver a complete reference genome of *L. perenne* in the near future (Ian Armstead, pers. comm.). Currently *de novo* sequencing of the *Festuca* mapping parents B14/16 and HF2/7 using Illumina HiSeq 2000 is ongoing in a collaborative effort between the University of Life Sciences, Ås, Norway and University of Aarhus, Denmark, using the *Lolium perenne* sequence assembly as a reference for assembly of the meadow fescue genome sequences.

5.8 Future Prospects

With advancement in next generation sequencing (NGS) technologies and bioinformatics tools, there is a lot of scope to further extend the *Festuca pratensis* genomic resources. The application of NGS will yield reference genomes for crop species including those with very large genomes. Re-sequencing of cultivated and wild variants will provide insights into domestication and adaptation to diverse environments and climates. Resequencing of multiple cultivars will facilitate the dissection of the genetic architecture of quantitative phenotypic traits; this will allow marker-assisted selection on a genome scale, an approached termed "genomic selection". It will also facilitate the incorporation of useful allelic variants from landraces and ecotypes into cultivars, thus helping to maintain the influx of novel genetic variation into elite breeding material. NGS-enabled predictive breeding is expected to advance efforts to adapt crops to changing environments. Large-scale genotyping of thousands of individual plants is already utilized in commercial breeding programs of major crops, and the need for bioinformatic resources and support will increase dramatically. There are scope for significant socio-economic impacts from NGS-powered advances in breeding.

References

Alm V, Busso CS, Ergon Å, Rudi H, Larsen A, Humphreys MW, Rognli OA (2011) QTL analyses and comparative genetic mapping of frost tolerance, winter survival and drought tolerance in meadow fescue (*Festuca pratensis* Huds.). Theor Appl Genet 123: 369–382.

Alm V, Fang C, Busso CS, Devos KM, Vollan K, Grieg Z, Rognli OA (2003) A linkage map of meadow fescue (*Festuca pratensis* Huds.) and comparative mapping with other Poaceae species. Theor Appl Genet 108: 25–40.

Armstead IP, Harper JA, Turner LB, Skøt L, King IP, Humphreys MO, Morgan WG, Thomas HM, Roderick HW (2006) Introgression of crown rust (*Puccinia coronata*) resistance from meadow fescue (*Festuca pratensis*) into Italian ryegrass (*Lolium multiflorum*): genetic mapping and identification of associated molecular markers. Plant Pathol 55: 62–67.

Armstead IP, Skøt L, Turner LB, Skøt K, Donnison IS, Humphreys MO, King IP (2005) Identification of perennial ryegrass (*Lolium perenne* L.) and meadow fescue (*Festuca pratensis* Huds.) candidate orthologous sequences to the rice Hd1(Se1) and barley HvCO1 CONSTANS-like genes through comparative mapping and microsynteny. New Phytol 167: 239–247.

Balfourier F, Imbert C, Charmet G (2000) Evidence for phylogeographic structure in *Lolium* species related to the spread of agriculture in Europe. A cpDNA study. Theor Appl Genet 101: 131–138.

Bartoš J, Sandve SR, Kölliker R, Kopecký D, Christelová P, Stočes Š, Østrem L, Larsen A, Kilian A, Rognli OA, Doležel J (2011) Genetic mapping of DArT markers in the *Festuca-Lolium* complex and their use in freezing tolerance association analysis. Theor Appl Genet 122: 1133–1147.

Bennett MD, Leitch IJ (2012) Angiosperm DNA C-values database (release 8.0, Dec. 2012) http://www.kew.org/cvalues/.

Boller B, Tanner P, Schubiger FX, Streckeisen P (2001) Selecting meadow fescue ecotypes for reduced susceptibility to bacterial wilt. In: P. Monjardino et al. (eds) Breeding for Stress Tolerance in Fodder Crops and Amenity Grasses. Proceedings of the 23rd Meeting of the Fodder Crops and Amenity Grass Section of EUCARPIA, Azores, Portugal. University of Azores, Terceira Island, pp 103–107.

Bonos SA, Clarke BB, Meyer WA (2006) Breeding for disease resistance in major cool-season turfgrasses. Annu Rev Phytopathol 44: 213–234.

Borrill M, Tyler BF, Morgan WG (1976) Studies in Festuca VII. Chromosome atlas (Part 2). An appraisal of chromosome race distribution and ecology, including *F. pratensis* var. *apennina* (De Not.) Hack,-tetraploid. Cytologia 41: 219–236.

Brink GE, Casler MD (2009) Meadow fescue, tall fescue, and orchardgrass response to nitrogen application rate. Forage and Grazinglands. doi:10.1094/FG-2009-0130-01-RS.

Brink GE, Casler MD, Martin NP (2010) Meadow fescue, tall fescue and orchardgrass response to defoliation management. Agron J 102: 667–674.

Brown AHD (1992) Human impact on plant gene pools and sampling for their conservation. Oikos 63: 109–118.

Casler MD, Brummer EC (2008) Theoretical expected genetic gains for among- and within-family selection methods in perennial forage crops. Crop Sci 48: 890–902.

Casler MD, Undersander DJ, Fredericks C, Combs DK, Reed JD (1998) An on-farm test of perennial forage grass varieties under management intensive grazing. J Prod Agri 11: 92–99.

Casler MD, van Santen E (2000) Patterns of variation in a collection of meadow fescue accessions. Crop Sci 40: 248–255.

Casler MD, van Santen E (2001) Performance of meadow fescue accessions under management-intensive grazing. Crop Sci 41: 1946–1953.

Charmet G, Ravel C, Balfourier F (1997) Phylogenetic analysis in the Festuca-Lolium complex using molecular markers and ITS rDNA. Theor Appl Genet 94: 1038–1046.

Darbyshire SJ (1993) Realignment of *Festuca* subgenus *Schedonorus* with the genus *Lolium*. Novon 3: 239–243.

Daugstad K (2012) On-farm conservation of the forage species timothy, meadow fescue and red clover: Generation of new landraces in Norway. In: Maxted N et al. (eds) Agrobiodiversity Conservation—Securing the Diversity of Crop Wild Relatives and Landraces. CAB International, Wallingford, UK, pp 125–130.

Donnison IS, O'Sullivan DM, Thomas A, Canter P, Moore B, Armstead I, Thomas H, Edwards KJ, King IP (2005) Construction of a *Festuca pratensis* BAC library for map-based cloning in *Festulolium* substitution lines. Theor Appl Genet 110: 846–851.

Dzyubenko NI, Dzyubenko EA (2009) Interactive agricultural ecological atlas of Russia and neighboring countries (http://www.agroatlas.ru/en/content/cultural/Festuca_pratensis_K/).

Ergon Å, Hamland H, Rognli OA (2012) Differential expression of VRN1 and other MADS-box genes in *Festuca pratensis* selections with different vernalization requirements. Biol Plant: 1–10, DOI: 10.1007/s10535-012-0283-z.

Ergon Å, Fang C, Jørgensen Ø, Aamlid TS, Rognli OA (2006) Quantitative trait loci controlling vernalisation requirement, heading time, and number of panicles in meadow fescue (*Festuca pratensis* Huds.). Theor Appl Genet 112: 232–242.

Fang C (2003) Comparative genome analyses, QTL mapping and genetic analyses of seed yield and related traits in meadow fescue (*Festuca pratensis* Huds.). Agricultural University of Norway. Doctor Scientiarum Thesis 2003: 10.

FAO (2005) Global Forest Resources Assessment 2005: Progress towards Sustainable Forest Management, Food and Agriculture Organization of the United Nations, Rome.

Fjellheim S, Blomlie ÅB, Marum P, Rognli OA (2007) Phenotypic variation in local populations and cultivars of meadow fescue–potential for improving cultivars by utilizing wild germplasm. Plant Breed 126: 279–286.

Fjellheim S, Rognli OA (2005a) Genetic diversity within and among Nordic meadow fescue (*Festuca pratensis* Huds.) cultivars based on AFLP markers. Crop Sci 45: 2081–2086.

Fjellheim S, Rognli OA (2005b) Molecular diversity of local Norwegian meadow fescue (*Festuca pratensis* Huds.) populations and Nordic cultivars—consequences for management and utilisation. Theor Appl Genet 111: 640–650.

Fjellheim S, Rognli OA, Fosnes K, Brochmann C (2006) Phylogeographical history of the widespread meadow fescue (*Festuca pratensis* Huds.) inferred from chloroplast DNA sequences. J Biogeogr 33: 1470–1478.

Frankel OH, Brown AHD, Burdon JJ (1995) The Conservation of Plant Biodiversity. Cambridge University Press, Cambridge.

Gao C, Liu J, Nielsen K (2009) *Agrobacterium*-mediated transformation of meadow fescue (*Festuca pratensis* Huds.). Plant Cell Rep 28: 1431–1437.

Ghesquière M, Humphreys MW, Zwierzykowski Z (2010) Festulolium. In: Boller B et al. (eds) Fodder crops and amenity grasses, Handbook of Plant Breeding 5, pp 293–316, Springer New York. DOI 10.1007/978-1-4419-0760-8_12.

Hand ML, Cogan NOI, Forster JW (2012) Molecular characterisation and interpretation of genetic diversity within globally distributed germplasm collections of tall fescue (*Festuca arundinacea* Schreb.) and meadow fescue (*F. pratensis* Huds.). Theor Appl Genet 124: 1127–1137.

Hansson M, Fogelfors H (2000) Management of a semi-natural grassland; results from a 15-year-old experiment in southern Sweden. J Veg Sci 11: 31–38.

Harper J, Armstead I, Thomas A, James C, Gasior D, Bisaga M, Roberts L, King I, King J (2011) Alien introgression in the grasses *Lolium perenne* (perennial ryegrass) and *Festuca pratensis* (meadow fescue): the development of seven monosomic substitution lines and their molecular and cytological characterization. Ann Bot 107: 1313–1321.

Havstad LT (2009) Frøavl av engsvingel. Dyrkingsveiledning 2009 (Seed production of meadow fescue. Growers guidance). Bioforsk Øst Landvik, p 10.

Hultén E, Fries, M (1986) Atlas of North European vascular plants: north of the Tropic of Cancer I-III. Koeltz Scientific Books, Königstein.

Humphreys J, Harper JA, Armstead IP, Humphreys MW (2005) Introgression-mapping of genes for drought resistance transferred from *Festuca arundinacea* var. *glaucescens* into *Lolium multiflorum*. Theor Appl Genet 110: 579–589.

Hopkins AA, Saha MC, Wang Z-Y (2007) Tall fescue breeding, genetics, and cultivars. In: Fribourg HA et al. (eds) Tall Fescue On-line Monograph (http://forages.oregonstate.edu/tallfescuemonograph/breeding_genetics/abstract).

Jauhar PP (1993) Cytogenetics of the Festuca–Lolium complex. Springer, Heidelberg.

King IP, MorganWG, Armstead IP, Harper JA, Hayward MD, Bollard A, Nash JV, Forster JW, Thomas HM (1998) Introgression mapping in the grasses. I. Introgression of *Festuca pratensis* chromosomes and chromosome segments into *Lolium perenne*. Heredity 81: 462–467.

King J, Armstead I, Donnison I, Harper J, Roberts L, Thomas H, Ougham H, Thomas A, Huang L, King IP (2007) Introgression mapping in the grasses. Chrom Res 15: 105–113.

King J, Armstead I, Harper J, Ramsey L, Snape J, Waugh R, James C, Thomas A, Gasior D, Kelly R, Roberts L, Gustafson P, King I (2013) Exploitation of interspecific diversity for monocot crop improvement. Heredity doi: 10.1038/hdy.2012.116.

Kosmala A, Bocian A, Rapacz M, Jurczyk B, Zwierzykowski Z (2009) Identification of leaf proteins differentially accumulated during cold acclimation between *Festuca pratensis* plants with distinct levels of frost tolerance. J Exp Bot 60: 3595–3609.

Kostoff D (1949) The application of cytology to grass and clover breeding. In: Proceedings of the 5th International Grassland Congress, The Hague, The Netherlands, pp 84–85.

Kopecký D, Bartoš J, Christelová P, Černoch V, Kilian A, Doležel J (2011) Genomic constitution of *Festuca* × *Lolium* hybrids revealed by the DArTFest array. Theor Appl Genet 122: 355–363.

Kopecký D, Bartoš J, Lukaszewski AJ, Baird JH, Černoch V, Kölliker R, Rognli OA, Blois H, Caig V, Lübberstedt T, Studer B, Doležel J, Kilian A (2009) Development and mapping of DArT markers within the *Festuca-Lolium* complex. BMC Genomics 10: 473.

Kopecký D, Lukaszewski AJ, Doležel J (2008) Meiotic behavior of individual chromosomes of *Festuca pratensis* in tetraploid *Lolium multiflorum*. Chrom Res 16: 987–998.

Kölliker R (1998) Genetic variability in *Festuca pratensis* Huds.; effect of management on natural populations and comparison of cultivars to other species. PhD thesis Diss. ETH No. 12753, Swiss Federal Institute of Technology, Zurich, Switzerland, pp 82.

Kölliker R, Stadelmann FJ, Reidy B, Nösberger J (1998) Fertilization and defoliation frequency affect genetic diversity of *Festuca pratensis* Huds. in permanent grasslands. Mol Ecol 7: 1557–1567.

Kölliker R, Stadelmann FJ, Reidy B, Nösberger J (1999) Genetic variability of forage grass cultivars: a comparison of *Festuca pratensis* Huds., *Lolium perenne* L. and *Dactylis glomerata* L. Euphytica 106: 261–270.

Maxted N, Ford-Lloyd BV, Hawkes JG (eds) (1997) Plant Genetic Conservation: the *in situ* approach. Chapman & Hall, London.

McGrath S, Hodkinson TR, Barth S (2007) Extremely high cytoplasmic diversity in natural and breeding populations of *Lolium* (Poaceae). Heredity 99: 531–544.

Michel VV (2001) Interactions between *Xanthomonas campestris* pv. *graminis* strains and meadow fescue and Italian ryegrass cultivars. Plant Dis 85: 538–542.

Momotaz A, Forster JW, Yamada T (2004) Identification of cultivars and accessions of cultivars and accessions of *Lolium*, *Festuca* and *Festulolium* hybrids through the detection of simple sequence repeat (SSR) polymorphism. Plant Breed 123: 370–376.

Nurminiemi M, Tufto J, Nilsson NO, Rognli OA (1998) Spatial models of pollen dispersal in the forage grass meadow fescue. Evol Ecol 12: 487–502.

Peeters A (2004) Wild and Sown Grasses. Blackwell Pub, Rome.

Peter-Schmid MKI, Boller B, Kölliker R (2008a) Habitat and management affect genetic structure of *Festuca pratensis* but not *Lolium multiflorum* ecotype populations. Plant Breed 127: 510–517.

Peter-Schmid MKI, Kölliker R, Boller B (2008b) Value of permanent grassland habitats as reservoirs of *Festuca pratensis* Huds. and *Lolium multiflorum* Lam. populations for breeding and conservation. Euphytica 164: 239–253.

Rognli OA, Fjellheim S, Pecetti L, Boller B (2013) Semi-natural grasslands as source of genetic diversity. Grassland Science in Europe 18: 303–313.

Rognli OA, Nilsson NO, Nurminiemi M (2000) Effects of distance and pollen competition on gene flow in the wind-pollinated grass *Festuca pratensis* Huds. Heredity 85: 550–560.

Rognli OA, Saha MC, Bhamidimarri S, Heijden SV (2010) In: Boller B et al. (eds) Fodder crops and amenity grasses, Handbook of Plant Breeding 5, pp 261–292, Springer New York. DOI 10.1007/978-1-4419-0760-8_12.

Rudi H, Sandve SR, Opseth LM, Larsen A, Rognli OA (2011) Identification of candidate genes important for frost tolerance in *Festuca pratensis* Huds. by transcriptional profiling. Plant Sci 180: 78–85.

Smarda P, Bures P, Horova L, Foggi B, Rossi G (2008) Genome size and GC content evolution of *Festuca*: Ancestral expansion and subsequent reduction. Ann Bot 101: 421–433.

Saha MC, Mian MAR, Eujayl I, Zwonitzer JC, Wang L, May GD (2004) Tall fescue EST-SSR markers with transferability across several grass species. Theor Appl Genet 109: 783–791.

Simonsen Ø (1975) Cytogenetic investigations in diploid and autotetraploid populations of Festuca pratensis Huds. Hereditas 79: 73–108.

Spangenberg G, Wang ZY, Potrykus I (1998) Biotechnology in Forage and Turf Grass Improvement. Vol. 23, Springer Berlin Heidelberg.

Spangenberg G, Wang Z-Y, Vallés MP, Potrykus I (1995) Genetic transformation in *Festuca arundinacea* Schreb. (tall fescue) and *Festuca pratensis* Huds. (meadow fescue). In: Bajaj YPS (ed) Biotechnology in agriculture and forestry, vol 34. Springer, Berlin, pp 183–203.

Studer B, Boller B, Herrmann D, Bauer E, Posselt U, Widmer F, Kölliker R (2006) Genetic mapping reveals a single major QTL for bacterial wilt resistance in Italian ryegrass (*Lolium multiflorum* Lam.). Theor Appl Genet 113: 661–671.

Takai T, Sadao N, Yasumichi T, Sadao H, Hisaaki D, Hiroshi A, Kazuhiko M, ShinIchi S, Koichi I (2001) Breeding of 'Harusakae' meadow fescue and its characteristics. Res Bull Hokkaido Nat Agri Exp Stn 173: 47–62.

Tamura KI, Yonemaru JI, Hisano H, Kanamori H, King J, King I, Tase K, Sanada Y, Komatsu T, Yamada T (2009) Development of intron-flanking EST markers for the *Lolium/Festuca* complex using rice genomic information. Theor Appl Genet 118: 1549–1560.

Tscharntke T, Klein AM, Kruess A, Steffan-Dewenter I, Thies C (2005) Landscape perspectives on agricultural intensification and biodiversity ecosystem service management. Ecol Lett 8: 857–874.

Valay R, Van Santen E (1999) Grazing induces a patterned selection response in tall fescue. Crop Sci 39: 44–51.

Wang ZY, Vallés MP, Montavon P, Potrykus I, Spangenberg G (1993) Fertile plant regeneration from protoplasts of meadow fescue (*Festuca pratensis* Huds.). Plant Cell Rep 12: 95–100.

Wilkins PW (1991) Breeding perennial ryegrass for agriculture. Euphytica 52: 201–214.

Yamada T, Kishida T (2003) Genetic analysis of forage grasses based on heterologous RFLP markers detected by rice cDNAs. Plant Breed 122: 57–60.

Yamada T (2011) Wild crop relatives: Genomic and Breeding Resources, Millets and Grasses, Springer-Verlag, Berlin, Heidelberg.

Dactylis and *Phleum*

Alan V. Stewart,[1,]* Nicholas W. Ellison[2] and B. Shaun Bushman[3]

ABSTRACT

Dactylis and *Phleum* are both represented by polyploid series, from $2x$ to $6x$ in *Dactylis* and from $2x$ to $8x$ in *Phleum*. In *Dactylis* (cocksfoot or orchard grass) the commercial cultivars are mainly tetraploids while in *Phleum* (Timothy) the commercial cultivars are mainly hexaploids.

Molecular data show that both genera arose in central Asia as diploids and underwent adaptive radiation and polyploidization to cover the extensive region today known from Asia to western Europe.

In the case of *Dactylis* the adaptive radiation of the diploids occurred prior to the last glaciation which subsequently reduced its distribution to disjunct glacial refugia in southern Europe and Asia but also opened up in North Africa and nearby Atlantic Islands for expansion. Post-glacial expansion was dominated by a few limited tetraploid forms such that they cover almost the entire range of *Dactylis* today.

In the case of *Phleum pratense* the modern hexaploids formed from a limited meeting of an ancestral diploid and a tetraploid in a southern European glacial refuge, to expand northwards as the climate warmed and cover the range known today.

There is now considerable potential to introgress and to re-synthesize very diverse new forms of both species, using wild relatives from other ploidy levels and the knowledge gained about their relationships from molecular studies.

[1]PGG Wrightson, PO Box 175, Lincoln 7640, New Zealand.
[2]AgResearch, Private Bag 11008, Palmerston North, New Zealand.
[3]USDA-ARS FRRL, 695 N 1100 E, Logan, UT, USA.
*Corresponding author: *astewart@pggwrightsonseeds.co.nz*

Unfortunately, genebank collections of both genera have a serious lack of diploid and tetraploid ancestral populations from glacial refugia, yet these forms can be expected to contain much more genetic diversity than those in commerce.

Collection of such resources must become a priority for storage in *ex situ* genebanks, as many are under threat from climate warming. In both genera genomic resources are only just beginning to be developed, but funding of breeding programs has diminished internationally so that few breeding groups will be in a strong position to use either the germplasm or genomic resources effectively.

Key words: *Dactylis*, *Phleum*, Polyploids, Glacial refugia, Genetic diversity, Germplasm, Genebanks, Genomic resources

Part 1: *Dactylis*

6.1 Introduction

Dactylis glomerata L., cocksfoot, or orchardgrass as it is commonly known, is widely used as a forage grass in the temperate world. As a genus *Dactylis* is subject to differing taxonomic interpretations in different regions of its natural range (Europe, Asia, North Africa and the Canary Islands). There are no modern taxonomic treatments which interpret all forms on the same basis. Here we follow Stebbins's interpretation that *Dactylis glomerata* L. is monotypic, consisting of one diverse species complex (Stebbins and Zohary 1959; Stebbins 1971). The genus however, is clearly on the cusp of speciation with many overlapping diploid, tetraploid and a hexaploid subspecies and some reduction in fertility when different diploid forms are crossed (Parker and Borrill 1968; Borrill 1977).

The basic chromosome number is $2n = 14$ and, while the species consists of diploids, tetraploids and at least one hexaploid population, the commonly used forage types are tetraploid.

6.2 Economic Importance

Dactylis glomerata is commonly used in drier and lower fertility pastures in most temperate regions of the world. It is possible to divide cultivars into at least three categories based on seasonal growth; winter-hardy types for most of Europe, Japan, the USA and Canada; winter active types for mild winter zones of Australia, New Zealand, South America and South Africa and summer dormant types for dry summer climates of the Mediterranean, Australia and California. Many further subdivisions can be made on flowering time, ploidy level and other agronomic features.

Each year approximately 14,000 tons of seeds are harvested making *Dactylis* the fourth most widely used grass genus with 3.3% of the world's temperate grass seed, following *Lolium, Festuca* and *Phleum* (Bondesen 2007).

6.3 Academic Importance

Dactylis has gained considerable academic significance through the evolutionary studies initiated by Stebbins in the 1950s (Stebbins and Zohary 1959). After Stebbins showed that *Dactylis* was a good example of adaptive radiation and polyploid development many studies have attempted to clarify the history of this species. Since then there have been a range of studies using different techniques to further clarify its evolution. These include studies with taxonomical and morphological traits (Borrill 1961, 1977), isozymes (Lumaret 1988), phenolic flavonoids (Ardouin et al. 1985; Fiasson et al. 1987), karyological (Wetschnig 1991), and finally molecular techniques, which have explained the history when appropriate populations are available for study (Sahuquillo and Lumaret 1999; Stewart and Ellison 2010).

As part of the evolutionary studies it has become clear that tetraploidy through unreduced gametes has been a significant factor in evolution and *Dactylis* has become a model species for this process (Lumaret et al. 1989; Lumaret and Barrientos 1990; Lumaret et al. 1992). Unreduced gametes occur at a moderately high frequency within diploid populations, often up to 10% or more. Breeders have used this mechanism to cross a diploid plant known to develop a few percent of unreduced gametes with tetraploids to obtain tetraploid progeny, as the triploids abort (Lewis 1975).

6.4 Botanical Origin and Evolution

As Stebbins postulated *Dactylis* is a good example of adaptive radiation combined with the formation of polyploids. A process which we now know has been driven by glaciation events forcing migration into and out of glacial refugia (Hewitt 1999). This has allowed divergent forms to meet and hybridize and for their more vigorous progeny to recolonize large areas of new territory as the glaciers recede.

Molecular studies by Stewart and Ellison (2010) have shown that during the interglacial period prior to the last glaciation a successful Central Asian diploid progenitor, similar to today's *altaica,* expanded its range to cover the region from the Himalayas to Portugal. This region was largely temperate rainforest and *Dactylis* could be expected to occur on the forest margins and in drier parts of the range as it does today.

As the last glaciation proceeded the northern range of *Dactylis* contracted into a series of scattered refugia in southern Europe. Glaciation also allowed a southern expansion into the increasingly moist North African grasslands where *Dactylis* appears to have been quite widespread. From there expansion to the Canary and Cape Verde Islands also occurred, most probably via the major annual bird migrations along this route. There is no evidence of spread into highland tropical regions of Africa, despite some European and North African species doing so.

Tetraploidy through unreduced gametes would likely have occurred throughout the history of *Dactylis* but autotetraploids from a single population appear little better adapted than the original diploid populations and remain sporadic and restricted in distribution. However, the migration of divergent diploids into, and out of glacial refugia allowed inter-population tetraploids to develop with potential hybrid vigor, so much so that tetraploids have an almost continuous distribution over the full range of the species today (Lumaret 1988).

The basis for this hybrid vigor in tetraploid *Dactylis* is due to their having more polymorphic loci than diploids, 0.80 compared to 0.70, a higher heterozygosity of 0.43 compared to 0.17, and a greater number of alleles per locus of 2.36 compared to 1.51 (Soltis and Soltis 1993). This diversity is also reflected within *Dactylis* cultivars (Kölliker et al. 1999).

Post-glacial warming allowed the northern expansion but favored inter-population tetraploid forms, leaving most diploids in very restricted ranges and the tetraploids over almost the entire range of the genus. A xeromorphic hexaploid also developed in North Africa of which little is known.

Since the glaciation the *Dactylis* range in North Africa has been restricted by post-glacial desertification, while climate warming is expected to further reduce this range.

Table 6-1 lists the various subspecies or common names at each ploidy level while the migration pattern is summarized in Fig. 6-1.

Table 6-1 Populations or subspecies with *Dactylis glomerata.*

Diploids		Tetraploids
altaica	*lusitanica*	*glomerata*
aschersoniana	*mairei*	*hispanica*
castellata	*metlesicsii*	*hylodes*
himalayensis	*parthiana*	*marina*
hyrcana	*reichenbachii*	*oceanica*
ibizensis = nestorii	*santai*	*slovenica*
izcoi = Galician	*sinensis*	
judaica	*smithii*	**Hexaploid**
juncinella	*woronowii*	*hispanica*

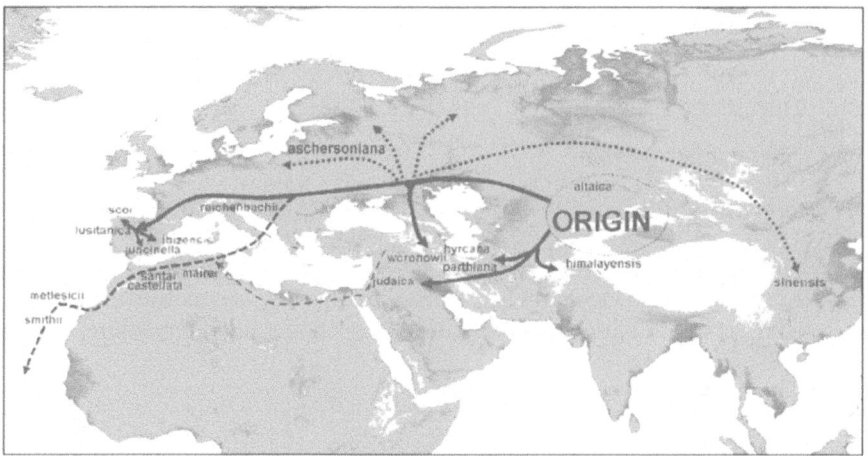

Figure 6-1 Probable migration routes of diploid *Dactylis* based on molecular results of Stewart and Ellison (2010) (black lines—before glaciation, dashed lines—North Africa during the glaciation; dotted lines—post-glacial, northern Europe and China).

6.5 Domestication and Dissemination

The northern European tetraploids were the first to be domesticated and used in agriculture. These have been disseminated around the world to all major temperate regions, particularly Asia, North and South America, Australia, New Zealand and South Africa.

The process of dissemination around the world has been driven by human migration and trade. Indeed, an interesting case is the Chinese diploid *sinensis* which has a disjunct distribution from all other *Dactylis*. Although originally thought to be a disjunct extension of diploid *himalayensis* of the Himalayas (Stebbins and Zohary 1959; Borrill 1977) it has became apparent that it is quite different from *himalayensis* (Lumaret 1987). Recent molecular studies of material from Sichuan Province of China show that it exhibits its greatest similarity to European *aschersoniana* than to *himalayensis* (Stewart and Ellison 2010). This suggests that they are derived from European forms that had migrated to China, perhaps naturally or via the "silk road" in historic times.

Breeders have by and large selected within the regional ecotypes from their region and supplemented these with new tetraploid germplasm from the original range of the species. Occasionally diploid material has been used to introduce new genes into tetraploid cultivars, and at least two diploid cultivars exist.

6.6 Germplasm and Gene Pools

Genetic resources for *Dactylis* breeding can be divided into four gene pools on the following basis (Stewart and Ellison 2010).

6.6.1 Primary Gene Pool

These can be defined as cultivars and elite breeding lines adapted to the region of agricultural use. This is the primary activity of *Dactylis* breeders and consists almost exclusively of tetraploids, although the diploids izcoi and aschersoniana have been used commercially. In general, these resources are well used by breeders and their "working" collections largely represent this gene pool.

6.6.2 Secondary Gene Pool

This can be defined as those tetraploid populations outside the regions of main agricultural use of the species, Europe, Asia and North Africa and the many islands. These resources are often explored by plant breeders but they are of lesser significance than the primary pool simply because adaptation to agricultural environments is usually inferior. In general, these resources still occur in nature and in genebank collections. However, many of the wild populations are under threat with climate warming and human-induced habitat changes, while even those in genebanks can only be maintained if adequate resources are available.

The secondary gene pool includes tetraploid forms sympatric with diploid populations. The Australian cultivar Porto and the Spanish cultivar Adac 1 are based on natural tetraploids with spp. *lusitanica* influence. A number of cultivars have been based on tetraploid forms of spp. *izcoi* from Galicia, such as Grasslands Wana (New Zealand), Cambria (UK) and Artibro (Spain). The tetraploid forms of many of these diploid populations represent an enormous resource for breeders and representative collections are required. Although many have been collected, few are identified as tetraploid forms of the remnant diploid population.

6.6.3 Tertiary Gene Pool

These can be defined as those populations from other ploidy levels where introgression into commercial breeding programs is possible. The tertiary genepool would largely consist of diploids, but potentially the hexaploid could also be used for tetraploid breeding if more was known about it.

Breeders have crossed germplasm from diploid spp. *lusitanica* into tetraploid material, with examples being the two UK commercial cultivars Saborto and Calder, as well as the NZ cultivar Grasslands Kara. Tetraploidy has been induced from the diploids by colchicine treatment as in Saborto, and via unreduced gametes as in Calder (Lewis 1975).

In general, these diploid resources are poorly represented in genebanks and are seldom used by breeders. Yet as the foundation upon which modern tetraploids are based, the remnant diploid populations represent an enormous resource. Many are under threat with climate warming and human induced habitat changes. Even those in genebanks will only survive if adequate funding is provided.

6.6.4 Quaternary Gene Pool

This can be defined as those species outside *Dactylis*, which may be a source of genes for breeding. However, only a few species have ever been hybridized with *Dactylis*, such as *Lolium multiflorum* (Oertel et al. 1996), *Festuca arundinacea* (Matzk 1981), and *Phleum pratense* (Nakazumi et al. 1997), and successful embryos have been formed in cereals (wheat, barley, rye) pollinated with tetraploid *Dactylis glomerata* (Zenkteler and Nitzsche 1984). As fertility is extremely low these hybrids have never been fully explored in breeding programs and their development does not appear to be a priority.

6.7 Occasional Mutants

At least three mutants have been reported in the literature. In studies of unreduced gametes Falistocco et al. (1996) discovered a desynaptic mutant. A glabrous mutant without leaf dentations were discovered after x-ray treatment (Zijp 1960), although such types exist naturally (Van Dijk GE 1961) and may be valuable for breeding. A chlorophyll mutant was discovered in subspecies judaica by Apiron and Zohary (1961) and found to act as a balanced polymorphism.

6.8 Karyotype

Dactylis has forms with $2n = 14$, 28 and 42 chromosomes, with Wetschnig (1991) providing an excellent review of the karyotypes of many of these forms. There has been no research on cytogenetic stocks, such as addition or substitution lines. Williams and Barclay (1972) described the transmission of B chromosomes in *Dactylis*.

6.9 Genome Size

The genome size of *Dactylis* has been determined in a number of studies for both diploid and tetraploid forms (Creber et al. 1994; Reeves et al. 1998; Vilhar 2002; Tuna et al. 2004, 2007) and has been shown to vary between populations. Genome size has been associated with environment factors in some situations, with Reeves et al. (1998) reporting a decrease of up to 30%

associated with increasing altitude but other authors found no association with altitude in different studies (Vilhar 2002). Tuna et al. (2007) showed that in the diploids from Mediterranean climates had approximately 15% lower DNA mass than northern European forms.

6.10 Traditional Breeding

There are over 700 named cultivars of *Dactylis glomerata* in genebanks and over 250 in current seed certification schemes.

Breeding objectives for *Dactylis* represent the usual forage grass objectives of forage yield, seasonal yield, disease and pest resistance, forage quality, seed yield, tolerance to adverse conditions such as cold winters or dry summers and drought, as appropriate to the region where it is bred. Breeding methodologies are those common to population breeding of forage grasses.

Classical breeding methodology has achieved a range of cultivars well adapted to their particular zones. Yields though have reached a plateau with genetic improvements of less than 0.5% per year and probably much less in most situations. The tetrasomic inheritance observed in *Dactylis* probably limits gain as does the very limited scale of breeding programs.

6.11 Genetic Diversity Studies

There have been a number of genetic diversity studies undertaken with *Dactylis* using molecular techniques both within the species and within breeding populations. Early molecular marker-based diversity studies utilized isozymes (Lumaret 1986; Lumaret et al. 1987; Sahuquillo and Lumaret 1995; Tosun et al. 2002), followed by RAPDs and other dominant markers (Kolliker et al. 1999; Tuna et al. 2004; Zeng et al. 2006; Peng et al. 2008; Zeng et al. 2008). Recently, heterologous and genomic-derived SSR markers have been used (Litrico et al. 2009; Xie et al. 2010), and an EST-based SSR resource of 1,100 SSR markers was developed and used to assess relationships of orchardgrass subspecies (Bushman et al. 2010).

6.11.1 Genomics Resources

Little genomic sequence is currently available for *Dactylis*. Trejo-Calzada and O'Connell (2005) conducted gene expression studies for drought-response, and as a side effect released a number of gene sequences. Bushman et al. (2011) generated a library of over 17,000 *Dactylis* expressed sequence tags (EST) using four tissues: cold acclimated crowns, etiolated seedlings, salt/drought stressed shoots, and salt/drought stressed roots. Simple sequence repeats (SSRs) identified from these libraries can be aligned to rice chromosomes to determine predicted locations of the SSR markers. PCR primers designed from 3' UTR regions from other species contain high degrees of polymorphism, and the same is expected for this library. Three genetic mapping populations have also been developed, including an F_1 map of Chinese diploid plants (Xie et al. 2011), an F_1 map of tetraploid Japanese plants (Song et al. 2011), and an F_1 map of tetraploid *himalayensis* and *aschersoniana* (Xie et al. 2012). As a *"near autotetraploid"*, genetic mapping in orchardgrass is not straightforward, and requires modeling of parental genotypes based on offspring gel-band phenotypes. Of these three maps, Xie et al. (2012) also mapped quantitative trait loci (QTL) associated with flowering time. They are also using genetic association mapping populations, and trait evaluation field plots to help overcome winter hardiness and to provide tolerance to dry summers.

6.12 Genetic Modification

Transformation systems to develop transgenic plants have been developed for *Dactylis* (Cho et al. 2001), but at this stage no transgenic cultivars have been released.

Genetic modification should provide a valuable mechanism for improvement in *Dactylis* but probably only for genes that are proven in major forage grass species first.

6.12.1 The Future Improvement of Dactylis

In order that molecular resources can be applied in an effective and balanced manner, it is important to ensure pragmatic field breeding programs continue in all major regions. This is of major concern for *Dactylis glomerata* as it is a species with limited current investment.

Germplasm of many of the diploid forms is under serious threat from habitat degradation and climate warming *in situ* and unfortunately collections of many forms are poorly represented in *ex situ* genebanks. It is critical that a wide range of these forms be collected for storage.

It is crucial that viable large-scale breeding programs are maintained internationally to allow adequate cultivar development, germplasm collection, introgressions of wild germplasm and application of molecular resources.

Part 2: *Phleum*

6.13 Introduction

The genus *Phleum* contains one important commercial species, Timothy, *Phleum pratense* and two minor commercial species, turf Timothy, *Phleum pratense* subsp. *bertolonii* and Alpine Timothy, *Phleum alpinum*.

6.14 Taxonomy

The genus *Phleum* contains 14 species in four sections over a polyploid series from diploid to octoploid as outlined by Joachimiak (2005) and Stewart et al. (2010).

The three commercial species are in section *Phleum*. In this chapter we use the widely accepted nomenclature of Humphries (1978, 1980) for European species and of Barkworth et al. (2007) for American species, as followed by Stewart et al. (2010).

6.14.1 ***P. alpinum*** L. or Alpine Timothy, is an alpine species differentiated into three different diploid or tetraploid cytotypes with ciliate or glabrous awns:

    6.14.1.1 A glabrous awned tetraploid; *P. alpinum* L. ≡ *P. commutatum* Gaudin. with a circumpolar northern hemisphere and South American distribution.

    6.14.1.2 A ciliate awned diploid form, known from the Rhaetic Alps of Italy and the Balkans; *P. alpinum* ssp. *rhaeticum* Humphries, ≡ *P. rhaeticum* (Humphries) Rauschert.

    6.14.1.3 A glabrous awned diploid form also known as *P. alpinum* L., currently referred to by the informal name "commutatum" following Joachimiak and Kula (1993). This form grows among the snow-bed vegetation at high altitudes (Zernig 2005) and occurs in the mountains of central Europe from the Alps north to the Carpathian Mountains.

6.14.2 ***P. pratense*** L. is a lowland species represented by a diploid to octoploid polyploid series:

    6.14.2.1 The diploids occur throughout much of Europe and parts of North Africa; *P. pratense* ssp. *bertolonii* (DC.) Bornm., ≡ *P. bertolonii* DC., ≡ *P. nodosum* L.

    6.14.2.2 Less common tetraploid forms in southern Europe; *P. pratense* ssp. *pratense*.

    6.14.2.3 Widespread agricultural hexaploid forms; *P. pratense* ssp. *pratense*.

6.14.2.4   An octoploid form restricted to southern Italy; *P. pratense* ssp. *pratense*.

6.14.3 *P. echinatum* Host. is a winter active annual grass of eastern Mediterranean mountains which is not used commercially.

The other three sections, *Chilochloa, Achnodon and Maillea* contain 11 species but none of these are commercial so will not be described here (Stewart et al. 2010).

6.15 Economic Importance

Timothy, *P. pratense* ssp. *pratense,* is commonly used in cold winter pastures for high quality hay. It is widely used especially in Scandinavia, northern Europe, northern Japan, Canada and northern USA but there is also a small amount used in Patagonia and in the south of New Zealand. Cultivars are all hexaploid and frequently divided into early and late flowering types.

The diploid Turf Timothy, *P. pratense* ssp. *bertolonii,* is used occasionally in northern Europe for turf purposes, often in mixtures with other turf grass species.

Alpine Timothy, *P. alpinum,* is used to a limited extent in Europe for high altitude revegetation plantings.

Each year approximately 34,000 tons of Timothy seeds are harvested, a few hundred tons of turf Timothy and only a few tons of Alpine Timothy. This makes *Phleum* the third most widely sown grass genus with 8% of the world's temperate grass seed, following *Lolium,* and *Festuca* (Bondesen 2007).

6.16 Academic Significance

Timothy has achieved academic significance in studies following the discovery that commercial forms have minimal vernalization requirement for flowering, with numerous studies being undertaken to further understand this flowering response. As a consequence, Timothy has no strong mechanism to prevent reproductive development except during summer and the growing point becomes elevated and vulnerable to removal by grazing, making it susceptible to grazing damage. However, a vernalization response is known in diploid germplasm (Cooper and Calder 1964) and more recent studies have shown considerable variation in vernalization requirement within Timothy (Fiil et al. 2011) but breeders have not introduced this feature into commercial cultivars.

6.17 Botanical Origin and Evolution

The molecular results of Stewart et al. (2010) show an Asian origin for the section *Phleum* and identify two separate migrations into Europe.

The first migration into Europe was of an ancestor of diploid *P. alpinum* subsp. *rhaeticum*. The penultimate Riss glaciation 130,000–150,000 year B.P. provided ample opportunity for this subalpine species to migrate vast distances through lowland areas to eventually become isolated on the Alps during the subsequent warmer interglacial period. Subsequent migration along mountain ranges has occurred so that today *rhaeticum* occurs in the Alps, Pyrenees, Apennines and the Balkans.

Migration also occurred onto the colder mountain ranges to the north into Germany and to the Carpathian Mountains of Poland and Romania but was associated with micro-evolutionary changes in morphology and cytology to develop into diploid "commutatum". The overlap of the range of *rhaeticum* and "commutatum" has since allowed considerable hybridization so that a swarm of hybrids overlaps the range of "commutatum" and part of the *rhaeticum* range. Occasional tetraploid hybrids have developed and these have migrated back east, at least as far as Kazakhstan.

Migration of *rhaeticum* populations back into the lowlands as a result of climate cooling eventually resulted in the first lowland species of this group, *P. pratense* subsp. *bertolonii*. This was also accompanied by micro-evolutionary changes in cytology, morphology and adaptation. As the climate cooled during the last glaciation (the Würm 22,000 to 13,000 yr B.P.), this lowland species retreated into southern European glacial refugia. Upon warming these subsequently reinvaded northern Europe from the Balkan/Italy refugia. Those in the Spanish/Portuguese glacial refuge remained restricted to that region. Hybridization occurred when these two forms met at the interface in France resulting in the generation of a recent autotetraploid.

Hybrids formed in the Italian Alps where subsp. *bertolonii* and the Balkans *rhaeticum* overlapped resulting in an allotetraploid *pratense*. It is probable that a further hybridization with the adjacent northern European subsp. *bertolonii* lead to the formation of the agricultural hexaploid *pratense*. Upon warming in the holocene these subsequently reinvaded northern Europe from the Balkan/Italy refugium, a refugium common to a wide range of European biota (Hewitt 1996, 1999) (Fig. 6-2).

Hexaploid and octoploid forms occur today within known glacial refugia, with two different hexaploids in southern Italy and Morocco and an octoploid in the mountains of southern Italy.

The very widespread allotetraploid *P. alpinum* formed over 300,000 yr B.P. in Asia from hybridization of an ancestral *rhaeticum* with another unknown genome. This form remained in Asia until eventually migrating

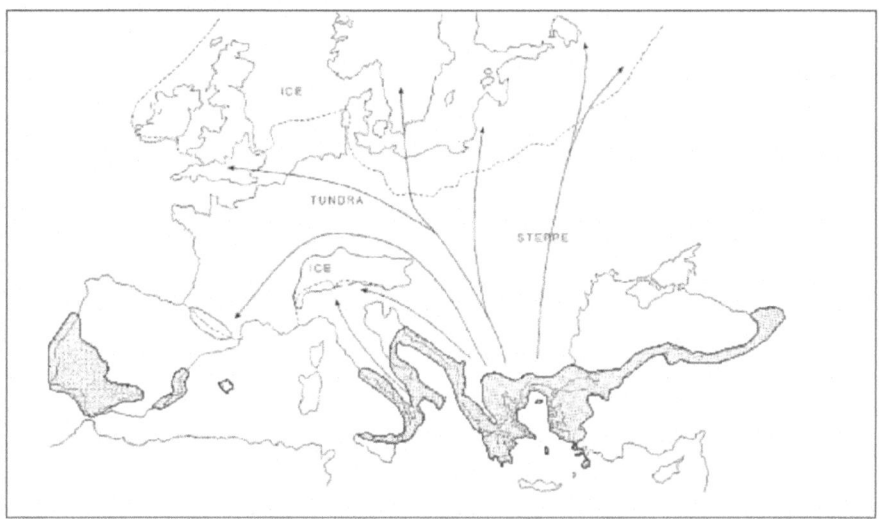

Figure 6-2 Generic glacial refugia of southern Europe (shaded areas) (after Hewitt 1999) and potential post-glacial migration route of diploid ssp. *bertolonii* and agricultural hexaploid *pratense.*

into Europe during the last glaciation (the Würm 22,000 to 13,000 yr B.P.), when conditions became suitable. At the same time many species, including this one, were able to migrate into the Americas via the Bering/Aleutian route, although probably not completing their entry into America until this route became open around 8000 yr (Hong et al. 1999; Weber 2003). This divergent migration has led to a divergence in molecular forms, one in northern Europe and the other in Japan and the Americas. The circumpolar migration was completed in Iceland where derivatives of both forms now occur.

Dogan (1991) describes the Mediterranean and western Asia as the center of origin for the genus *Phleum.* This remains true for *P. pratense* even today, although it may be considered surprising that a species so common in high latitude cold temperate zones originated within the southern European glacial refugia and that these areas still retain high genetic diversity today. Although it has been suggested that northern Europe is a center of diversity (Guo et al. 2003), this appears unlikely as it has only been free of glaciation in the last 12,000 yr and any diversity must be recent, or of migratory origin.

6.18 Domestication and Dissemination

The northern European hexaploids were the first to be domesticated and used in agriculture largely as a result of their natural distribution and value for hay. These have been disseminated around the world to all major cold temperate regions, particularly Asia, Japan, North and South America, Australia and New Zealand. This process of dissemination around the world has been driven by human migration and trade.

6.19 Germplasm and Gene Pools

Genetic resources for commercial Timothy breeding at the hexaploid level can be divided into four genepools on the following basis (Stewart et al. 2010).

6.19.1 Primary Gene Pool

These can be defined as cultivars and elite breeding lines adapted to the region of agricultural use. This is largely northern Europe, northern Asia and North America and to a much lesser extent New Zealand. This is the primary source of material for Timothy breeders and consists almost exclusively of hexaploids, although the diploids subsp. *bertolonii* are used occasionally for turf. In general, these resources are frequently used by breeders and their "working" collections largely represent this genepool.

6.19.2 Secondary Gene Pool

This pool includes hexaploid germplasm of *P. pratense* from the center of origin in the Balkans/Italy region of southern Europe, the Mediterranean mountains and North Africa, a region largely outside the region of commercial use. These regions have never been targeted for genebank collection because this material is not sufficiently winter hardy for use in more northern European climates. However, such material deserves special attention as it represents unique genomic hexaploid constitutions, and genetic erosion in these southern regions is occurring at an alarming rate as many *in situ* populations are under threat from climate warming and human-induced habitat degradation. Genebank material from these southern regions is very limited and we know of only one sample available from North Africa and only two from southern Italy.

6.19.3 Tertiary Gene Pool

This consists of germplasm of ploidy levels other than hexaploid. Ploidy remains the major barrier to hybridization between *Phleum* species but apart from ploidy difficulties most forms cross readily (Nath 1967). The one exception to this is the widespread paleo-allotetraploid *P. alpinum* which is difficult to cross and here we would classify this into the quaternary genepool.

With the latest knowledge of the genomic constitution of *P. pratense* (Stewart et al. 2010) it should now be possible to either resynthesize

P. pratense from different forms of the same genomes, or use genomes from other forms and ploidy levels of section *Phleum* for introgression into *P. pratense* hexaploids.

Crosses between diploid *bertolonii* and hexaploid *P. pratense* are easy to produce artificially (Nordenskiold 1945), and although they are not always tetraploid (Løhde 1978), they are found in nature (Müntzing 1935; Foerster 1968, 2005). These readily cross back to hexaploid *P. pratense*.

Hybrids between tetraploid *P. pratense* forms and the octoploids have been made by the senior author and these form fertile hexaploid plants. These plants have been crossed with regular agricultural hexaploids to generate fertile progeny.

Hybrids of tetraploid and hexaploid forms are pentaploid as expected (Nielsen and Nath 1961) and these may be backcrossed to hexaploids quite easily.

This tertiary germplasm pool represents an enormous untapped pool of unexplored material for breeders, but collections and molecular characterization will be necessary.

6.19.4 Quaternary Genepool

By definition these would consist of the more difficult to cross material such as widespread allotetraploid *P. alpinum*, as well as species in other sections. Other crosses which appear to be possible with hexaploid *P. pratense* include *P. phleoides* in section *Chilochloa*, of which there is a herbarium sample in the Leiden University National Herbarium in the Netherlands; *P. hirsutum* in section *Chilochloa*, which the senior author has crossed to obtain tetraploids; *P. subulatum* in section Achnodon, which Myers (1941) crossed to obtain a male sterile tetraploid progeny.

There is also a report of a sterile cross between tetraploid *Dactylis glomerata* and hexaploid *P. pratense* (Nakazumi et al. 1997).

At this stage the quaternary genepool offers much less potential for breeders than the secondary and tertiary genepools and resources would be much better targeted at the latter.

6.19.5 Cytology

There have been numerous cytological studies in *Phleum*, largely in an effort to understand the relationship between the genomes in *P. pratense*, for example as described by Joachimiak (2005). The B chromosomes have also been studied (Bosemark 1967; Fröst 1969; Joachimiak 1982, 1986). There have been no attempts to develop chromosome substitution or deletion lines.

Anther culture methodology has been developed and triploid and hexaploid regenerants have been obtained (Abdullah et al. 1994).

6.20 Genomic and Molecular Resources in *Phleum*

Molecular techniques have been developed for phylogenetic and genomic identification (Stewart et al. 2010). A study on genetic resources of Nordic Timothy has been undertaken in Norway using molecular techniques (Fjellheim et al. 2007).

Only a few studies for molecular marker development were reported. Cai and Bullen (1994) has isolated genome-specific sequences for *Phleum* species from *Sau3AI* genomic libraries. Cai et al. (2003) developed a set of simple sequence repeat (SSR) markers from SSR-enriched genomic libraries, which were used to assess diversity in Japanese Timothy top-crosses to predict yield potential (Tanaka et al. 2011). Jonavičienė et al. (2009) applied inter-simple sequence repeat (ISSR) markers to study the genetic diversity in Lithuanian Timothy and a team in Finland announced its intention in 2006 (Manninen et al. 2006) to use a candidate gene approach, bulked segregant analysis and gene expression chips to find modern genetic tools applicable to Timothy breeding. This has resulted in a publication on the development of markers for feeding quality and gray snow mold resistance (Tanhuanpää et al. 2007, 2008).

Other genetic studies by groups in Japan, Lithuania and Denmark, have identified DNA sequences of Timothy genes involved in fructan biosynthesis (Tamura et al. 2009, 2011; Jonaviciene et al. 2012). In molecular linkage map construction, Cai et al. (2009) has reported the construction of hexaploid AFLP-SSR based linkage map and SSR-based diploid linkage map.

Although no genetically modified Timothy cultivar has been developed, Timothy has been included in a number of patents for genes for disease resistance, herbicide tolerance and allergy studies.

6.20.1 Recommendations for Future Actions

The current understanding of the genomic constitution within *Phleum* should allow breeders to utilize the genetic resources more effectively than before. It should now be possible to resynthesize or introgress a much wider range of diverse *P. pratense* than has occurred naturally.

However, many of the genetic resources of wild relatives are under threat from climate warming and human induced habitat degradation ('t Mannetje 2007).

There is an urgent need to collect hexaploid *P. pratense* germplasm from Mediterranean mountain glacial refuge areas as well as a wide range of genetically diverse diploid, tetraploid and octoploid *P. pratense* and the readily crossable forms of *P. alpinum*. These include diploid subspecies *rhaeticum* and "commutatum" as well as their diploid and tetraploid hybrids.

These collections should be integrated into core collections to maximize molecular diversity of the available genomes.

It is also important that each of the major regions where Timothy is used maintains strong functional field breeding programs to allow adequate cultivar development, germplasm collection, introgressions of wild germplasm and exploration of molecular resources.

References

Abdullah AA, Pedersen S, Andersen SB (1994) Triploid and hexaploid regenerants from hexaploid Timothy (*Phleum pratense*) via anther culture. Plant Breed 112: 342–345.

Apirion D, Zohary D (1961) Chlorophyll lethal in Natural populations of the Orchard Grass (*Dactylis glomerata* L.). A case of balanced polymorphism in plants. Genetics 46: 393–399.

Ardouin P, Fiasson JL, Jay M, Lumaret R, Hubac JM (1985) Chemical diversification within *Dactylis glomerata* L. polyploid complex (*Graminaceae*). In: Jacquard P et al. (eds) Proc NATO Symp on Genetic Differentiation and Dispersal in Plants. Vol G5, Springer, Basel, Switzerland.

Barkworth ME (2007) Flora of North America, North of Mexico; v. 24: *Magnoliophyta*; *Commelinidae (in part): Poaceae*, part 1. Barkworth ME, Capels KM, Long S, Anderton LK, Piep MB (eds) New York: Oxford U. Press. 911 pp.

Bondesen OB (2007) Seed production and seed trade in a globalised world. Seed production in the northern light. Proc 6th Int Herbage Seed Conf., Norway, pp 9–12.

Borrill M (1961) Patterns of morphological variation in diploid and tetraploid *Dactylis*. Bot J Linn Soc 56: 441–439.

Borrill M (1977) Evolution and genetic resources in Cocksfoot. Annu Rep Welsh Plant Breed. Stat, pp 90–209.

Bosemark NO (1967) Edaphic factors and the geographic distribution of accessory chromosomes in *Phleum phleoides*. Hereditas 57: 239–262.

Bushman BS, Larson SR (2010) Transferability of orchardgrass (*Dactylis glomerata*) SSR primers and genetic diversity of orchardgrass subspecies. Poster presentation, 6th Int. Symp. Mol Breed. of Forage and Turf, 15–19 March, 2010, Buenos Aires, Argentina.

Cai Q, Bullen MR (1994) Analysis of genome specific sequences in *Phleum* species: identification and use for study of genomic relationships. Theor Appl Genet 88: 831–837.

Cai HW, Inoue M, Yuyama N, Hirata M (2009) Genome mapping in cool-season forage grass. In: Yamada T, Spangenberg G (eds) Molecular Breeding of Forage and Turf. Springer, pp 173–183.

Cai HW, Yuyama N, Tamaki H, Yoshizawa A (2003) Isolation and characterization of simple sequence repeat markers in the hexaploid forage grass timothy (*Phleum pratense* L.). Theor Appl Genet 107: 1337–1349.

Cho MJ, Choi HW, Lemaux PG (2001) Transformed T0 orchardgrass (*Dactylis glomerata* L.) plants produced from highly regenerative tissues derived from mature seeds. Plant Cell Rep 20: 318–324.

Congera BV, Lowe KW, Carabaia JV (1980) Relative DNA content of cells in *Festuca arundinacea* and *Dactylis glomerata* calli of different ages. Envir Exp Botany 20: 401–408.

Cooper JP, Calder DM (1964) The inductive requirements for flowering of some temperate grasses. Grass Forage Sci 19: 6–14.

Creber HMC, Davies MS, Francis D, Walker DHD (1994) Variation in DNA C Value in Natural Populations of *Dactylis glomerata* L. New Phytol 128: 555–561.

Dogan M (1991) A taxonomical revision of the genus *Phleum* L. (Gramineae). Karaca Arbor Mag 1: 53–70.

Falistocco E, Lorenzetti S, Falcinelli M (1996) Microsporogenesis in a desynaptic mutant of diploid *Dactylis*. Cytologia 59: 309–316.

Fiasson JL, Ardouin P, Jay M (1987) A phylogenetic groundplan of the specific complex *Dactylis glomerata*. Biochem Syst Ecol 15: 225–230.

Fiil A, Jensen LB, Fjellheim S, Lübberstedt T, Andersen JR (2011) Variation in the vernalization response of a geographically diverse collection of Timothy genotypes. Crop Sci 51: 2689–2697.

Fjellheim S, Pedersen AJ, Andersen JR, Antonius-Klemola K, Bondo L, Brantestam AK, Dafgård L, Helgadottir A, Isolahti M, Jensen LFB, Lübberstedt T, Mannien O, Marum P, Merker A, Tanuanpää P, Weibull J, Weibull P, Rognli OA (2007) Phenotypic and molecular characterization of genetic resources of Nordic timothy (*Phleum pratense* L.). In: Abstract Book of XXVIIth EUCARPIA Symposium on Improvement of Fodder Crops and Amenity Grasses, held 19th August to 23rd August 2007 Copenhagen.

Foerster E (1968) Ein beitrag zur untercheidung von *Phleum pratense* und *Phleum nodosum*. Göttinger Flor Rundbreife 1: 9.

Foerster E (2005) Natürliche Hybriden zwischen *Phleum pratense* und *Phleum bertolonii* (Natural hybrids between *Phleum pratense* and *Phleum bertolonii*). Gründungstagung der Vereinigung zur Erforschung der Flora Deutschlands, 29 und 30 Oktober 2005 in Vechta.

Fröst S (1969) The inheritance of accessary chromosomes in plants, especially in *Ranunculus acris* and *Phleum nodosum*. Hereditas 61: 317–326.

Guo Y-D, Yli-Matilla T, Pulli S (2003) Assessment of genetic variation in timothy (*Phleum pratense* L.) using RAPD and UP-PCR. Hereditas 138: 101–113.

Hewitt GM (1996) Some genetic consequences of ice ages, and their role in divergence and speciation. Biol J Linn Soc 58: 247–276.

Hewitt GM (1999) Post glacial recolonisation of European biota. Biol J Linn Soc 68: 87–112.

Hong Q, White P, Klinka K, Chourmouzis C (1999) Phytogeographical and community similarities of alpine tundras of Changbaishan Summit, and Indian Peaks, USA J Veg Sci 10: 869–882.

Humphries CJ (1978) Notes on the genus *Phleum* L. Bot J Linn Soc 76: 337–340.

Humphries CJ (1980) *Phleum*. In: Tutin TC et al. (eds) Flora Europeaea 5. *Alismataceae to Orchidaceae (Monocotyledones)*, Cambridge University Press, Cambridge, pp 239–241.

Joachimiak A (1982) Cyto-genetics of standard B-chromosomes in *Phleum boehmeri* from Poland. Acta Biol Cracov Ser Bot 24: 63–77.

Joachimiak A (1986) B-chromosome condensation in *Phleum* pollen grains. Genetica 68: 169–174.

Joachimiak A (2005) Heterochromatin and microevolution in *Phleum*. In: Sharma AK, Sharma A (eds) Plant Genome: Biodiversity and Evolution. Vol 1, Part B: Phanerogams. Science Publishers Inc. Enfield, New Hampshire, pp 89–117.

Joachimiak A, Kula A (1993) Cytotaxonomy and karyotype evolution in *Phleum* sect. *Phleum* (Poaceae) in Poland. Plant Syst Evol 188: 11–25.

Jonavičiené K, Paplauskiené V, Brazauskas G (2009) Isozymes and ISSR markers as a tool for the assessment of genetic diversity in *Phleum* spp. Zemdirbyste-Agriculture 96: 47–57.

Jonavičiené K, Studer B, Asp T, Jensen LB, Paplauskiene V, Lazauskas S, Brazauskas G (2012) Identification of genes involved in a water stress response in timothy and mapping of orthologous loci in perennial ryegrass. Biologia Plantarum 56: 473–483.

Kölliker R, Stadelmann FJ, Reidy B, Nösberger J (1999) Genetic variability of forage grass cultivars: A comparison of *Festuca pratensis* Huds., *Lolium perenne* L., and *Dactylis glomerata* L. Euphytica 106: 261–270.

Lewis EJ (1975) An alternate technique for the production of amphidiploids. Ann Rep Welsh Plant Breed Stn 1974: 15.

Litrico I, Bech N., Flajoulot S, Cadier D, Talon C, Gibelin C, Barre P (2009) Cross-species amplification tests and diversity analysis using 56 PCR markers in *Dactylis glomerata* and *Lolium perenne*. Mol Ecol Resources 9: 159–164.

Løhde JJH (1978) *Phleum pratense* and *Phleum bertolonii* hybridisation, morphology and ecology in Denmark. Thesis, Kongelige veterinaer-og Landbohoejskole, Denmark (1977) 80 pp From Årsskrift, Kongelige veterinaer- og Landbohoejskole 1978 p 167.

Lumaret R (1986) Doubled duplication of the structural gene for cytosolic phosphoglucose isomerase in the *Dactylis glomerata* L. polyploidy complex. Mol Biol Evol 3: 499–521.

Lumaret R (1987) Differential degree in genetic divergence as a consequence of a long isolation in a diploid entity of *Dactylis glomerata* L. from the Guizhou region (China). Presentation to the 2nd Symp Paleoenvironment East Asia, Hong Kong, Jan 9–14, 1987.

Lumaret R (1988) Cytology, genetics and evolution in the genus *Dactylis*. Crit. Rev. Plant Sci 7: 55–91.

Lumaret R, Bowman CM, Dyer TA (1989) Autotetraploidy in *Dactylis glomerata* L.: further evidence from studies of chloroplast DNA variation. Theor Appl Genet 78: 393–399.

Lumaret R, Barrientos E (1990) Phylogenetic relationships and gene flow between sympatric diploid and tetraploid plants of *Dactylis glomerata* (Gramineae). Plant Syst Evol 169: 1615–6110.

Lumerat R, Bretagnolle F, Maceira NO (1992) 2n gamete frequency and bilateral polyploidization in *Dactylis glomerata*. In: Mariana A, Tavoletti S. Gametes with somatic chromosome number in the evolution and breeding of polyploid polysomic species: achievements and perspectives, Perugia: Forage Plant Breeding Institute 15–21.

Manninen O, Erkkilä M, Isolahti M, Nissinen O, Pärssinen P, Rinne M, Tanhuanpää P (2006) Biotechnological tools for breeding feeding quality and optimal growth rhythm in timothy, *Phleum pratense*. Timothy productivity and forage quality—possibilities and limitations. NJF Seminar 384, 10–12 August 2006 Akureyri, Iceland 119–120.

Matzk F (1981) Successful crosses between *Festuca arundinacea* Schreb. and *Dactylis glomerata* L. Theor Appl Genet 60: 119–122.

Müntzing A (1935) Cyto-genetic studies on hybrids between two *Phleum*-species Hereditas 20: 103–136.

Myers, WM (1941) Meiotic behaviour of *Phleum pratense*, *Phleum subulatum* and their F1 hybrid. J Ag Res 63: 649–655.

Nakazumi H, Furuya M, Shimokouji H, Fujii H (1997) Wide hybridization between timothy (*Phleum pratense* L.) and orchardgrass (*Dactylis glomerata* L.) Bull Hokkaido Prefectural Agric Expt Stat (Japan) 72: 11–16.

Nath J (1967) Cytogenetical and related studies in the genus *Phleum* L. Euphytica 16: 267–282.

Nielsen EL, Nath J (1961) Cytogenetics of a tetraploid form of *Phleum pratense* L. Euphytica 10: 343–350.

Nördenskiold H (1945) Cyto-genetic studies in the genus *Phleum*. Acta Agr Suec 1: 1–138.

Oertel C, Fuchs J, Matzk F (1996) Successful hybridization between *Lolium* and *Dactylis*. Plant Breed 115: 101–105.

Parker PF, Borrill M (1968) Studies in *Dactylis*. 1 Fertility relationships in some diploid subspecies. New Phytol 67: 649–662.

Peng Y, Zhang X-Q, Deng Y, Ma X (2008) Evaluation of genetic diversity in wild orchardgrass (*Dactylis glomerata* L.) based on AFLP markers. Hereditas 145: 174–181.

Reeves G, Francis D, Davies MS, Rogers HJ, Hodkinson TR (1998) Genome size is negatively correlated with altitude in natural populations of *Dactylis glomerata*. Ann Bot (Lond) 82 (Suppl A): 99–105.

Robins JG, Bushman BS, Jensen KB (2008) New genomic resources for orchardgrass. Proc 27th EUCARPIA Symp. Improvement of Fodder Crops and Amenity Grasses. Copenhagen, Denmark, 19th to 23rd August 2007, pp 202–203.

Sahuquillo E, Lumaret R (1995) Variation in the subtropical group of *Dactylis glomerata* L. Evidence from allozyme polymorphism. Biochem Syst Ecol 23: 407–418.

Sahuquillo E, Lumaret R (1999) Chloroplast DNA variation in *Dactylis glomerata* L. taxa endemic to the Macaronesian islands. Mol Ecol 8: 1797–1803.

Soltis DE, Soltis PS (1993) Molecular data and the dynamic nature of polyploidy. Crit Rev Plant Sci 12: 243–273.

Song Y, Liu F, Zhu Z, Tan L, Fu Y, Sun C, Cai H (2011) Construction of a simple sequence repeat marker-based genetic linkage map in the autotetraploid forage grass *Dactylis glomerata* L. Grassland Sci 57: 158–167.

Stebbins GL, Zohary D (1959) Cytogenetic and evolutionary studies in the genus *Dactylis*. I Morphological, distribution and inter relationships of the diploid subspecies. Univ Calif Pub Bot 31: 1–40.

Stebbins GL (1971) Chromosomal Evolution in Higher Plants. Edward Arnold, London.

Stewart AV, Ellison N (2010) The genus *Dactylis*; Wealth of Wild Species: Role in Plant Genome Elucidation and Improvement. (Springer: New York).

Stewart AV, Ellison N, Joachimiak A (2010) The Genus *Phleum*; Wealth of Wild Species: Role in Plant Genome Elucidation and Improvement. Springer, New York.

Tamura K, Kawakami A, Sanada Y, Tase K, Komatsu T, Yoshida M (2009) Cloning and functional analysis of a fructosyltransferase cDNA from synthesis of highly polymerized levans in timothy (*Phleum pratense* L.). J Exp Bot 60: 893–905.

Tamura K, Sanada Y, Tase K, Komatsu T, Yoshida M (2011) Pp6-FEH1 encodes an enzyme for degradation of highly polymerized levan and is transcriptionally induced by defoliation in timothy (*Phleum pratense* L.). J Exp Bot 62: 3421–3431.

Tanaka T, Tamaki H, Cai HW, Ashikaga K, Fujii H, Yamada T (2011) DNA profiling of seed parents and a topcross tester and its application for yield improvement in Timothy (*Phleum pratense* L.). Crop Sci 51: 612–620.

Tanhuanpää P, Erkkilä M, Nissinen O, Rinne M, Manninen O, Isolahti M, Pärssinen P (2007) Developing DNA markers for feeding quality and gray snow mould resistance in timothy, *Phleum pratense*. In: Plant GEM 6: plant genomics European meeting, 3–6 October 2007, Tenerife, p 94.

Tanhuanpää P, Isolahti M., Nissinen O, Pärssinen P, Kalendar R, Schulman A, Manninen O (2008) Identification of DNA markers for gray snow mold resistance in timothy, *Phleum pratense*, using bulked segregant analysis. In: Molecular mapping and marker assisted selection in plants, Vienna, Austria 3–6 February 2008, p 68.

Tosun M, Akgun I, Taspinar MS (2002) Determination of variations of some enzymes in orchardgrass (*Dactylis glomerata* L.) ecotypes. Soil Plant Sci 52: 110–115.

Trejo-Calzada R, O'Connell MA (2005) Genetic diversity of drought-responsive genes in populations of the desert forage *Dactylis glomerata*. Plant Sci 168: 1327–1335.

Tuna M, Teykin E, Buyukbaser A, Budak H (2007) Nuclear DNA variation in the grass genus *Dactylis* L. Poster presentation, 5th Int Symp Mol Breed of Forage and Turf, 1–6th July, 2007, Sapporo, Japan.

Tuna M, Khadka DK, Shrestha MK, Arumuganathan K, Golan-Goldhirsh A (2004) Characterization of natural orchardgrass (*Dactylis glomerata* L.) populations of the Thrace Region of Turkey based on ploidy and DNA polymorphisms. Euphytica 135: 39–46.

't Mannetje L (2007) Climate change and grasslands through the ages: an overview. Grass Forage Sci 62: 113–117.

Van Dijk GE (1961) The inheritance of harsh leaves in tetraploid cocksfoot. Euphytica 13: 305–313.

Vilhar B, Vidic T, Jogan N, Dermastia M (2002) Genome size and the nucleolar number as estimators of ploidy level in *Dactylis glomerata* in the Slovenian Alps. Plant Syst Evol 234: 1–13.

Weber WA (2003) The Middle Asian element in the Southern Rocky Mountain flora of the western United States: a critical biogeographical review. J Biogeog 30: 649–688.

Weibull J, Ottosson F, Kolodinska Brantestam A, Dafgard L, Weibull P, Merker A (2007) Vanishing variation—the diversity of Timothy (*Phleum pratense* L.) in historical grasslands. 18th EUCARPIA Genetic Resource Section Meeting, Plant Genetic Resources and their Exploitation in the Plant Breeding for Food and Agriculture, Piešťany, Slovak Republic 23–27 May 2007.

Wetschnig W (1991) Karyotype Morphology and Some Diploid Subspecies of *Dactylis glomerata* L. (Poaceae). Phyton Horn 31: 35–55.

Williams E, Barclay PC (1972) Transmission of B-chromosomes in *Dactylis*. NZ J Bot 10: 573–584.

Xie W-G, Zhang X-Q, Ma X, Cai H-W, Huang L-K, Peng Y, Zeng B (2010) Diversity comparison and phylogenetic relationships of cocksfoot (*Dactylis glomerata* L.) germplasm as revealed by SSR markers. Can J Plant Sci 90: 13–21.

Xie W, Zhang X, Cai H, Huang L, Peng Y, Ma X (2011) Genetic maps of SSR and SRAG markers in diploid orchardgrass (*Dactylis glomerata* L.) using the pseudo-testcross strategy. Genome 54: 212–221.

Xie W, Robins JG, Bushman BS (2012) A genetic linkage map of tetraploid orchardgrass (*Dactylis glomerata* L.) and quantitative trait loci for heading date. Genome 55: 360–369.

Zeng B, Zhang X-Q, Lan Y (2006) Genetic diversity of *Dactylis glomerata* germplasm resources detected by intersimple sequence repeats (ISSR) molecular markers. Hereditas (Beijing) 28: 1093–1100.

Zeng B, Zhang X-Q, Lan Y, Yang W-Y (2008) Evaluation of genetic diversity and relationships in orchardgrass (*Dactylis glomerata* L.) germplasm based on SRAP markers. Can J Plant Sci 88: 53–60.

Zenkteler M, Nitzsche W (1984) Wide hybridization experiments in cereals Theor Appl Genet 68: 311–315.

Zernig K (2005) *Phleum commutatum* and *Phleum rhaeticum* (*Poaceae*) in the Eastern Alps: characteristics and distribution. Phyton 45: 65–79.

Zijp MJ (1960) Some observations on a possible x-ray mutant in cocksfoot (*Dactylis glomerata*) Euphytica 9: 222–224.

7

Bentgrasses and Bluegrasses

Keenan Amundsen and *Scott Warnke**

ABSTRACT

Species in the genus *Agrostis* are widely utilized as highly managed turfgrasses for recreation throughout the world and species in the genus *Poa* are widely adapted for use on sports fields and lawns. The progress related to genomics of these species has been limited by their complex genetic systems. The genus *Agrostis* is a polyploidy series with high levels of outcrossing that has created a very complex speciation history and limits the progress that can be made using genomics technologies. The genus *Poa* is widespread throughout the world with the most widely cultivated species *Poa pratensis* being apomictic that limits the ability to apply genomics technologies. The progress to date has primarily resulted in an improved understanding of the genetic diversity of these important species. Future progress will require careful planning to overcome the complex breeding systems of these important turfgrass species.

Key words: *Agrostis, Poa,* Polyploidy, Genetic diversity, Miniature inverted repeat transposable elements, Genetic mapping, Apomixis

7.1 Introduction

Species in the genera *Agrostis* and *Poa* are often used as turfgrasses throughout the world as they are capable of spreading by rhizomes and stolons creating an interconnected sward that is capable of repairing damage caused by biotic and abiotic stresses. The genetics of the species in these genera are complex with frequent polyploidy and apomixis operating in

USDA-ARS, Beltsville, MD 20705, USA.
*Corresponding author: *Scott.Warnke@ars.usda.gov*

Poa. In addition, these genera are not primary human food sources and in most cases are used as ornamentals or on recreational fields. Therefore, funding for research to improve these species through breeding is limited. However, turfgrasses play a significant role in the lives of many people and the importance of developing new species and cultivars that require fewer inputs while maintaining high quality is being recognized by research funding agencies. The combination of higher funding levels and lower cost technologies such as high-throughput sequencing will hopefully lead to significant improvements in these genera in the near future.

7.1.1 Agrostis

There are more than 150 species of *Agrostis* (Harvey 2007), important because several *Agrostis* species (the bentgrasses) can form interspecific hybrids and wild germplasm may be sources of novel stress tolerance. Bentgrass does not make a significant contribution to forage, but some of the finer textured species with high shoot density have aesthetic qualities that make them ideally suited for use as a turf. Creeping bentgrass (*Agrostis stolonifera* L.) is the most widely utilized turf species for highly managed playing surfaces such as golf course greens, tees and fairways (Turgeon 1996). It outperforms other cool season turfgrass species because of its strong stoloniferous growth, ability to maintain a high level of uniformity after mowing, fine leaf texture, high shoot density, and tolerance of mowing down to heights of 3 mm (Warnke 2003). Creeping bentgrass is well adapted to periodically flooded, well-drained, fine-textured, fertile soils (Beard 1973). It is believed to have originated in Eurasia and is found throughout the world in cool, humid temperate climates (Harvey 2007). While it is one of the most hardy of the cool season grasses used as turf, the quality of creeping bentgrass is adversely affected by wear, soil compaction and a number of pathogens such as dollar spot (*Sclerotinia homoeocarpa* F.T. Benn.) and brown patch (*Rhizoctonia solani* Kühn). Modern cultivars of creeping bentgrass have superior quality over accessions used just a few decades ago (2003 National Bentgrass Test, National Turfgrass Evaluation Program, http://www.ntep. org/), but even currently used varieties are adversely affected by stress. The focus of many *Agrostis* breeding programs is on developing improved biotic and abiotic stress tolerant germplasm (Bonos et al. 2006) while making selections based on morphological characters that contribute to high turf quality at low mowing heights (Warnke 2003; Phillips 2007). The efficiency of new cultivar development could be improved by identifying molecular markers linked to traits of interest and making selections based on those molecular markers (Warnke 2003).

Colonial bentgrass (*Agrostis capillaris* L.) and velvet bentgrass (*A. canina* L.) are also commonly used as turf, but differ from creeping bentgrass since they are not stoloniferous. Hybrids generated from wide crosses to incorporate wild germplasm into cultivated material have been a successful method for introducing novel genes for plant improvement in breeding programs (Stalker 1980; Kalloo 1992). Even though many of the species used as turf have evolved from naturally occurring interspecific hybrids (Casler and Duncan 2003), Belanger et al. (2003) was the first to investigate interspecific hybridization as a means of improving *Agrostis* species. Belanger et al. demonstrated that successful hybrids were formed between *A. stolonifera* and *A. canina, A. capillaris, A. gigantea,* and *A. castellana.* Belanger et al. (2004) generated hybrids between *A. stolonifera* and *A. capillaris* to test if interspecific hybridization could be used to improve dollar spot resistance in *A. stolonifera.* Dollar spot is an economically important disease for the turfgrass industry and *A. stolonifera* is considered susceptible to the disease while *A. capillaris* is tolerant. Hybrid plants were recovered with significantly more dollar spot resistance than the *A. stolonifera* parent suggesting that interspecific hybridization could be an effective tool to improve *Agrostis* species.

7.1.2 Polyploidy and Genome Size Variation

A number of *Agrostis* species including those used as turf are polyploids, with the common turf-type *Agrostis* species being either diploid (*A. canina*) or tetraploid (*A. stolonifera* and *A. capillaris*). Jones (1956a,b,c) directed the first studies to examine the genome constitution of the species *A. canina, A. vinealis, A. capillaris, A. stolonifera* and *A. gigantea.* Jones' (1956a,b,c) data on chromosome pairing behavior during metaphase I of meiosis suggests that *A. vinealis* is an autotetraploidization of *A. canina,* the tetraploids *A. capillaris* and *A. stolonifera* share the A_2A_2 subgenome in common and the hexaploid *A. gigantea* ($A_1A_1A_2A_2A_3A_3$) has two sub-genomes in common with both *A. capillaris* ($A_1A_1A_2A_2$) and *A. stolonifera* ($A_2A_2A_3A_3$). Jones warned that this naming convention was used to describe the chromosome set interactions of the hybrids, but it should not imply that the A_2 genome is identical between the species because the *A. stolonifera* X *A. capillaris* and *A. stolonifera* X *A. gigantea* hybrids had increased sterility and lower seed set. Zhao et al. (2007) conducted similar cytology experiments on *Agrostis* hybrids and found chromosome associations between *A. stolonifera* and *A. capillaris,* and *A. stolonifera* and *A. gigantea.* Zhao et al. also found chromosome associations in hybrids of *A. stolonifera* crossed with *A. idahoensis* ($2n = 4x = 28$), *Polypogon monspeliensis* ($2n = 4x = 28$), *P. fugax* ($2n = 6x = 42$), and *P. viridis* ($2n = 4x = 28$). The findings of Zhao et al. (2007) were consistent with Jones' earlier work. The observations by Jones on chromosome associations during metaphase I of meiosis are still cited as the most comprehensive investigations into the subject (Warnke 2003; Belanger et al. 2003a; Zhao et al. 2007).

There is increased interest, in revisiting the early cytology experiments of Jones with modern molecular biology techniques to distinguish *Agrostis* species. Bonos et al. (2002) proposed using flow cytometry DNA content measures as a means of differentiating certain species of *Agrostis*. In the Bonos et al. study, the mean 2C DNA content (pg) of *A. canina*, *A. stolonifera*, *A. capillaris*, *A. vinealis*, *A. gigantea*, and *A. castellana* was determined to be 3.42, 5.27, 5.87, 6.31, 8.18, and 8.71 respectively. Additionally Bonos et al. counted the chromosomes from metaphase cells of *A. canina* (2n = 2x = 14), *A. capillaris* (2n = 4x = 28), *A. stolonifera* (2n = 4x = 28), *A. castellana* (2n = 6x = 42), and *A. alba* (2n = 6x = 42). A strong correlation between 2C DNA content and ploidy level was observed. The DNA content values reported by Bonos et al. (2002) were consistent with other studies by Hollman et al. (2005), Arumuganathan et al. (1999), and Arumuganathan and Earle (1991). Similarly, Hollman et al. (2005) were able to identify putative *A. canina* accessions based on random amplified polymorphic DNA (RAPD) markers and flow cytometry. The advantage of predicting ploidy by flow cytometry over chromosome counts is the speed at which the DNA content can be measured. Amundsen et al. (2010) used flow cytometry as a rapid tool to classify more than 300 *Agrostis* accessions by ploidy but found a near continuous distribution of DNA content measures making ploidy predictions difficult (Fig. 7-1). In cases where DNA content does not fall into a clearly defined ploidy level group, chromosome counts could be done to determine ploidy (Jones 1956a,b,c; Ahloowalia 1965; Bonos et al. 2002; Zhao et al. 2006).

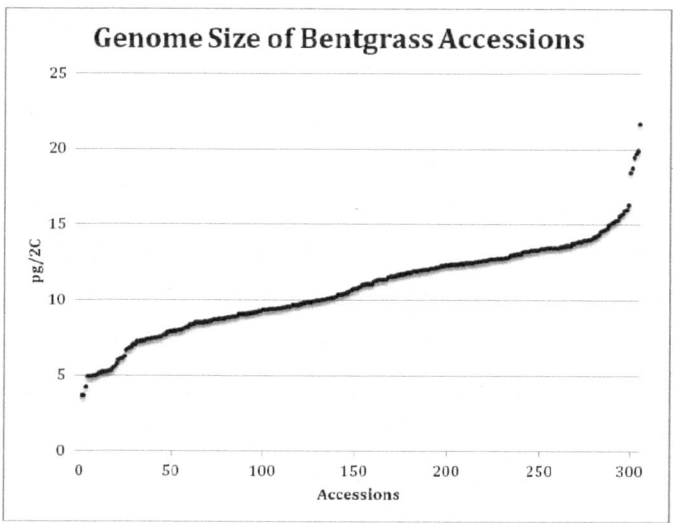

Figure 7-1 DNA content variation based on Flow cytometry data from 300 *Agrostis* accessions maintained by the US National Plant Germplasm System.

7.1.3 Genetic Segregation

The chromosome pairing behavior of tetraploid creeping bentgrass has been studied using isozyme markers (Warnke et al. 1998) and by looking at coupling vs repulsion phase linkage prevalence with dominant Amplified Fragment Length Polymorphism (AFLP) markers (Chakraborty et al. 2007). Isozyme markers were studied in controlled crosses and the presence of fixed heterozygosity as well as segregation ratios that could only be explained by disomic inheritance strongly supported that creeping bentgrass is a bivalent forming allotetraploid. Chromosome pairing behavior can also be studied on a genome-wide basis by looking at the prevalence of repulsion phase linkages in the segregating progeny of controlled crosses. Random chromosome pairing or autopolyploidy results in the number of repulsion phase linkages being significantly less than coupling phase linkages and the repulsion phase linkages are located at the ends of linkage groups. The analysis using AFLP markers resulted in an equal number of coupling and repulsion phase linkages and were evenly spaced along the linkage groups supporting disomic inheritance or allopolyploidy.

7.1.4 Marker Development and Genetic Diversity

Genetic marker studies in *Agrostis* are complicated by a number of factors: 1) The major turf-type *Agrostis* species are highly outcrossing and exhibit high rates of heterozygosity. 2) Polyploidy is common and can result in poor chromosome pairing and reduced recombination. 3) Intermediate morphologies exist between taxa making taxonomic classification difficult. 4) The species in the genus *Agrostis* are long lived perennials with good asexual reproduction and in some cases clones do not even go through a sexual cycle.

    Molecular markers have been used extensively in studies of species differentiation and genetic diversity in *Agrostis*. For example, Warnke et al. (1997) examined isozyme markers to assess the relationships between 18 *A. stolonifera* lines including both cultivated and non-cultivated accessions. Polymorphic allozymes were scored for presence or absence and the frequency of each marker calculated. Nei's distance formula (Nei 1972) was used to calculate genetic distances, which were used to cluster the accessions. The European germplasm was found to be distinct from the United States cultivated material suggesting that the European material might be a source of novel germplasm to improve US cultivars. Zhao et al. (2006) investigated the genetic diversity among 39 *A. capillaris* accessions by AFLP. They compared genetic similarity of AFLP markers between each pair of accessions and found a considerable amount of genetic variation. Vergara and Bughrara (2003) used 400 AFLP markers to evaluate genetic diversity of 40 *Agrostis* accessions representing 14 different species. Based on chromosome counts they found that *A. transcaspica* is a diploid and based on clustering of AFLP markers this species may have contributed the A_3A_3 genome to the evolution of *A. stolonifera* and *A. gigantea*. Karlsen and Steiner (2007) used RAPD markers to describe the genetic diversity of *A. capillaris* plants in the Scandinavian region and found that collected material from the northern Scandinavian region was distinct from accessions present in the National Plant Germplasm System. Hollman et al. (2005) performed RAPD analysis to characterize *Agrostis* germplasm and found markers that were able to differentiate *A. canina, A. capillaris* and *A. stolonifera* germplasm. The DNA content and predicted ploidy level of each accession was validated by flow cytometry and supported the characterization of plants at the species level by RAPD markers. Caceres et al. (2000) identified seven restriction fragment length polymorphism (RFLP) markers that could differentiate five *A. stolonifera* accessions. Sequence characterized amplified region (SCAR) markers designed from RAPD fragments by Scheef et al. (2003) were found to co-segregate with either the A_1A_1 or A_3A_3 genomes of *A. stolonifera* and *A. capillaris* respectively and could therefore be used as sub-genome specific markers to differentiate the species. Amundsen et al. (2010) conducted the most thorough genetic marker study of *Agrostis* thus far, whereby 1,309 MITE-display genetic markers were applied to study the diversity of 75 *Agrostis* accessions along with six closely related species. MITE-display is a modified AFLP technique that anchors the amplified fragments to miniature inverted-repeat transposable element (MITE) sequences. MITE-display is a relatively new class of molecular markers that could target transcriptionally active regions of the genome because MITEs have a tendency to insert near genes (Casa et al. 2004). MITE-display may therefore be more informative than AFLP markers that map to non-coding regions. Amundsen et al. (2010) found eight distinct clusters of markers and combined with flow cytometry data found two distinct groups of diploid germplasm. The data presented by Amundsen et al. (2010) also demonstrates the early formation of a genetic bottleneck among cultivated *A. stolonifera* accessions. These studies further our understanding of *Agrostis* genomics and demonstrate that molecular marker techniques can be used for diversity studies and to distinguish closely related *Agrostis* species. The use of multi-allelic markers such as SSRs has not been widely adopted in *Agrostis* studies because they are complicated in outcrossing polyploids due to the complex banding profiles.

7.1.5 Map Construction and mapping Quantitative Trait Loci (QTLs)

Chakraborty et al. (2005) reported the first genetic linkage map of *Agrostis stolonifera*. This map includes 169 RAPD, 180 AFLP, and 75 cDNA RFLP markers for a total of 424 mapped loci covering 1,110 cM. Amundsen et al. (2010) report the estimated genome length of *A. stolonifera* to be 2,873 cM and therefore the current map of *A. stolonifera* covers only 39% of the genome. Rotter et al. (2009) developed the first *A. capillaris* genetic linkage map. The colonial bentgrass map was developed from 212 AFLP and 110 gene-based markers and covers approximately 1,156 cM. Chakraborty et al. (2006) conducted the only published quantitative trait loci (QTL) study in *Agrostis*. Chakraborty et al. (2006) compared dollar spot severity over a two-year period to genetic marker data in a Full-sib mapping population. Eight QTLs were identified by 15 markers that showed significant association with dollar spot resistance. Only the QTL on linkage group 7.1 had significant association with disease resistance over both years of the study. Since the creeping bentgrass map covers only 39% of the genome, QTL studies remain challenging in *Agrostis*. The development of more detailed genetic linkage maps will improve the effectiveness of QTL studies.

7.1.6 Expressed Sequence Tags (ESTs)

There is limited *Agrostis* sequence data available in publicly available databases, for example the National Center for Biotechnology Information had 21,656 *Agrostis* expressed sequence tag (EST) sequences as of September 21st, 2010. The majority (16,201 ESTs) of this sequence data was part of a collaborative EST sequencing project presented by Rotter et al. (2007). The *Agrostis* sequencing efforts of Rotter et al. (2007) produced 7528 and 8470 EST sequences of *A. capillaris* and *A. stolonifera* respectively. From an analysis of these sequences, Rotter et al. identified 177 and 161 *Oryza* COS (conserved ortholog set; Fulton et al. 2002) genes conserved in the *A. capillaris* and *A. stolonifera* DNA sequence databases respectively. Seven of the 161 *A. stolonifera*, and 11 of the 177 *A. capillaris* *Oryza* COS gene homologues were found to have conserved sequences in *Festuca arundinacea* Schreb., *Avena sativa* L., *Triticum aestivum* L., *Hordeum vulgare* L., *Brachypodium distachyon* (L.) Beauv., *Oryza sativa* L., and *Zea mays* L. These sequences were used in phylogenetic analyses which placed the *Agrostis* species closer to *F. arundinacea* than to *A. sativa*, which is interesting since *Agrostis* is in the same tribe as *A. sativa* while *F. arundinacea* is in the Poeae tribe. *Agrostis* was placed in the Aveneae tribe based on morphological characteristics and enzyme restriction patterns of chloroplast DNA (Watson 1990; Soreng and Davis 1998). The two tribes are closely related (Kellog 1998; Soreng et al. 2007) and share many common chloroplast DNA restriction sites (Soreng et al. 1990).

Another interesting analysis to come from the *Agrostis* sequence data is a predicted time of evolution for the species (Rotter et al. 2007). A number of the *Agrostis* COS genes were identified as having multiple distinct copies in the EST sequence data, which was expected since the two species sequenced were tetraploids. Four of the multiple copy COS sequences were found in both the *A. stolonifera* and *A. capillaris* sequence databases. Rotter et al. predicted which of the COS gene copies originated from the A_2A_2 subgenome by identifying which copy had the highest amount of sequence similarity between *A. stolonifera* and *A. capillaris*. By comparing the COS gene pairs to their corresponding *Oryza* ortholog and calculating the number of substitutions at synonymous sites and non-synonymous sites, it was determined that these COS genes were undergoing purifying selection. The rate of divergence of the sub-genomes for both *A. capillaris* and *A. stolonifera* were estimated. By calibrating to *Oryza* and comparing the rate of divergence of the *Agrostis* sub-genomes, Rotter et al. found the rate between *A. capillaris* A_1A_1 and A_2A_2 sub-genomes, *A. stolonifera* A_2A_2 and A3A3 sub-genomes, and *A. capillaris* and *A. stolonifera* A_2A_2 sub-genomes to be 8.9, 10.6, and 2.2 MYA (Million Years Ago) respectively, suggesting a recent origin of the tetraploid *Agrostis*.

7.2 *Poa*

Like the bentgrasses, the bluegrasses (*Poa* spp.) are difficult to categorize taxonomically because they are polyploid and hybridization events often result in intermediate morphologies. They are most easily distinguished from other grasses by their characteristic boat-shaped leaf tip (Beard 1973). Several species of *Poa* can be apomictic, either being derived from bisexual species or dioecious species. There are approximately 500 species of *Poa*, and several species used as both forage and turfgrasses (Soreng et al. 2007). Huff (2010) reports that *P. alpina* L., *P. annua* L., *P. arachnifera* Torr., *P. bulbosa* L., *P. compressa* L., *P. nemoralis* L., *P. palustris* L., *P. pratensis* L., *P. secunda* J. Prest., *P. supine* Schrad., and *P. trivialis* L. are the *Poa* species that currently have the most agronomic and economic importance.

Kentucky bluegrass (*Poa pratensis*) is an important cool season turfgrass species because it forms a high quality turf with dark green color, good uniformity and relatively high shoot density (Beard 1973). Kentucky bluegrass is vigorously rhizomatous, which helps its ability to recover from damage during the growing season, and also makes it a desirable species for sod production. Originating from Eurasia, it is found throughout the world and has adapted to cool, humid climates. Kentucky bluegrass is a facultative apomict that is believed to be an allopolyploid with a near continuous distribution of chromosome numbers from 24 to 124 (Huff 2003; Love and Love 1975). Annual bluegrass (*P. annua*) is another important species because it is widely distributed and often thought of as an invasive weed species. It is one of the few grass species that can compete with the bentgrasses on tightly mown golf course putting greens. A significant amount of research has been invested into the control of *P. annua*, but a second camp of research is centered on making use of the positive qualities of *P. annua* and developing it as a high quality turf (Huff 2003). Annual bluegrass can form a high quality turf with high shoot density and fine texture when properly managed (Beard 1973). Annual bluegrass is a self-pollinated tetraploid (2n = 4x = 28). Supina bluegrass (*P. supina*) is most commonly used on athletic fields and golf courses because of its ability to tolerate wear, shade and low temperature. Additionally, *P. supina* is strongly stoloniferous contributing to its use in athletic fields (Bughrara 2003). Supina bluegrass is a highly self-fertile diploid (2n = 2x = 14). Rough bluegrass (*Poa trivialis*) is also stoloniferous and well adapted to damp, shaded environments. Rough bluegrass has poor heat, drought and wear tolerance (Beard 1973). Rough bluegrass is a highly sexual diploid (2n = 2x = 14).

7.2.1 Marker Development and Genetic Diversity

Marker studies of *Poa* are often difficult to interpret due to the complexities of the genome. Several marker types have been used to study *Poa* spp. Frenot et al. (1999) used morphological characters and isozyme markers to study the variation among 30 randomly chosen individuals from six sub-Antarctic populations of *P. annua*. Morphological characters clearly differentiated Crozet and Kerguelen Island populations. The genetic diversity among these populations was difficult to assess because of the limited variation among the isozymes tested. The origins of Antarctic *P. annua* were also assessed in 48 individuals from each of three populations based on 133 polymorphic AFLP markers (Chwedorzewska 2008). The three populations were clearly differentiated from one another by a principal component analysis and analysis of molecular variance. Genetic diversity of AFLP and RAPD markers was assessed between Texas bluegrass (*P. arachnifera* Torr.) accessions representing distinct collection sites and having superior turf or forage characteristics (Renganayaki et al. 2001). A total of 2,541 polymorphic AFLP and 335 RAPD markers were scored. Genetic similarity of the accessions based on the AFLP markers ranged from 0.46 to 0.87, similar values were obtained with a separate analysis of the RAPD data. Each analysis divided the accessions into two groups based on UPGMA clustering and principal component analysis, but there was not a high degree of correlation between marker types ($r = 0.66$). Mengistu et al. (2000) also used 18 RAPD markers to test *P. annua* genetic diversity from 10 seed production fields in Oregon. The 1,357 individuals were divided into 47 populations and based on clustering the diversity between populations correlated well with herbicide treatments and season of germination. By screening 47 *Lolium* SSR markers in eight *Poa* species, Kindiger (2006) showed that cross-species amplification of SSRs was not an effective method for developing *Poa* markers. Five microsatellite markers were successfully used by Rudmann-Maurer et al. (2007) to assess genetic diversity of *P. alpina* accessions. There was limited genetic diversity within populations but an increased genetic diversity between populations at greater distances from one another, between grazed and mowed populations, and between sexual and vegetative reproducing populations.

7.2.2 Genetic Linkage Mapping

Porceddu et al. (2002) constructed the only genetic linkage map to date of *Poa pratensis* L. from 70 AFLP and 161 SAMPL (selectively amplified microsatellite polymorphic loci) markers in a population of 67 F_1 progeny of a cross between a highy apomictic clone (paternal parent) and a completely sexual clone (maternal parent). Paternal and maternal linkage maps were developed covering 367 and 338.4 cM respectively. The maps were based on 41 paternal and 47 maternal coupling phase markers.

7.2.3 Apomixis

Several reproduction systems are observed in *Poa*. For example *P. annua* is highly self-fertile, *P. trivialis* is highly outcrossing, and *P. wheeleri* is apomictic. *P. pratensis* is a facultative apomict and can produce both sexual and asexual seed. An ability to control apomixis in *P. pratensis* would allow for the development and perpetuation of elite cultivars with complex sets of genes (Albertini et al. 2004). Several studies have been conducted to identify genes involved in apomixis. Albertinie et al. (2004) used cDNA-AFLP to identify genes with differential expression between asexual and sexual reproductive genotypes. Of the 179 differentially expressed genes, 60% were similar to previously reported genes. Few genes were identified that were specific to either mode of reproduction. Three of the genes were analyzed in depth, *PpRAB1*, *PpARM*, and *PpAPK* because they were believed to be involved in signal transduction, but further work needs to be done to determine if they are involved in apomixis. The genes *PpSERK* (*somatic empryogenesis receptor-like kinase*) and *APOSTART* were isolated from the same cDNA-AFLP experiments (Albertini et al. 2005). The *SERK* genes are believed to play a role in somatic embryogenesis and apomixis. The expression of *PpSERK* was not detected in the megaspore mother cell of apomictic plants but was detectable in sexual genotypes. Pattern of *PpSERK* expression suggests it has a role in the development of the embryo sac. The *APOSTART* gene is similar to an Arabidopsis gene located next to a *MOB1*-like gene, an apomeiotic mutant in *Medicago fulcata*. The expression of *APOSTART* suggests that it has a role both in sporogenesis and apoptosis. Matzk et al. (2005) proposed a five-gene model involved in apomixis. They found that apospory and parthenogenesis segregated independently from one another and that multiple genes control each pathway. Their data suggests incomplete penetrance by the *Alt* gene, whereas *Pit*, *Apv*, and *Ppv* always showed complete penetrance. Fully understanding the genetic components of the apomictic pathway is complicated by a number of factors, but Matzk et al. (2005) suggested that it is best to study each component of the pathway separately giving consideration to expression levels.

7.2.4 Polyploidy and Genome Size Variation

Knowledge of the genome size, ploidy level and mode of reproduction is critical in order to apply an appropriate breeding strategy for improving the bluegrasses. Polyploidy and aneuploidy are common in *Poa* species; the apomictic nature of several of the species allows for the persistence of unbalanced chromosome numbers. Murray et al. (2005) surveyed the genome size and chromosome numbers of several grasses of New Zealand. This study included 50 voucher specimens of *Poa* representing 32 different species. Chromosome numbers ranged from 28 (42 specimens) to between 263 and 266 (*P. litorosa* Cheeseman). Genome size ranged from 4.24 (*P. breviglumis* Hook.f.) to 32.56 pg/2C (*P. litorosa* Cheeseman). Genome size has also been used as a tool to determine the reproductive mode in *P. pratensis* (Wieners et al. 2006). Wieners et al. (2006) examined 38 accessions from the USDA *P. pratensis* core collection to determine the reproductive pathway and estimate DNA content of each accession. The mean 2C DNA content of the *P. pratensis* accessions ranged from 4.85 (PI298096) to 16.56 pg (PI230120). The presence of different combinations of 1C, 2C, 3C, and 5C peaks in the flow cytometry data was used to predict the reproductive pathway of the accessions.

References

Ahloowalia BS (1965) A root tip squash technique for screening chromosome number in *Lolium*. Euphytica 14: 170–172.

Albertini E, Marconi G, Barcaccia G, Raggi L, Falcinelli M (2004) Isolation of candidate genes for apomixis in *Poa pratensis* L. Plant Mol Biol 56(6): 879–894.

Albertini E, Marconi G, Reale L, Barcaccia G, Porceddu A, Ferranti F, Falcinelli M (2005) SERK and APOSTART. Candidate genes for apomixis in *Poa pratensis*. Plant Physiol 138(4): 2185–2199.

Amundsen K, Rotter D, Li HM, Messing J, Jung G, Belanger F, Warnke S (2011a) Miniature inverted-repeat transposable element identification and genetic marker development in *Agrostis*. Crop Sci 51(2): 854–861.

Amundsen K, Warnke S (2011b) Species relationships in the genus *Agrostis* based on flow cytometry and MITE- display molecular markers. Crop Sci 51(3): 1224–1231.

Arumuganathan K, Earle ED (1991) Nuclear DNA content of some important plant species. Plant Mol Biol Rep 9: 208–218.

Arumuganathan K, Tallury SP, Fraser ML, Bruneau AH, Qu R (1999) Nuclear DNA content of thirteen turfgrass species by flow cytometry. Crop Sci 39: 1518–1521.

Beard JB (ed) (1973) Creeping bentgrass. In: Turfgrass: Science and Culture. Prentice-Hall, Englewood Cliffs, NJ, pp 71–78.

Belanger FC, Bonos S, Meyer WA (2004) Dollar spot resistant hybrids between creeping bentgrass and colonial bentgrass. Crop Sci 44: 581–586.

Belanger FC, Meagher TR, Day PR, Plumley K, Meyer WA (2003a) Interspecific hybridization between *Agrostis stolonifera* and related *Agrostis* species under field conditions. Crop Sci 43: 240–246.

Belanger FC, Plumley KA, Day PR, Meyer WA (2003b) Interspecific hybridization as a potential method for improvements of *Agrostis* species. Crop Sci 43: 2172–2176.

Bonos SA, Clarke BB, Meyer WA (2006) Breeding for disease resistance in the major cool-season turfgrasses. Annu Rev Phytopathol 44: 213–234.

Bonos SA, Plumley KA, Meyer WA (2002) Ploidy determination in *Agrostis* using flow cytometry and morphological traits. Crop Sci 42: 192–196.

Bughrara S (2003) Supina bluegrass (*Poa supine* Schrad.). In: Casler MD, Duncan RR (eds) Turfgrass Biology, Genetics and Breeding. John Wiley & Sons, Hoboken, NJ, pp 53–59.

Caceres ME, Pupilli F, Piano E, Arcioni S (2000) RFLP markers are an effective tool for the identification of creeping bentgrass (*Agrostis stolonifera* L.) cultivars. Genet Resour Crop Evol 47: 455–459.

Casa AM, Nagel A, Wessler SR (2004) MITE Display. In: Miller WJ, Capy P (eds) Methods in Molecular Biology, v. 260: Mobile Genetic Elements: Protocols and Genomic Applications. Humana Press, Totowa, NJ, pp 175–188.

Casler MD, Duncan RR (2003) Introduction. In: Casler MD, Duncan RR (eds) Turfgrass Biology, Genetics, and Breeding. John Wiley & Sons, Hoboken, NJ, pp 3–26.

Chakraborty N, Bae J, Warnke S, Chang T, Jung G (2005) Linkage map construction in allotetraploid creeping bentgrass (*Agrostis stolonifera* L.). Theor Appl Genet 111(4): 795–803.

Chakraborty N, Curley J, Warnke S, Casler MD, Jung G (2006) Mapping QTL for dollar spot resistance in creeping bentgrass (*Agrostis stolonifera* L.). Theor Appl Genet 113: 1421–1435.

Chwedorzewska KJ (2008) *Poa annua* L. in Antarctic: searching for the source of introduction. Polar Biol 31(3): 263–268.

Frenot Y, Aubry M, Misset MT, Gloaguen JC, Gourret JP, Lebouvier M (1999) Phenotypic plasticity and genetic diversity in *Poa annua* L. (Poaceae) at Crozet and Kerguelen Islands (subantarctic). Polar Biol 22(5): 302–310.

Fulton TM, van der Hoeven R, Eannetta NT, Tanksley SD (2002) Identification, analysis, and utilization of conserved ortholog set markers for comparative genomics in higher plants. Plant Cell 14(7): 1457–67.

Harvey MJ (2007) Agrostis. In Barkworth ME, Anderton LK, Capels KM, Long S, Piep MB (eds) Manual of Grasses for North America. Utah State University Press, Logan, Utah, USA, pp 148–154.

Hollman AB, Stier JC, Casler MD, Jung G, Brilman LA (2005) Identification of putative velvet bentgrass clones using RAPD markers. Crop Sci 45: 923–930.

Huff DR (2003) Kentucky bluegrass. In Casler MD, Duncan RR (eds) Turfgrass Biology, Genetics and Breeding. John Wiley & Sons, Hoboken, NJ, pp 27–38.

Huff DR (2010) Bluegrasses. In: Boller B et al. (eds) Fodder Crops and Amenity Grasses. Springer, New York, NY, pp 345–380.

Jones K (1956a) Species determination in *Agrostis*. Part I. Cytological relationships in *Agrostis canina* L. J Genet 54: 370–376.

Jones K (1956b) Species differentiation in *Agrostis* II. The significance of chromosome pairing in the tetraploid hybrids of *Agrostis canina* subsp. *montana* Hartm., *A. tenuis* Sibth. and *A. stolonifera* L. J Genet 54: 377–393.

Jones K (1956c) Species differentiation in *Agrostis* III. *Agrostis gigantea* Roth. and its hybrids with *A. tenuis* Sibth. and *A. stolonifera* L. J Genet 54: 394–399.

Kalloo G (1992) Utilization of wild species. In: Kalloo G, Chowdhury JB (eds) Distant Hybridization of Crop Plants. Springer-Verlag, New York, pp 149–167.

Karlsen ÅK, Steiner JJ (2007) Scandinavian colonial bentgrass diversity described by RAPD, variable chlorophyll fluorescence, and collecting site ecogeography. Acta Agriculturae Scandinavica, Section B—Plant Soil Science 57: 23–34.

Kellogg EA (1998) Relationships of cereal crops and other grasses. Proc Natl Acad Sci USA 95(5): 2005–2010.

Kindiger B (2006) Cross-species amplification of *Lolium* microsatellites in *Poa*. Grassland Sci 52(3): 105–115.

Matzk F, Prodanovic S, Bäumlein H, Schubert I (2005) The Inheritance of apomixis in *Poa pratensis* confirms a five locus model with differences in gene expressivity and penetrance. Plant Cell 17(1): 13–24.

Mengistu LW, Mueller-Warrant GW, Barker RE (2000) Genetic diversity of *Poa annua* in western Oregon grass seed crops. Theor Appl Genet 101(1-2): 70–79.

Murray BG, De Lange PJ, Ferguson AR (2005) Nuclear DNA variation, chromosome numbers and polyploidy in the endemic and indigenous grass flora of New Zealand. Ann Bot 96(7): 1293–1305.

Nei M (1972) Genetic distance between populations. Am Nat 106: 283–292.

Novak SJ, Welfley AY (1997) Genetic diversity in the introduced clonal grass *Poa bulbosa* (Bulbous bluegrass). Northwest Sci 71(4): 271–280.

Phillips TD (2007) Breeding Turfgrasses. In: Pessarakli P, Pessarakli M (eds) Handbook of Turfgrass Management and Physiology. CRC Press, Boca Raton, Fl, pp 203–211.

Porceddu A, Albertini E, Barcaccia G, Falistocco E, Falcinelli M (2002) Linkage mapping in apomictic and sexual Kentucky bluegrass (*Poa pratensis* L.) genotypes using a two way pseudo-testcross strategy based on AFLP and SAMPL markers. Theor Appl Genet 104(2-3): 273–280.

Renganayaki K, Read JC, Fritz AK (2001) Genetic diversity among Texas bluegrass genotypes (*Poa arachnifera* Torr.) revealed by AFLP and RAPD markers. Theor Appl Genet 102(6-7): 1037–1045.

Rotter D, Amundsen K, Bonos SA, Meyer WA, Warnke SE, Belanger FC (2009) Molecular genetic linkage map for allotetraploid colonial bentgrass. Crop Sci 49(5): 1609–1619.

Rotter D, Bharti AK, Li HM, Luo C, Bonos SA, Bughrara S, Jung G, Messing J, Meyer WA, Rudd S, Warnke SE, Belanger FC (2007) Analysis of EST sequences suggests recent origin of allotetraploid colonial and creeping bentgrasses. Mol Genet Genom 278: 197–209.

Rudmann-Maurer K, Weyand A, Fischer M, Stöcklin J (2007) Microsatellite diversity of the agriculturally important alpine grass *Poa alpina* in relation to land use and natural environment. Ann Bot 100 (6): 1249–1258.

Scheef EA, Casler M, Jung G (2003) Development of species-specific SCAR markers in bentgrass. Crop Sci 43: 345–349.

Soreng RJ (2007) *Poa*. In: Barkworth ME et al. (eds) Manual of Grasses for North America. Utah State University Press, Logan, Utah. USA.

Soreng RJ, Davis JI (1998) Phylogenetics and character evolution in the grass family (Poaceae): simultaneous analysis of morphological and chloroplast DNA restriction site character sets. Bot Rev 64: 1–85.

Soreng RJ, Davis JI, Doyle JJ (1990) A phylogenetic analysis of chloroplast DNA restriction site variation in Poaceae subfam. Pooideae. Plant Syst Evol 172: 83–97.

Soreng RJ, Davis JI, Voionmaa MA (2007) A phylogenetic analysis of Poaceae tribe Poeae sensu lato based on morphological characters and sequence data from three plastid-encoded genes: evidence for reticulation, and a new classification for the tribe. Kew Bull 62: 425–454.

Stalker HT (1980) Utilization of wild species for crop improvement. Adv Agron 33: 111–147.

Turgeon AJ (1996) Turf Quality (p4–12) and Turfgrass Species (p43–84). In: Turfgrass Management. Prentice Hall, Upper Saddle River, NJ.

Vergara GV, Bughrara SS (2003) AFLP analyses of genetic diversity in bentgrass. Crop Sci 43: 2162–2171.

Warnke S (2003) Creeping bentgrass (*Agrostis stolonifera* L.). In: Casler MD, Duncan RR (eds) Turfgrass Biology, Genetics and Breeding. John Wiley & Sons, Hoboken, NJ, pp 175–185.

Warnke SE, Douches DS, Branham BE (1997) Relationships among creeping bentgrass cultivars based on isozyme polymorphisms. Crop Sci 37(1): 203–207.

Watson L (1990) The grass family. Poaceae. In: Chapman GP (ed) Reproductive versatility in the grasses. Cambridge University Press, Cambridge, MA, pp 1–31.

Wieners RR, Fei SZ, Johnson RC (2006) Characterization of a USDA Kentucky bluegrass (*Poa pratensis* L.) core collection for reproductive mode and DNA content by flow cytometry. Genet Res Crop Evol 53(8): 1531–1541.

Zhao H, Bughrara SS, Oliveira JA (2006) Genetic diversity in colonial bentgrass (*Agrostis capillaris* L.) revealed by EcoRI-MseI and PstI-MseI AFLP markers. Genome 49: 328–335.

Zhao H, Bughrara SS, Wang Y (2007) Cytology and pollen grain fertility in creeping bentgrass interspecific and intergeneric hybrids. Euphytica 156: 227–235.

Warm-Season Grasses

Masumi Ebina,[1,] Shin-ichi Tsuruta[1]* and *Yukio Akiyama[2]*

ABSTRACT

In tropical regions, the needs of animal livestock and husbandry have rapidly increased along with increases in economic development, and the forage use of tropical grasses is becoming more important in these areas. The grass family (Poaceae) has been estimated to have evolved from taxa in the tropics and subtropics about 50 to 60 million years ago (Mya) based on fossilized material. The extant tropical grasses have C_4 photosynthesis mechanisms, and this system developed about 50 Mya, based on the time of appearance of the genes responsible for C_4 photosynthesis; thus, the progenitor of the tropical grasses appeared 50 Mya. A total of 372 genera of tropical grasses have a C_4 photosynthesis system, which is much more prevalent in tropical grasses than in other plant species. Apomixis and polyploidy are two other characteristics that distinguish tropical grasses from other plant species and other grasses. Apomixis is a mode of asexual reproduction which produces progeny as maternal clones through seed but without chromosomal recombination, and is one of the most intriguing subjects in plant reproductive biology. Apomixis is recognized in many tropical grasses regardless of their phyletic lines. The application of apomixis could be a tremendous tool for plant breeding, because it can maintain heterozygosity without segregation in hybrid progeny, providing genetically uniform seeds. The history of comprehensive breeding of the tropical grasses began in the 1940s and has successfully proceeded despite the presence of apomixis and complex polyploidy.

Key words: Apomixis, C_4 photosynthesis system, Forage, Polyploidy, Tropical grass

[1]National Agricultural Research Organization, National Institute of Livestock and Grassland Science, Nasushiobara, 329-2793 Tochigi, Japan.
[2]National Agricultural Research Organization, National Institute of Tohoku Region Agricultural Research Center, Morioka, 020-0198 Morioka, Japan.
*Corresponding author: *triticum@affrc.go.jp*

8.1 Basic Information on the Tropical Grasses

Although tropical grasses are grown as warm-season grasses in temperate regions, these grasses are native to tropical regions and were originally grown in these areas. Since the 1920s, several superior lines and numerous native accessions of tropical grasses have been introduced into temperate regions and used intensely in these areas; thus, the term "warm-season grass" might be preferable for these "tropical" grasses. On the other hand, in tropical regions, the needs of animal livestock and husbandry have rapidly increased along with increases in economic development, and the forage use of tropical grasses is becoming more important in these areas. For this reason, we have retained the original term "tropical grass" rather than "warm-season grass". The grass family (Poaceae) has been estimated to have evolved from taxa in the tropics and subtropics about 50 to 60 million years ago (Mya) based on fossilized grass spikelets and inflorescence fragments containing pollen (Crepet and Feldman 1991), and 100 to 115 Mya based on chloroplast genome sequences (Chaw et al. 2004). The extant tropical grasses have C_4 photosynthesis mechanisms, and this system developed about 50 Mya, based on the time of appearance of the genes responsible for C_4 photosynthesis (Giussani et al. 2001; Edwards and Still 2008); thus, the progenitor of the tropical grasses appeared 50 Mya. A total of 372 genera of tropical grasses have a C_4 photosynthesis system, which is much more prevalent in tropical grasses than in other plant species; for example, all of the species of subfamily Chlorideae of tropical grasses exhibit C_4 photosynthesis (Sage 2001).

Apomixis and polyploidy are two other characteristics that distinguish tropical grasses from other plant species and other grasses. Apomixis is a mode of asexual reproduction which produces progeny as maternal clones through seed but without chromosomal recombination, and is one of the most intriguing subjects in plant reproductive biology (Nogler 1984; Ozias-Akins and van Dijk 2007). Apomixis has been described in ~ 400 flowering plant taxa in 126 genera of 33 families among both monocots and eudicots (Carman 1997). Apomixis is recognized in many tropical grasses regardless of their phyletic lines (Carman 1997). The application of apomixis could be a tremendous tool for plant breeding, because it can maintain heterozygosity without segregation in hybrid progeny, providing genetically uniform seeds. However, apomixis has not been found in major commercial crops and is rare in wild crop relatives. Attempts to introduce apomixis into major crops have not been successful, except for *Pennisetum glaucum* (pearl millet). Apomixis can be subdivided into two categories, sporophytic apomixis (adventitious embryony) and gametophytic apomixis, which can be further subdivided into apospory and diplospory. Among tropical grasses, brachiaria grass (most of genus *Brachiaria*, except for *B. ruziziensis* and *B. decumbens*), buffelgrass (*Cenchrus ciliaris*), guinea grass (*Panicum maximum*), and many *Pennisetum* species have apospory which arises via an unreduced embryo sac directly from a somatic cell of the ovary. In Table 8-1, the major tropical grasses, methods of reproduction and ploidy levels are summarized (Moser et al. 2004). The history of comprehensive breeding of tropical grasses began in the 1940s and has successfully proceeded despite the presence of apomixis and complex polyploidy.

Because of the high productivity of C_4 species, many tropical grasses are utilized as forage for grazing and hay for beef cattle and also as coarse cereals. In a broad sense, the major cereal crops such as maize (*Zea mays*), rice (*Oryza sativa*), and sorghum (*Sorghum bicolor*) are also tropical grasses, based on their classification within the grass family.

8.2 Classical Genetics and Traditional Breeding

In tropical grasses, comprehensive breeding started more recently than in other forages grasses and turfgrasses, so classical mapping efforts have been limited. Early in the breeding of each grass, the heritability of several important traits, and especially the mode of reproduction, has been characterized. The mode of reproduction is one of the most important information as it affects the breeding strategy. For example, apomixis directly propagates the genotype of the maternal plant through seeds. Apomixis in tropical grasses and other apomictic species exhibits monogenic inheritance (Nogler 1984; Asker and Jerling 1992; Ozias-Akins and van Dijk 2007) in *Panicum maximum* (Savidan 1980; Nakajima 1990), *Brachiaria* spp. (Valle and Savidan 1996), and *Pennisetum* spp. (Ozias-Akins et al. 1993), whereas apospory in *Paspalum notatum* exhibits complex inheritance (Martínez et al. 2001). In these apomictic species, superior clone selections can be made directly from natural distributed populations and used to breed new apomictic cultivars (Vogel and Burson 2004). As funding has become available, sexual accessions and sexual lines from natural diploid sexual relatives have been used for crossing and construction of breeding populations in several apomictic tropical grass species, especially in brachiaria grass (Valle and Glienke 1991; Lutts et al. 1994; Ishigaki et al. 2009a) and guinea grass (Hanna et al. 1973; Nakagawa and Hanna 1992). In guinea grass, superior apomictic cultivars can be directly selected from crossing populations (Hanna 1986; Kouki et al. 2006).

Although valuable knowledge of apomixis has accumulated, the inheritance of other agronomic traits is not well understood because of the high ploidy levels of many tropical grasses. Extensive phylogenetic analysis has been performed between species and genera to examine the metabolic functions of C_4 photosynthesis (Giussani et al. 2001; Edwards and Still 2008); however, the intraspecies variation for this trait is insufficient for use in

Table 8-1 Propagation method, chromosome number and ploidy level of tropical grass species.

Academic name	English name	Apomixis or sexual		Chromosome no.					Ploidy level					Linkage analysis
		sexual	apomixis											
Bothriochloa and other	Old world bluegrass	sexual	apomixis											
Bouteloua spp.	Gramas	sexual (2x)				20	40	60			2x	4x	6x	
Brachiaria brizantha	Brachiariagrass		apomixis					36					4x	yes
Brachiaria decumbens	Brachiariagrass	(sexual)	apomixis					36					4x	
Brachiaria humidicola	Brachiariagrass	(sexual)	apomixis					36					4x	
Brachiaria ruziziensis	Brachiariagrass	sexual						18					2x	
Buchloë dactyloides	Buffalograss	sexual				20	40	60			2x	4x	6x	
Chloris gayana	Rhodesgrass	sexual					20	40				2x	4x	yes
Cynodon dactylon	Bermudagrass	sexual					18	(36)				2x	(4x)	
Cynodon nlemfuënsis	Stargrass	sexual					(18)	36				(2x)	4x	
Digitaria eriantha	Digtgrass	sexual	high sterility	18	27	36	45	54	2x	3x	4x	5x	6x	
Eleusine coracana	Finger millet							36					4x	
Eragrostis tef	Lovegrasses	sexual	apomixis		40	60	80	120	2x	4x	6x	8x	12x	
Hemarthria altissima	Limpograss	sexual				18	36	54			2x	4x	6x	
Panicum antidotale	Blue panic	sexual						18					2x	
Panicum coloratum	Kleingrass	sexual	(apomixis?)		(18)	36	45	54		(2x)	4x	5x	6x	
Panicum maximum	Guineagrass	(sexual)	apomixis				(16)	32				(2x)	4x	yes
Panicum miliaceum	Proso millet	sexual(self-pollinated)						36					4x	
Panicum obtusum	Vine mesquite	sexual	apomixis				20	40				2x	4x	
Panicum virgatum	Switchgrass	sexual					36	72				4x	8x	
Paspalum dilatatum	Dallisgrass	sexual	apomixis					50						
Paspalum notatum	Bahiagrass	sexual	apomixis				20	40				2x	4x	yes

Table 8-1 contd.....

Table 8-1 contd.

Academic name	English name	Apomixis or sexual		Chromosome no.						Ploidy level						Linkage analysis
				8	(27)	36	(45)	(54)	(63)	4x	(5x)	(6x)	(7x)			
Pennisetum ciliare	Buffelgrass	sexual	(apomixis)			36			(63)				4x	(7x)		
Pennisetum clandestinum	Kikuyugrass		apomixis			36							4x			
Pennisetum glaucum	Pearl millet	sexual		14						2x						
Pennisetum orientale	Flaccidgrass		apomixis	8	(27)	36	45	(54)		2x	(3x)	4x	5x	(6x)		
Pennisetum purpureum	Nepiagrass	sexual				28							4x			
Pennisetum squamulatum	–		apomixis					54						6x		yes
Setaria italica	Foxtail millet	sexual		18						2x						
Setaria sphacelata	Setaria	sexual					18–90					2x–10x				
Sorghastrum nutans	Indiangrass	sexual		20		40		60								
Tripsacum dactyloides	Eastern gamagrass	sexual	apomixis	36	54	72	90	108		2x	3x	4x	5x	6x		yes

() limited observation
From Moser et. al. (2004)

breeding. There are three main types of C_4 photosynthesis: the aspartate, alanine, and NADP versus NAD malate pathways. All three types of C_4 photosynthesis have been recognized among the 22 *Panicum* species based on variations in leaf anatomy (Osugi and Murata 1986). Crossing among species with these different types of C_4 photosynthesis could lead to useful genetic variation in economically important species such as *P. maximum*.

In species used for both grain and forage, such as pearl millet, artificial selection might have been performed during the course of human farming history (Burton et al. 1968); in contrast, the breeding of many tropical grasses did not begin until the mid-1930s (Vogel and Burson 2004). Almost all sexually propagated tropical grass species exhibit cross-pollination and highly complex polyploidy. For these grasses, recurrent population selection is the preferred breeding system. In the case of Rhodes grass (*Chloris gayana*), the cultivars "Finecut" and "Topcut" were constructed by using a well-designed recurrent selection system that optimized the characteristics of the breeding population (Loch et al. 2004). This system has also been applied to sexual populations of apomictic grasses. In brachiaria grass, intensive improvement has been performed by recurrent selection within sexual breeding populations (Miles et al. 2004). Final crossing with an apomictic accession followed by selection led to development of the superior apomictic lines "Mulato" and "Mulato II", which are well-known cultivars used throughout the tropical grasslands (Pedro et al. 2007). In contrast, only a few agronomically important traits have been accumulated in the available sexual populations of guinea grass; therefore, more intensive collection from superior natural lines of sexual guinea grass will be necessary.

The other type of breeding activity used for tropical grasses is direct isolation of vegetative clones from natural habitats; this is necessary for species in which crossing is difficult because of highly complex polyploidy, such as Pangola grass (*Digitaria eriantha*; Pitman et al. 2004), or because flowering takes many years, such as zoysia grass (*Zoysia* spp.; Tsuruta et al. 2011). Pangola grass exhibits complex polyploidy in natural populations (chromosome numbers 18, 27, 30, 35, 36, 40, 45, and 54) and individuals with different chromosome numbers are often found together in the same habitat; therefore, the seed is usually infertile. Pangola grass cultivar "Transvala" is a famous vegetative cultivar that has strong resistance against viral disease and water-logging and exhibits high yield and palatability (Pitman et al. 2004). In zoysia grass, cultivars "Asamoe" and "Asagake" have highly vigorous stoloniferous growth, a trait that exhibits polygenetic recessive inheritance: progeny with more vigorous stoloniferous growth have not been found among numerous self and cross progeny of these cultivars.

8.3 Diversity Analysis

Extensive phylogenic analysis of the grass family has been performed using various techniques because of the importance of grasses as major crops. A basic phylogeny using botanical morphological variation was constructed by Clayton and Renvoize (1986). Since then, phylogeny has been studied by using variation in the DNA sequences of the chloroplast *ndhF* gene (Clark et al. 1995), the internal transcribed spacers of nuclear ribosomal DNA (Hsiao et al. 1999), genes related to C_4 photosynthesis (Giussani et al. 2001), and the entire chloroplast DNA sequence (Matsuoka et al. 2002).

In tropical grasses, especially in vegetatively propagated and apomictic grasses, phylogenetic analysis using phenotypic and genotypic diversity is important to enable discrimination of accession from each other. In other words, phylogenetic analysis is the only the way to evaluate the uniqueness of the selected accession. Phylogenetic analysis could also be informative for selection of the best crossing combinations when constructing populations and developing breeding strategies. The levels and patterns of genetic diversity in natural populations depend upon many factors. The biological factors that may affect the amount of genetic diversity include the propagation system and the amount of time that has passed following the establishment of the species or natural population. Grass species that are propagated by asexual reproduction (such as apomixis) and vegetative propagation produce many individuals that are genetically identical or very closely related, and some accessions cannot be distinguished. Asexual species can expand their genetic diversity through the accumulation of mutations over long periods; thus, the genetic diversity of populations that have speciated recently or that have been geographically restricted would be narrow compared with that of older species and those spread out over a wide area such as a continent.

Many phylogenic analyses have been performed using amplified fragment length polymorphisms (AFLPs) and simple sequence repeat (SSR) polymorphisms, instead of comparatively high-cost whole-genome sequencing. In guinea grass, a considerable number of accessions had been collected mainly from East Africa (the center of origin) up until the 1970s; these accessions were evaluated and subjected to breeding selection (Muir and Jank 2004). SSRs were developed and phylogenetic analysis performed using these accessions and cultivars (Ebina et al. 2005a, 2007; Chandra and Tiwari 2010). The phylogenetic analysis revealed that the accessions from western Africa contained more variation than those from other regions that are also natural guinea grass habitats. One SSR had 39 alleles among 77 accessions (Ebina et al. 2007). Although guinea grass propagates apomictically, genetic variation is clearly maintained in its natural populations.

In Rhodes grass, phylogenetic analysis has been done by using AFLPs (Ubi et al. 2001, 2003). This grass propagates sexually and exhibits outcrossing. In this case, phylogenetic analysis revealed that geographic and breeding selection attributed the 12–13% of the total allelic diversity of the 13 tested diploid cultivars.

In addition to apomixis and outcrossing, vegetative propagation is also recognized in many tropical grasses. Although vegetative propagation allows for duplicating heterozygous genotypes, this propagation method can be difficult to use for maintenance and management of a collection. In napier grass (*Pennisetum purpureum*), phylogenetic studies based on agronomic classification have suggested that some similar or closely related genotypes may be duplicates, and reclassification of the accessions within the existing collections was performed based on analysis of physiological traits (Wouw et al. 1999). There has also been research on genetic diversity assessment for effect maintenance and cultivar identification in vegetatively propagated turfgrasses such as buffalograss (*Buchloe dactyloides*; Budak et al. 2005), bermudagrass (*Cynodon* spp.; Kamps et al. 2011) and zoysia grass (Tsuruta et al. 2011). In zoysia grass, many ecotypes have been collected in Japan and elsewhere. More than 1200 accessions of *Zoysia* species have been collected in Japan, some of which have been used for characterization of genetic diversity based on phenotypic variation (Ebina et al. 2000).

In addition to the analysis of phenotypic variation, isozyme patterns and genome-wide polymorphisms have also been studied to estimate phylogenetic relatedness and trace the evolutionary path of zoysia grass (Table 8-2). Yaneshita et al. (1997) used restriction fragment length polymorphism (RFLP) markers to estimate the genetic variation among five species of *Zoysia* (*Z. japonica*, *Z. matrella*, *Z. tenuifolia*, *Z. sinica*, and *Z. macrostachya*) collected from natural populations in Japan by analyzing nuclear and chloroplast DNA. AFLP variation in zoysia grass has also been investigated and used to characterize its genetic diversity and the geographical variation among natural populations (Tsuruta et al. 2011). Recently, random amplified polymorphic DNA (RAPD) analysis (Weng et al. 2007), and SSR or microsatellite analysis (Tsuruta et al. 2005; Ma et al. 2007) were applied to clarify the phylogenetic relatedness among *Zoysia* species. Despite these analyses, the phylogeny and taxonomy of this genus remain unclear. Furthermore, leaf width, which was traditionally used as an indicator for species discrimination, shows continuous variation among species, making species identification difficult.

We explored a classification system for three species of *Zoysia* (*Z. japonica*, *Z. matrella*, and *Z. tenuifolia*) without assuming predefined taxonomic and geographic groups. Instead a population structure based on genotypes at 20 microsatellite markers and a model-based Bayesian clustering approach implemented in the STRUCTURE software (Pritchard et al. 2000) to explain

Table 8-2 Molecular marker-based isozyme and DNA profiling studies of *Zoysia* species.

Target	Marker system[a]	No. of markers	No. of accessions	Species[b]	Reference
Protein	Isozyme	1	24	Zj	Yamada and Fukuoka (1984)
	Isozyme	2	182	Zm, Zs	Weng (2002)
	Isozyme	2	131	Zm, Zt, Zs	Weng et al. (2007)
Nuclear	RFLP	20	17	Zj, Zm, Zt, Zs, Zmac	Yaneshita et al. (1993, 1997)
	RAPD	12	131	Zm, Zt, Zs	Weng et al. (2007)
	AFLP	4	46	Zj, Zm, Zt, Zs, Zmac	Ebina et al. (2000)
	AFLP	10	20	Zj	Hong et al. (2008)
	SSR	12	24	Zj, Zm, Zt, Zs, Zmac	Tsuruta et al. (2005)
	SSR	12	41	Zj, Zm, Zt	Hashiguchi et al. (2007)
	SSR	30	20	Zj	Ma et al. (2007)
Chloroplast	RFLP	10	17	Zj, Zm, Zt, Zs, Zmac	Yaneshita et al. (1997)
	Sequence	6	3	Zj, Zm, Zt	Tsuruta et al. (2008)
	SSR	21	3	Zj, Zm, Zt	Tsuruta et al. (2008)

[a]AFLP, amplified fragment length polymorphism; RFLP, restriction fragment length polymorphism; SSR, simple sequence repeat
[b]Zj, *Z. japonica*; Zm, *Z. matrella*; Zt, *Z. tenuifolia*; Zmac, *Z. macrostachya*; Zs, *Z. sinica*

incomplete lineage sorting and clarify species boundaries was used. The patterns of genetic structure within *Zoysia* revealed a clear differentiation of populations following phenotypically recognized species boundaries (Fig. 8-1). The average of the admixture proportions for individuals of a species to be assigned to an appropriate species cluster was 0.986–0.990, whereas that for individuals assigned to another species cluster was 0.004–0.008. Twenty eight (14.5%) of the 193 accessions used in this study were inferred to be admixed individuals with other accessions. Most of these individuals were collected in the southwestern islands of Kyushu and in the Okinawa archipelago in Japan, places where the three *Zoysia* species occur together, suggesting that admixed individuals may be produced by interspecific hybridization. Pairwise Divergence value (F_{st}) values between different species had significant P values, indicating that the clusters identified by STRUCTURE were well resolved as a result of redefining groups and removing admixed individuals (Table 8-3).

Tsuruta et al. (2008) investigated sequence variation in several tropical turfgrasses, including *Zoysia* species, in six chloroplast non-coding regions that have been extensively used for estimation of genetic diversity and

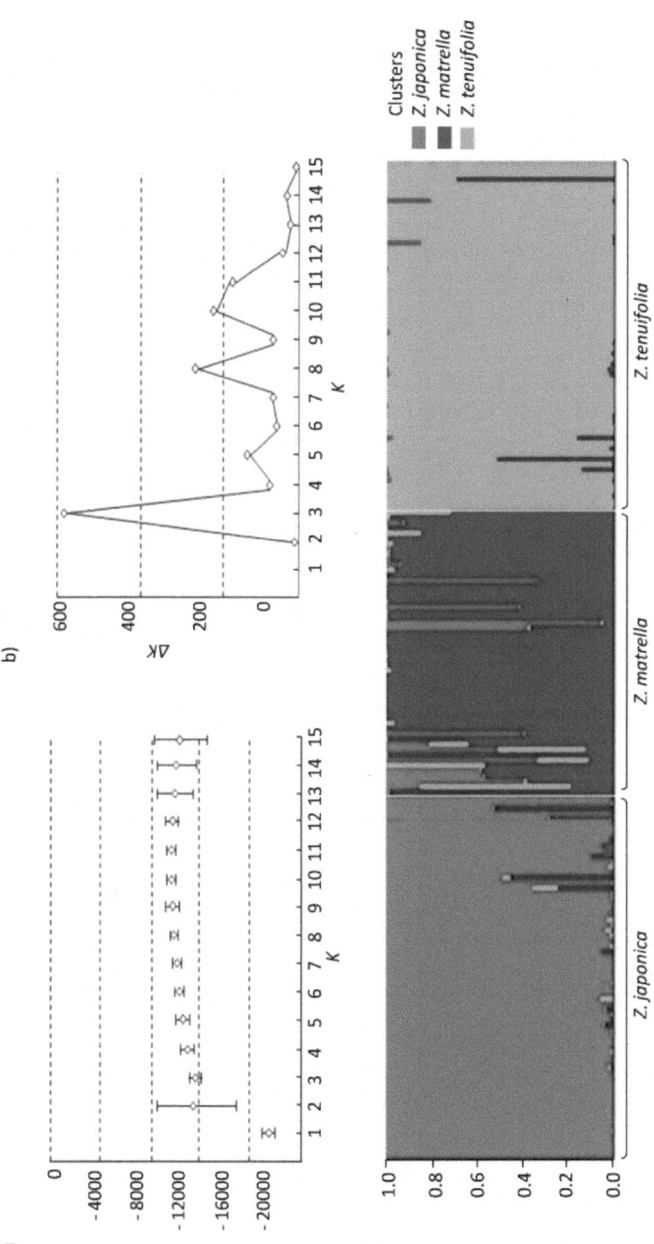

Figure 8-1 Genetic structure of 193 zoysia grass accessions identified by analysis using the STRUCTURE software based on 20 microsatellite polymorphisms. (a) Plot of the mean values (±SD) of the log likelihood of the genotyping data, for 10 runs at values of K from 1 to 15. (b) Plot of ΔK as a function of the number of clusters (K) for $K = 1$ to 15. Following calculation of the rate of change in the likelihood distribution ($L'(K) = L(K) - L(K-1)$) and absolute values of the second-order rate of change of the likelihood distribution ($|L''(K)| = |L'(K+1) - L'(K)|$), ΔK was calculated as $\Delta K = m|L''(K)|/s[L(K)]$. A maximum ΔK value at $K = 3$ indicated that a model with three clusters best explained the highest hierarchical level of genetic structure. (c) Hierarchical organization of 193 individuals from three *Zoysia* species identified by the STRUCTURE analysis for the $K = 3$ model.

Color image of this figure appears in the color plate section at the end of the book.

Table 8-3 Average admixture proportion for each *Zoysia* species and pairwise F_{st} among the *Zoysia* species.

Species	N[a]	Cluster (K)[b]				F_{st}[c]	
		1	2	3	$n_{admixed}$	*Z. japonica*	*Z. matrella*
Z. japonica	73	0.987 (0.01)	0.008 (0.01)	0.005 (0.01)	8	-	-
Z. matrella	54	0.007 (0.01)	0.986 (0.02)	0.007 (0.01)	14	0.122	-
Z. tenuifolia	66	0.004 (0.00)	0.006 (0.01)	0.990 (0.01)	6	0.173	0.074

[a]Total number of individuals
[b]The average proportion of individuals was calculated in each of the $K = 3$ clusters as inferred by the STRUCTURE software. Individuals were estimated a single-species if proportion was ≥90 % for one population. Figures in parentheses indicate the standard deviation of the average. $n_{admixed}$: number of admixed individuals ($q < 0.9$ in each of the three clusters)
[c]All pairwise F_{st} values were significant at 0.05 ($P < 0.05$).

phylogenetic relationships. The result showed that the tropical turfgrasses were clearly separated from temperate (cool-season) turfgrasses and cereal crops, and the results were generally in good agreement with conventional taxonomy (Fig. 8-2). By using the base-substitution rate in

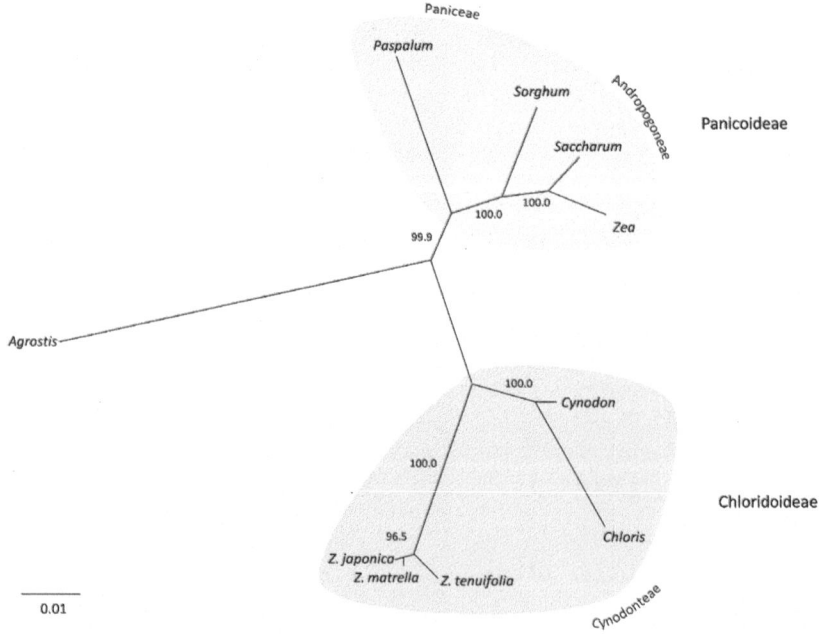

Figure 8-2 Phylogenetic analysis of five turfgrass species and three cereal crops using the neighbor-joining method based on the sequences of six non-coding chloroplast DNA regions (*trnK* intron, *trnD-psbM*, *atpA-rps*14, *atpB-rbcL*, *petE-psaJ* and *rps*3-*rps*12). Numerals next to the branches indicate bootstrap values (%) from 1000 replicates. Scales indicate the evolutionary distances of the base substitutions at each site, estimated by Kimura's two-parameter method.

these six chloroplast regions between *Sorghum bicolor* (Saski et al. 2007) and *Saccharum officinarum* (Calsa Júnior et al. 2004), we estimated the divergence times among the *Zoysia* species and between *Zoysia* species and *S. officinarum*. Pairwise comparison of the combined data sets in the six corresponding regions between *S. bicolor* and *S. officinarum* gave 1.20×10^{-3} substitutions/site. Since the divergence time of *S. bicolor* and *S. officinarum* was previously estimated at 5 Mya (Munkacsi et al. 2007), the substitution rate in the region was estimated to be 2.403×10^{-9} substitutions/site/year. Based on the substitution rate calculated here, the time of divergence calculated from the base-substitution rate between *Zoysia* and *Saccharum* was estimated at approximately 28 to 34 Mya. The divergence time calculated from the substitution rates (Table 8-4) between *Z. tenuifolia* and the other *Zoysia* species (*Z. japonica* and *Z. matrella*) was estimated at approximately 1.4 to 2.3 Mya. In contrast, the divergence time between *Z. japonica* and *Z. matrella* was estimated at approximately 0.8 to 1.0 Mya, suggesting that these species diverged relatively recently. Classification of *Z. matrella* has many ambiguities and this species tends to be grouped with the cluster containing *Z. japonica* accessions; consequently, the species boundary remains uncertain. In the structural analysis, F_{st} in *Z. matrella* compared to the other two *Zoysia* species was low (0.074 to 0.122), although the values indicate a significant difference between species. The unclear classification of *Z. matrella* has been assumed to be due to genetic introgression through interspecific hybridization, because *Z. japonica*, *Z. matrella*, and *Z. tenuifolia* can be used to generate interspecific hybrids in all combinations (Forbs 1952), and it appears that based on morphological observation, many naturally occurring interspecific hybrids can be found in several ecotypes (Fukuoka 2000). In addition to this phenomenon, the relatively short time after speciation may also be one of the factors contributing to the unclear classification of *Z. matrella*.

Generally, individuals within populations that naturally undergo vegetative propagation are closely related. Nevertheless, analysis of genetic diversity using DNA markers has indicated that ecotypes of zoysia grasses have relatively high levels of variability. As described previously,

Table 8-4 Pairwise comparisons of total mutations (above the diagonal) and rate of base substitution (below the diagonal) among three *Zoysia* species.

	Z. japonica	*Z. matrella*	*Z. tenuifolia*
Z. japonica	-	8	17
Z. matrella	1.97 [a]	-	14
Z. tenuifolia	4.19	3.50	-

[a] Rates of base substitution were estimated from the total sequence of six chloroplast noncoding regions, not including insertion/deletion events (indels) or gaps. Each value represents the number of base substitutions per 1000 sites.

the genus *Zoysia* is estimated to have diverged from an ancestral monocot approximately 28 to 34 Mya based on the base-substitution rate of chloroplast sequences. The divergence time of the genus *Zoysia* is the same or slightly earlier than the estimated divergence of *Z. mays* (11 to 24 Mya; Munkacsi et al. 2007). Given the estimated time of divergence of *Zoysia*, the high degree of genetic diversity found within *Zoysia* species may be correlated with its long history after speciation and population establishment. Recently, zoysia grass ecotypes have been gradually disappearing because of changes in their natural habitats; for example, habitats of endemic species including *Zoysia* are being lost as human activities expand and invasive foreign species encroach (Esler 1991; Fukuoka 2000). For the conservation of genetic diversity in the genus *Zoysia*, every effort should be made to maintain populations of native plants, which are useful breeding materials, through habitat preservation and plant collection.

8.4 Mapping and Tagging of Simply Inherited Traits

8.4.1 Apomixis

A number of molecular linkage maps have been constructed mainly for the purpose of mapping apomixis-related genes. In several apomictic tropical grasses, linkage between apomixis and molecular markers has been established. The genetic region responsible for apomixis has been mapped in detail in *Pennisetum* (Ozias-Akins 1998). Partial linkage maps focused on apomixis have been constructed in *Paspalum* (Oritz et al. 1997; Pupilli et al. 2001) and *Brachiaria* (Pessino et al. 1997, 1998), and unsaturated but complete linkage maps have been constructed in *Panicum* (Ebina et al. 2005b) and *Paspalum* spp. (Labombarda 2002). The details of tropical grasses apomixis linkages research and the other apomictic plants described and reviewed in Ozias-Akins and van Dijk (2007) and Ozias-Akins (2006).

Aposporous *Pennisetum squamulatum* has been used extensively to study apomixis (Ozias-Akins et al. 2003). This trait has been transferred to a sexual relative (*P. glaucum*) by crossing *P. glaucum* and *P. squamulatum* followed by backcrossing to *P. glaucum*. Apospory appears to follow simple dominant Mendelian inheritance in the F_1 progeny. Twelve molecular markers linked with apospory were obtained as sequence-characterized amplified regions (SCARs) in *P. squamulatum* by RFLP analysis (Ozias-Akins et al. 1998). Genetic linkage analysis of 397 individuals using these 12 markers showed no recombination between these markers and the apospory trait, and this linkage block has been termed the apospory-specific genomic region (ASGR). Similar segregation distortion to that seen in *P. squamulatum* is also observed in *C. ciliaris* (Sherwood et al. 1994). A set of markers tightly linked to apomixis indicated a DNA region referred to as the apomixis-specific genomic region (ASGR). The chromosomal structure of the ASGR was revealed by fluorescence *in situ* hybridization (FISH) using bacterial artificial chromosome (BAC) libraries of *P. squamulatum* screened using the SCAR markers (Goel et al. 2003). In this analysis, the ASGR-carrier chromosome was detected as a hemizygous region. The entire hemizygous region was estimated as ~ 50 Mbp by image analysis (Akiyama et al. 2004, 2006). In *Panicum*, Ebina et al. (2005b) found more than 21 tightly linked AFLP markers in the ASGR using 56 AFLP primer combinations; subsequently, more than 200 markers consistent with the ASGR structure have been found in the ASGR of *Panicum* using further AFLP primer combinations. In *Paspalum*, the segregation ratio is often distorted by the influence of a few recessive genes, and the inheritance of apomixis fits a model of tetrasomic inheritance of a single dominant gene with pleiotropic effects and incomplete penetrance (Martínez et al. 2001). Both the structure of the ASGR and the complexity of apomixis inheritance prevent map-based cloning of the gene or genes responsible for apomixis. Other attempts to discover the apomixis gene or genes have used BAC contig construction and its precise sequencing (Conner et al. 2008). Despite the difficulties inherent in identifying the actual gene or genes responsible for apomixis, the apomixis-linked markers that have been identified are already being used for selection of apomixis in breeding programs, as an alternative to progeny testing or embryo sac analysis.

8.4.2 Lignin Biosynthesis Genes

Digestibility is one of major traits of forage grasses and is negatively correlated with lignin content (Gabrielsen et al. 1990). The genes encoding caffeic acid *O*-methyltransferase (COMT) and cinnamyl alcohol dehydrogenase (CAD), the key enzymes in the lignin biosynthesis pathway, have been isolated and proven to affect digestibility in maize (Piquemal et al. 2002; Halpin 1998). Cherney et al. (1991) proposed the use of this gene for breeding to improve forage quality. In forage-type sorghum breeding, the *brown-midrib* mutants (*bmr*) have been widely used for studies of the regulation of lignin biosynthesis in grasses. These mutants show a reddish-brown pigment in the leaf midrib and stem. In addition, this trait is inherited as a simple recessive trait and is closely associated with improved forage digestibility (Barrière and Argillier 1993; Bittinger et al. 1981). For these reasons, the *bmr* mutants were integrated into conventional breeding programs using the visible brown midrib phenotype to follow the trait (Kasuga 2002). Biochemical analysis has suggested that the alteration of lignin content or composition in these mutants is caused by reductions in the activities of some enzymes involved in lignin biosynthesis during the middle stages of lignification, 40 to 60 d after sowing (Bucholtz et al.

1980; Palmer et al. 2008; Pillonel et al. 1991; Tsuruta et al. 2003). Several *bmr* mutants in sorghum contain a nonsense mutation within the open reading frame of a gene encoding a lignin biosynthesis enzyme such as COMT (*bmr*-12 and -18; Bout and Vermerris 2003) or CAD (*bmr*-6; Saballos et al. 2009; Sattler et al. 2009; Tsuruta et al. 2010). These mutations are considered to activate an mRNA degradation system (nonsense-mediated mRNA decay) which plays a role in mRNA quality control in eukaryotic cells (Conti and Izaurralde 2005); consequently, OMT and CAD activity were depressed by an almost complete absence of these proteins (Saballos et al. 2009; Sattler et al. 2009; Tsuruta et al. 2010).

8.4.3 Mapping in Rhodes Grass

In some other economically important tropical grasses, genetic linkage maps have been constructed. Rhodes grass is grown worldwide, especially in Australia (tropical Queensland) and warm temperate zones such as southwestern Japan. The linkage map consists of 72 AFLP and 12 RFLP markers covering 443.3 cM with 12 linkage groups (Ubi et al. 2004). Interestingly, Rhodes grass has one of the smallest genome sizes (350 Mbp) in the grass family. This linkage map could facilitate not only breeding of Rhodes grass but also gene isolation using comparative mapping with grasses that have larger genomes.

8.5 Molecular Mapping of Complex Traits

Little molecular mapping of complex traits has been done in tropical grasses except in the case of zoysia grass. For this species, breeding is only now entering the clonal selection stage. Many cultivars have been directly selected from vegetative clones in natural populations; however, zoysia exhibits both inbreeding and outcrossing species with allotetraploid inheritance. This mode of reproduction can allow one to study some of complex traits in zoysia grass by using simple molecular approaches, as in diploid crops (Tsuruta et al. 2011).

Zoysia grass is unique for a C_4 plant in that it exhibits freezing tolerance. This trait may have originated in *Zoysia japonica*, because this species evolved within the Asian cool temperate zone in the pan-Pacific rim. This freezing tolerance is controlled by a major QTL (for which the nearest marker, B1-227, had a LOD score of 6.49 and explained 29.4% of the observed variation) and several minor genes (Tsuruta et al. 2011). Zoysia grass is also salt-tolerant (Oishi and Ebina 2001), which is a common trait in tropical grasses, with strong tolerance to even sea-water salt strength. An attempt was made to characterize and analyze this trait in zoysia (Tsuruta et al. 2011); however, the strong salt tolerance appeared to be controlled by many minor genes, and no major QTL was detected.

8.6 Molecular Breeding and Transformation

8.6.1 Molecular Breeding

Since the ASGR, which harbors the gene or genes responsible for apomixis (see earlier), covers a large genome region and the inheritance of apomixis behaves as a single major gene, following this region with molecular markers is relatively simple. Without marker-assisted selection (MAS), the only other ways to follow the trait are embryo sac analysis, which requires the plant to mature to the flowering stage, and progeny segregation testing, which requires generating progeny. Application of these conventional techniques, therefore, requires that each plant be maintained for years. On the other hand, MAS for apomixis requires only a small DNA sample for each plant. Thus, the gain in efficiency realized by using MAS for apomixis is considered to be greater than that for other traits (Ebina et al. 2005b). In the relatively new brachiaria breeding program, the same technique has been applied for the progeny of crosses between sexual tetraploid lines (Ishigaki et al. 2009a) and apomictic cultivars. Applied in combination with selection for other traits that can be checked early in plant development (e.g., initial growth rate), MAS for apomixis is an efficient method to discriminate superior clones from among a large number of initial breeding progeny.

8.6.2 Transformation

Takahashi et al. (2010) developed a plant regeneration system for the giant reed (*Arundo donax*), a biofuel plant. Tropical grasses can yield as much as 20 to 50 tDM/ha/yr; therefore, transformation systems are being developed for several of the species. The transformation systems for tropical grasses are being developed not only to make it possible to improve biomass yield for biofuel production, but also for so-called "cascade use" for isolation of valuable biomaterials. Plant regeneration or transformation systems have been constructed for guinea grass (Akashi et al. 1995), rhodes grass (Gondo et al. 2007), bahia grass (*Paspalum notatum*; Gondo et al. 2005), and brachiaria grass (Pinheiro et al. 2000; Ishigaki et al. 2009b). The brachiaria grass system is being used to construct tetraploid sexual maternal plants from obligate diploid plants via a colchicine chromosome-doubling method. The constructed tetraploid brachiaria grasses are used directly in conventional breeding and are crossed with apomictic *Brachiaria* species.

8.7 Map-based Cloning and Genome Sequencing Initiatives

Several attempts to isolate and clone genes related to apomixis have been made with *P. squamulatum* and *C. ciliaris*. As discussed in earlier, these two interesting apomictic species have highly conserved ASGRs, as observed by FISH analysis (Goel et al. 2003; Akiyama et al. 2011), and constructed considerable large covering BAC of 2.7 Mb contigs together with both *P. squamulatum* and *C. ciliaris*. The entire 2.7-Mb contig was used in predicted protein sequence comparisons with rice and sorghum (Conner et al. 2008). Notably, the predicted protein sequence of an *ASGR-BBM-like* gene (*ASGR-BABY BOOM*) revealed 98.6 to 99.8% similarity between the contig sequence and the sequences from rice and sorghum. *BBM* was originally discovered as a gene responsible for initiation and maintenance of the embryonic pathway of the AP2/ERF family of transcription factors. Transformation experiments revealed that *BBM* led to the spontaneous formation of somatic embryos and cotyledon-like structures on seedlings (Boutilier et al. 2002).

8.8 Structural Genomics

Because of the importance of apomixis in tropical grasses, several EST analyses and functional gene discovery experiments related to apomictic embryo formation have been performed. In guinea grass, genes differentially expressed between apomictic and sexual plants were precisely analyzed by differential display analysis (Chen et al. 2005) and by integration of microarray and EST data (Yamada-Akiyama et al. 2009). A specific gene for the apomictic embryo initial stage named *ASG-1* (Chen et al. 2005) was clearly identified as being specifically expressed in several apomictic lines and cultivars, such as in guinea grass and in a apomictic cultivar of *Paspalum notatum*. *ASG-1* gene homologues include *RD22*, a seed-specific gene in *Arabidopsis thaliana*, which has 49 to 61% similarity, and *EA30, -92, and -87*, which are abundant genes in *Vicia faba* embryos and which have 48 to 70% similarity.

To identify genes associated with apomixis, a set of ESTs was constructed from a total of 4608 genes. These ESTs were obtained from the apomictic progeny of a cross between apomictic cultivar "Natsukaze" and obligate sexual line "Noh PL-1", and from apomictic cultivar "Ku5954". All these genes were subjected to microarray analysis to compare their expression to that of another apomictic accession, "Oki 64". Only 394 of the 4608 genes clearly showed gene expression in the immature pistils of all three genotypes (i.e., the apomictic progeny of "Natsukaze" and "Noh PL-1", "Ku 5954", and "Oki 64"). Of these 394 candidate apomixis-specific genes, 196 genes were sequenced and compared with sequences in public databases. Out of the 196 genes that were sequenced, 12 genes showed significant homology with genes in the EST database of the apomictic plant *C. ciliaris*. When the 12 genes were subjected to quantitative PCR, 3 of the 12 genes were expressed early in the appearance of apospory, suggesting a tight relationship to apomixis (Yamada-Akiyama et al. 2009).

8.9 Comparative Genomics

Cenchrus ciliaris is represented by both aposporous and obligate sexual accessions and is used extensively to study apomixis. BAC clones of *C. ciliaris* screened with SCAR markers of *P. squamulatum* were mapped in *C. ciliaris* by FISH (Goel et al. 2003). The ASGR-carrier chromosome was not observed in obligate sexual accessions and it was larger than other homo(eo)logous chromosomes (Akiyama et al. 2005). Interestingly, the BACs of *C. ciliaris* and *P. squamulatum* derived with same SCAR markers were mapped to positions that are syntenic between the two species (Goel et al. 2006). The order of the ASGR-linked BAC clone sequences on the chromosome was similar between the species, which provides evidence for extensive macrosynteny between the ASGRs of the two species.

Recently, three ASGR-specific BACs of *P. squamulatum* were used to probe other *Cenchrus* and *Pennisetum* species by FISH. The aposporous species were *C. setigres*, *P. flaccidum*, *P. macrourum*, *P. massaicum*, *P. orientale*, *P. pedicellatum*, *P. polystachion*, *P. setaceum*, *P. subangustum*, and *P. villosum*, and the sexual species were *P. alopecuroides*, *P. basedowii*, *P. mezianum*, *P. nervosum*, *P. ramosum*, and *P. schweinfurthii* (Akiyama et al. 2011). The BAC sequences did not hybridize to chromosomes of the sexual species but did hybridize to chromosomes in the aposporous species, providing further evidence that apospory in these species is associated with the ASGR.

Because the ability to perform positional cloning for the gene or genes related to apospory is limited by the segregation distortion of the ASGR, information regarding the ASGR-carrier chromosome is necessary to isolate the gene(s). Therefore, chromosome research is considered to be essential for the study of apomixis.

8.10 Future Prospects

In tropical regions, the requirements of animal livestock and husbandry have rapidly increased in parallel with economic development, and the forage use of tropical grasses is gaining more importance in these areas. In addition, the need for renewable energy crops is a pressing issue all over the world. Tropical grasses provide a potential solution for both problems because of their distinctive characteristics such as C_4 photosynthesis and apomixis. Comprehensive breeding programs are still in their infancy compared with direct selection from natural populations; however, future breeding activity in tropical grasses holds great promise for the development of new desirable cultivars.

References

Akashi R, Sachmann S, Hoffmann F, Adachi T (1995) Embryogenic callus formation from protoplasts derived from suspension cells of apomictic guineagrass (*Panicum maximum* Jacq.). Jpn J Breed 45: 445–448.

Akiyama Y, Conner JA, Goel S, Morishige DT, Mullet JE, Hanna WW, Ozias-Akins P (2004) High-resolution physical mapping in *Pennisetum squamulatum* reveals extensive chromosomal heteromorphism of the genomic region associated with apomixis. Plant Physiol 134: 1733–1741.

Akiyama Y, Hanna WW, Ozias-Akins P (2005) High-resolution physical mapping reveals that the apospory-specific genomic region (ASGR) in *Cenchrus ciliaris* is located on a heterochromatic and hemizygous region of a single chromosome. Theor Appl Genet 111: 1042–1051.

Akiyama Y, Goel S, Chen Z, Hanna WW, Ozias-Akins P (2006) *Pennisetum squamulatum*: Is the predominant cytotype hexaploid or octaploid? J Hered 97: 521–524.

Akiyama Y, Goel S, Conner JA, Hanna WW, Yamada-Akiyama H, Ozias-Akins P (2011) Evolution of the apomixis transmitting chromosome in *Pennisetum*. BMC Evol Biol 11: 289.

Asker SE, Jerling L (1992) Apomixis in plants. CRC Press, Boca Raton.

Barrière, Y, Argillier O (1993) Brown-midrib genes of maize: A Review. Agronomie 13: 865–876.

Bittinger TS, Cantrell RP, Axtell JD (1981) Allelism tests of the brown midrib mutants of sorghum. J Hered 72: 147–148.

Bout S, Vermerris W (2003) A candidate-gene approach to clone the sorghum brown midrib gene encoding caffeic acid O methyltransferase. Mol Genet Genom 269: 205–214.

Boutilier K, Offringa R, Sharma VK, Kieft H, Ouellet KT, Zang L, Hattori J, Liu CM, van Lammeren AAM, Miki BLA, Custers JBM, van Lookeren-Campagne MM (2002) Ectopic expression of BABY BOOM triggers a conversion from vegetative to embryonic growth. Plant Cell 14: 1737–1749.

Bucholtz DL, Cantrell RP, Axtell JD, Lechtenberg VL (1980) Lignin biochemistry of normal and brown midrib mutant sorghum. J Agri Food Chem 28: 1239–1241.

Budak H, Shearman R, Gulsen O, Dweikat I (2005) Understanding ploidy complex and geographic origin of the *Buchloe dactyloides* genome using cytoplasmic and nuclear marker systems. Theor Appl Genet 111: 1545–1552.

Burton GW, Powell JB (1968) Pearl millet breeding and cytogenetics. In: Norman AG (ed) Advance in Agronomy Vol 20. Elsevier, New York, pp 49–89.

Calsa Júnior T, Carraro DM, Benatti MR, Barbosa AC, Kitajima JP, Carrer H (2004) Structural features and transcript-editing analysis of sugarcane (*Saccharum officinarum* L.) chloroplast genome. Curr Genet 46: 366–373.

Carman JG (1997) Asynchronous expression of duplicate genes in angiosperms may cause apomixis, bispory, tetraspory, and polyembryony. Biol J Linn Soc 61: 51–94.

Chandra A, Tiwari KK (2010) Isolation and characterization of microsatellite markers from guineagrass (*Panicum maximum*) for genetic diversity estimate and cross-species amplification. Plant Breed 129: 120–124.

Chaw SM, Chang CC, Chen HL, Li WH (2004) Dating the monocot-dicot divergence and the origin of core eudicots using whole chloroplast genomes. J Mol Evol 58: 424–441.

Chen L, Guan L, Seo M, Hoffmann F, Adachi T (2005) Developmental expression of ASG-1 during gametogenesis in apomictic guinea grass (*Panicum maximum*). J Plant Physiol 162: 1141–1148.

Cherney JH, Cherney KJR, Akin DE, Axtell JD (1991) Potential of brown-midrib, low-lignin mutants for improving forage quality. In: Sparks DL (ed) Advances in Agronomy, vol 46. Academic Press, San Diego, CA, pp 158–198.

Clark LG, Zhang W, Wendel JF (1995) A phylogeny of the grass family (Poaceae) based on *ndh*F sequence data. Syst Bot 20: 436–460.

Clayton WD, Renvoize SA (1986) Genera Graminum: Grasses of the world. In: Cope TA (ed) Kew Bulletin Additional Series 13, Royal Botanic Gardens, Kew, London, pp 1–389.

Conner JA, Goel S, Gunawan G, Cordonnier-Pratt MM, Johnson VE, Liang C, Wang H, Pratt LH, Mullet JE, DeBarry J, Yang L, Bennetzen JL, Klein PE, Ozias-Akins P (2008) Sequence analysis of bacterial artificial chromosome clones from the apospory-specific genomic region of *Pennisetum* and *Cenchrus*. Plant Physiol 147: 1396–1411.

Conti E, Izaurralde E (2005) Nonsense-mediated mRNA decay: molecular insights and mechanistic variations across species. Curr Opin Cell Biol 17: 316–325.

Crepet WL, Feldman GD (1991) The earliest remains of grass in the fossil record. Am J Bot 78: 1010–1014.

Ebina M, Abe A, Kobayashi M, Kasuga S, Araya H, Nakagawa H (2000) Phylogenetic analysis of genus *Zoysia* for improvement of indigenous grazing grassland. Proc International Workshop Integration of Biodiversity and Genome Technology for Crop Improvement. Tsukuba, Japan, pp 133–134.

Ebina M, Kouki K, Tsuruta S, Takahara M, Kobayashi M, Yamamoto T, Nakajima K, Nakagawa H (2005a) Development of simple sequence repeat (SSR) markers and their use to assess genetic diversity in apomictic Guineagrass (*Panicum maximum* Jacq.). In: Humphreys MO (ed) Molecular breeding for the genetic improvement of forage crops and turf, Proc 4th International Symposium on the Molecular Breeding of Forage and Turf, a satellite workshop of the XXth International Grassland Congress, July 2005, Aberystwyth, Wales, pp 127.

Ebina M, Nakagawa H, Yamamoto T, Araya H, Tsuruta S, Takahara M, Nakajima K (2005b) Co-segregation of AFLP and RAPD markers to apospory in Guineagrass (*Panicum maximum* Jacq.). Grassl Sci 51: 71–78.

Ebina M, Kouki K, Tsuruta S, Akashi R, Yamamoto T, Takahara M, Inafuku M, Okumura K, Nakagawa H, Nakajima K (2007) Genetic relationship estimation in guineagrass (*Panicum maximum* Jacq.) assessed on the basis of simple sequence repeat markers. Grassl Sci 53: 155–164.

Edwards EJ, Still CJ (2008) Climate, phylogeny and the ecological distribution of C_4 grasses. Ecol Lett 11: 266–276.

Esler AE (1991) Changes in the native plant cover of urban Auckland, New Zealand. NZ J Bot 29: 177–196.

Forbs IJ (1952) Chromosome numbers and hybrids in *Zoysia*. Agron J 44: 147–151.

Fukuoka H (2000) Breeding of zoysia grass. 1. Collection of genetic resources and general view of their characteristics. J Jpn Soc Turfgrass Sci 29: 11–21 (In Japanese with English summary).

Gabrielsen BC, Vogel KP, Anderson BE, Ward JK (1990) Alkali-labile cell-wall phenolics and forage quality in switchgrasses selected for differing digestibility. Crop Sci 30: 1313–1320.

Giussani LM, Cota-Sánchez JH, Zuloaga FO, Kellogg EA (2001) A molecular phylogeny of the grass subfamily Panicoideae (Poaceae) shows multiple origins of C_4 photosynthesis. Am J Bot 88: 1993–2012.

Goel S, Chen Z, Conner JA, Akiyama Y, Hanna WW, Ozias-Akins P (2003) Delineation by fluorescence *in situ* hybridization of a single hemizygous chromosomal region associated with aposporous embryo sac formation in *Pennisetum squamulatum* and *Cenchrus ciliaris*. Genetics 163: 1069–1082.

Goel S, Chen Z, Akiyama Y, Conner JA, Basu M, Gualtieri G, Hanna WW, Ozias-Akins P (2006) Comparative physical mapping of the apospory-specific genomic region in two apomictic grasses: *Pennisetum squamulatum* and *Cenchrus ciliaris*. Genetics 173: 389–400.

Gondo T, Tsuruta S, Akashi R, Kawamura O, Hoffmann F (2005) Green, herbicide-resistant plants by particle inflow gun-mediated gene transfer to diploid bahiagrass (*Paspalum notatum*). J Plant Physiol 162: 1367–1375.

Gondo T, Matsumoto J, Yamakawa K, Tsuruta S, Ebina M, Akashi R (2007) Somatic embryogenesis and multiple-shoot formation from seed-derived shoot apical meristems of rhodesgrass (*Chloris gayana* Kunth). Grassl Sci 53: 138–142.

Halpin C, Holt K, Chojecki J, Oliver D, Chabbert B, Monties B, Edwards K, Barakate A, Foxon GA (1998) *Brown-midrib* maize (*bm1*)—a mutation affecting the cinnamyl alcohol dehydrogenase gene. Plant J 14: 545–553.

Hanna WW, Powell JB, Millot JC, Burton GW (1973) Cytology of obligate sexual plants in *Panicum maximum* Jacq. and their use in controlled hybrids. Crop Sci 13: 695–697.

Hanna WW (1986) Registration of 12 winter-hardy *Panicum maximum* germplasm clones. Crop Sci 26: 389–390.

Hashiguchi M, Tsuruta S, Matsuo T, Ebina M, Kobayashi M, Akamine H, Akashi R (2007) Analysis of genetic resources in *Zoysia* spp. 2. Evaluation of genetic diversity in zoysiagrass indigenous to southwest islands of Japan based on simple sequence repeat markers. Jpn J Grassl Sci 53: 133–137 (In Japanese with English summary).

Hong J, Liebao H, Zhang Y (2008) AFLP analysis on genetic diversity of *Zoysia japonica*. Acta Hort 783: 265–272.

Hsiao C, Jacobs SWL, Chatterton JH, Asay KH (1999) A molecular phylogeny of the grass family (*Poaceae*) based on the sequences of nuclear ribosomal DNA (ITS). Aust Syst Bot 11: 667–688.

Ishigaki G, Gondo T, Suenaga K, Akashi R (2009a) Induction of tetraploid ruzigrass (*Brachiaria ruziziensis*) plants by colchicine treatment of *in vitro* multiple-shoot clumps and seedlings. Grassl Sci 55: 164–170.

Ishigaki G, Gondo T, Suenaga K, Akashi R (2009b) Multiple shoot formation, somatic embryogenesis and plant regeneration from seed-derived shoot apical meristems in ruzigrass (*Brachiaria ruziziensis*). Grassl Sci 55: 46–51.

Kamps TL, Williams NR, Ortega VM, Chamusco KC, Harris-Shultz Karen, Scully BT, Chase CD (2011) DNA polymorphisms at bermudagrass microsatellite loci and their use in genotype fingerprinting. Crop Sci 21: 1122–1131.

Kasuga S (2002) Growing of high digestibility sorghum variety and utilization as feed (2). 2. Growing of sorghum new breed "Hazuki" using high digestibility gene. Anim Husb 56: 465–469 (In Japanese).

Kouki K, Ebina M, Hayasaka J, Inafuku M, Okumura K (2006) Characterization of new registered cultivar 'Paikaji' guineagrass. Bull of Okinawa Livestock Research Center 44: 95–101 (in Japanese).

Labombarda P, Busti A, Caceres ME, Pupilli ME, Arcioni S (2002) An AFLP marker tightly linked to apomixis reveals hemizygosity in a portion of the apomixis-controlling locus in Paspalum simplex. Genome 45: 513–519.

Loch DS, Rethman NFG, van Niekerk WA (2004) Rhodesgrass. In: Moser LE et al. (eds) Warm-Season (C₄) Grasses. Agron Monogr 45. ASA, CSSA, SSSA, Madison, WI, pp 833–872.

Lutts S, Ndikumana J, Louant BP (1994) Male and female sporogenesis and gametogenesis in apomictic *Brachiaria brizantha*, *Brachiaria decumbens* and F1 hybrids with sexual colchicines induced tetraploid *Brachiaria ruziziensis*. Euphytica 78: 19–25.

Ma KH, Jang DH, Dixit A, Chung JW, Lee SY, Lee JR, Kang HK, Kim SM, Park YJ (2007) Characterization of 30 new microsatellite markers, developed from enriched genomic DNA library of zoysiagrass, *Zoysia japonica* Steud. Mol Ecol Notes 7: 1323–1325.

Martínez EJ, Urbani MH, Quarin CL, Oritz JPA (2001) Inheritance of apospory in bahiagrass, *Paspalum notatum*. Hereditas 135: 19–25.

Matsuoka Y, Yamazaki Y, Ogihara Y, Tsunewaki K (2002) Whole chloroplast genome comparison of rice, maize, and wheat: implications for chloroplast gene diversification and phylogeny of cereals. Mol Biol Evol 19: 2084–2091.

Miles JW, Valle CB do, Rao IM, Euclides VPD (2004) Brachiariagrass. In: Moser LE et al. (eds) Warm-Season (C$_4$) Grasses. Agron Monogr 45. ASA, CSSA, SSSA, Madison, WI, pp 745–783.

Moser LE, Burson BL, Sollenberger LE (2004) Warm-season (C$_4$) grasses. Agron Monogr 45. ASA, CSSA, SSSA, Madison, WI.

Muir JP, Jank L (2004) Guineagrass. In: Moser LE et al. (eds) Warm-season (C$_4$) grasses. Agron Monogr 45. ASA, CSSA, SSSA, Madison, WI, pp 589–621.

Munkacsi AB, Stoxen S, May G (2007) Domestication of maize, sorghum and sugarcane did not drive the divergence of their smut pathogens. Evolution 61: 388–403.

Nakagawa H, Hanna WW (1992) Induced sexual tetraploids for breeding guineagrass (*Panicum maximum* Jacq.). J Jpn Soc Grassl Sci 38: 152–159.

Nakajima K (1990) Apomixis and its application to plant breeding. In: Gamma Field Symposia No. 29, Institute RadBreed, NIAR, MAFF, Ibaraki, Japan, pp 228–238.

Nogler GA (1984) Gametophytic apomixis. In: Johri BM (ed) Embryology of Angiosperms. Springer-Verlag, Berlin, pp 475–518.

Oishi H, Ebina M (2005) Isolation of cDNA and enzymatic properties of betaine aldehyde dehydrogenase from *Zoysia tenuifolia*. J Plant Physiol 162: 1077–1086.

Oritz JPA, Pessino SC, Leblanc O, Hayward MD, Quarin CL (1997) Genetic fingerprinting for determining the mode of reproduction in *Paspalum notatum*, a subtropical apomictic forage grass. Theor Appl Genet 95: 850–856.

Osugi R, Murata T (1986) Variations in the leaf anatomy among some C$_4$ *Panicum* species. Ann Bot 58: 443–453.

Ozias-Akins P, Lubbers EL, Hannna WW, McNay JW (1993) Transmission of the apomictic mode of reproduction in *Pennisetum*: co-inheritance of the trait and molecular markers. Theor Appl Genet 85: 632–638.

Ozias-Akins P, Roche D, Hanna WW (1998) Tight clustering and hemizygosity of apomixis-linked molecular markers in Pennisetum squamulatum implies genetic control of apospory by a divergent locus that may have no allelic form in sexual genotypes. Proc Natl Acad Sci USA 95: 5127–5132.

Ozias-Akins P, Akiyama Y, Hanna WW (2003) Molecular characterization of the genomic region linked with apomixis in *Pennisetum/Cenchrus*. Funct Integr Genom 3: 94–104.

Ozias-Akins P, van Dijk PJ (2007) Mendelian genetics of apomixis in plant. Annu Rev Genet 41: 509–537.

Ozias-Akins P (2006) Apomixis: developmental characteristics and genetics. Crit Rev Plant Sci 25: 199–214.

Palmer NA, Sattler SE, Saathoff AJ, Funnell D, Pedersen JF, Sarath G (2008) Genetic background impacts soluble and cell wall-bound aromatics in brown midrib mutants of sorghum. Planta 229: 115–127.

Pedro MAJ, Miles JW, Guiot JD, Cuadrado H, Lascano CE (2007) Cultivar Mulato II (Brachiaria hybrid CIAT 36087): A high-quality forage grass, resistant to spittlebugs and adapted to well-drained, acid tropical soils, CIAT, Cali, Colombia.

Pessino SC, Oritz JPA, Valle CB do, Evans C, Hayward MD (1997) Identification of a maize linkage group related to apomixis in *Brachiaria*. Theor Appl Genet 94: 439–444.

Pessino SC, Evans C, Oritz JPA, Armstead I, Valle CB do, Hayword MD (1998) A genetic map of the apospory-region in *Brachiaria* hybrids: identification of two markers closely associated with the trait. Hereditas 128: 153–158..

Pillonel C, Mulder MMJ, Boon J, Forstert B, Binder A (1991) Improvement of cinnamyl-alcohol dehydrogenase in the control of lignin formation in *Sorghum bicolor* L. Moench. Planta 185: 538–544.

Pinheiro AA, Pozzobon MT, Valle CB do, Penteado IO, Carneiro VTC (2000) Duplication of the chromosome number of diploid *Brachiaria brizantha* plants using colchicines. Plant Cell Rep 19: 274–278.

Piquemal J, Chamayou S, Nadaud I, Beckert M, Barrière Y, Mila I, Lapierre C, Rigau J, Puigdomenech P, Jauneau A, Digonnet C, Boudet A-M, Goffner D, Pichon M (2002)

Down-regulation of caffeic acid O-methyltransferase in maize revisited using a transgenic approach. Plant Physiol 130: 1675–1685.

Pitman WD, Chambliss CG, Hacker JB (2004) Digitgrass and other species of digitaria. In: Moser LE et al. (eds) Warm-Season (C_4) Grasses. Agron Monogr 45. ASA, CSSA, SSSA, Madison, WI, pp 715–743.

Pritchard JK, Stephens M, Donnelly P (2000) Inference of population structure using multilocus genotype data. Genetics 155: 945–959.

Pupilli F, Labombarda P, Caceres ME, Quarín, Arcioni S (2001) The chromosome segment related to apomixis in *Paspalum simplex* is homoeologous to the telomeric region of the long arm of rice chromosome 12. Mol Breed 8: 53–61.

Saballos A, Ejeta G, Sanchez E, Kang C, Vermerris W (2009) A genomewide analysis of the cinnamyl alcohol dehydrogenase family in sorghum [*Sorghum bicolor* (L.) Moench] identifies *SbCAD2* as the *Brown midrib6* gene. Genetics 181: 783–795.

Sage RF (2001) Environmental and evolutionary preconditions for the origin and diversification of the C_4 photosynthetic syndrome. Plant Biol 3: 202–213.

Saski C, Lee S-B, Fjellheim S, Guda S, Jansen RK, Luo H, Tomkins J, Rognli OA, Daniell H, Clarke JL (2007) Complete chloroplast genome sequences of *Hordeum vulgare*, *Sorghum bicolor* and *Agrostis stolonifera*, and comparative analyses with other grass genomes. Theor Appl Genet 115: 571–590.

Sattler SE, Saathoff AJ, Haas EJ, Palmer NA, Funnell-Harris DL, Sarath G, Pedersen JF (2009) A nonsense mutation in a cinnamyl alcohol dehydrogenase gene is responsible for the sorghum *brown midrib6* phenotype. Plant Physiol 150: 584–595.

Savidan Y (1980) Chromosomal and embryological analysis in sexual x apomictic hybrids of *Panicum maximum* Jacq. Theor Appl Genet 57: 153–156.

Sherwood RT, Berg CC, Young BA (1994) Inheritance of apospory in buffelgrass. Crop Sci 34: 1490–1494.

Takahashi W, Takamizo T, Kobayashi M, Ebina M (2010) Plant regeneration from calli in giant reed (*Arundo donax* L.). Grassl Sci 56: 224–229.

Tsuruta S, Ebina M, Kobayashi M, Akashi R, Kawamura O (2010) Structure and expression profile of the cinnamyl alcohol dehydrogenase gene and its association with lignification in the sorghum (*Sorghum bicolor* (L.) Moench) *bmr-6* mutant. Breed Sci 60: 314–323.

Tsuruta S, Hashiguchi M, Ebina M, Matsuo T, Yamamoto T, Kobayashi M, Takahara M, Nakagawa H, Akashi R (2005) Development and characterization of simple sequence repeat markers in *Zoysia japonica* Steud. Grassl Sci 51: 249–257.

Tsuruta S, Hosaka F, Otabara T, Hashiguchi M, Yamamoto T, Akashi R (2008) Genetic diversity of chloroplast DNA in *Zoysia* and other warm-season turfgrasses. Grassl Sci 54: 151–159.

Tsuruta S, Kobayashi M, Ebina M (2011) *Zoysia*. In: Kole C (ed) Wild Crop Relatives: Genomic and Breeding Resources. Vol Millets and Grasses. Springer-Verlag, Berlin, Heidelberg, pp 297–309.

Tsuruta S, Sakatani Y, Kawabe T, Kawano T, Ebina M, Arashi R, Kawamura O (2003) Enzymatic activities on the phenylpropanoid metabolism and their relationships with an accumulation of lignin in brown-midrib-6 (bmr-6) mutant of sorghum (*Sorghum bicolor* L. Moench). Glassl Sci 49: 379–383.

Ubi BE, Fujimori M, Ebina M (2001) Amplified fragment length polymorphism analysis in diploid cultivars of rhodesgrass. Plant Breed 120: 85–87.

Ubi BE, Kölliker R, Fujimori M, Komatsu T (2003) Genetic diversity in diploid cultivars of rhodesgrass determined on the basis of amplified fragment length polymorphism markers. Crop Sci 43: 1516–1522.

Ubi BE, Fujimori M, Mano Y, Komatsu T (2004) A genetic linkage map of rhodesgrass based on an F_1 pseudo-testcross population. Plant Breed 123: 247–253.

Valle CB do, Glienke C (1991) new sexual accessions in *Brachiaria*. Apomixis Newsl 3: 11–13.

Valle CB do, Savidan YH (1996) Genetics, cytogenetics and reproductive biology of *Brachiaria*. In: Miles JW et al. (ed) *Brachiaria*: Biology, agronomy, and improvement. CIAT, Cali, Colombia, and CNPGC/EMBRATA, Campo Grande, MS, Brazil, pp 147–163.

Vogel KP, Burson BL (2004) Breeding and Genetics In: Moser LE et al. (eds) Warm-Season (C_4) Grasses. Agron Monogr 45. ASA, CSSA, SSSA, Madison, WI, pp 51–94.

Weng JH (2002) Genetic variation of *Zoysia* in Taiwan as analysed by isozyme patterns and salinity tolerance. Plant Prod Sci 5: 236–241.

Weng JH, Fan MJ, Lin CY, Liu YH, Huang SY (2007) Genetic variation of *Zoysia* as revealed by random amplified polymorphic DNA (RAPD) and isozyme pattern. Plant Prod Sci 10: 80–85.

Wouw MVD, Hanson J, Luethi S (1999) Morphological and characterization of a collection of napiergrass (*Pennisetum purpureum*) and *P. purpureum* × *P. glaucum*. Trop Grassl 33: 150–158.

Yamada T, Fukuoka H (1984) Variations in peroxidase isozyme of Japanese lawn grass (*Zoysia japonica* Steud.) population in Japan. Jpn J Breed 34: 431–438.

Yamada-Akiyama H, Akiyama Y, Ebina M, Xu Q, Tsuruta S, Yazaki J, Kishimoto N, Kikuchi S, Takahara M, Takamizo T, Sugita S, Nakagawa H (2009) Analysis of expressed sequence tags in apomictic guineagrass (*Panicum maximum*). J Plant Physiol 166: 750–761.

Yaneshita M, Nagasawa R, Engelke MC, Sasakuma T (1997) Genetic variation and interspecific hybridization among natural populations of zoysiagrasses detected by RFLP analyses of chloroplast and nuclear DNA. Genes Genet Syst 72: 173–179.

Yaneshita M, Ohmura T, Sasakuma T, Ogihara Y (1993) Phylogenetic relationships of turfgrasses as revealed by restriction fragment analysis of chloroplast DNA. Theor Appl Genet 87: 129–135.

Zoysiagrass

Hongwei Cai,[1,2,]* Manli Li,[3] Xun Wang,[4] Nana Yuyama[2]
and Mariko Hirata[2]

ABSTRACT

The genus *Zoysia* consisting of eight species, are found in coastal areas or grasslands of southeastern and eastern Asia (north to China and Japan), Australia and New Zealand. *Zoysia* species are perennial grasses that have both rhizome roots and stolons and usually grow to no more than 30 cm high. *Zoysia japonica* and *Z. matrella* have been utilized extensively as turf in Japan and other countries in eastern Asia. These species also are used on golf courses to create fairways and teeing areas, as well as for grazing livestock. Here, we first review the brief history of the crop and classical breeding efforts, and then describe in detail the results for diversity analysis both based on phonotypical and molecular markers including RAPD, RFLP, SRAP and SSR molecular genetic map construction for both *Z. japonica* and *Z. matrella* based on RFLP, AFLP and SSR markers and QTL analysis, finally, genetically modified *Zoysia* were also reviewed.

Key words: Zoysiagrass, Breeding, Diversity, Molecular map, Genetically modified *Zoysia*

[1]Department of Plant Genetics and Breeding, College of Agronomy and Biotechnology, China Agricultural University, 2 Yuanmingyuan West Road, Beijing 100193, China.
[2]Forage Crop Research Institute, Japan Grassland Agricultural and Forage Seed Association, 388-5 Higashiakada, Nasushiobara, Tochigi 329-2742, Japan.
[3]Department of Grassland Science, College of Animal Science and Technology, China Agricultural University, 2 Yuanmingyuan West Road, Beijing 100193, China.
[4]Chengdu Institute of Biology, Chinese Academy of Sciences, No.9 Section 4, Renmin Nan Road, Chengdu, Sichuan, 610041 P.R. China.
*Corresponding author: *caihw@cau.edu.cn; hcai@jfsass.or.jp*

9.1 *Zoysia* Natural History and Breeding Achievements

The genus *Zoysia* (Poaceae, Poales, Liliopsida, Magnoliophyta) was named after the Austrian botanist Karl von Zois. The eight species of the genus, commonly known as zoysia or zoysiagrass, are found in coastal areas or grasslands of southeastern and eastern Asia (north to China and Japan), Australia, and New Zealand. The eight species are *Z. japonica*, *Z. macrantha*, *Z. macrostachya*, *Z. matrella*, *Z. minima*, *Z. pauciflora*, *Z. sinica*, and *Z. tenuifolia*. *Z. macrantha* is found only in Australia (Loch et al. 2005), *Z. minima* and *Z. pauciflora* are found only in New Zealand (Stewart 2005), and the other five species usually appear in eastern Asia, including China and Japan (Fu 1995; Dong and Gong 2001).

Zoysia species are perennial grasses that have both rhizome roots and stolons and usually grow to no more than 30 cm (some *Z. japonica* cultivars grow taller than 30 cm). The leaves of zoysiagrasses are smooth, wiry, stiff and become brown after the first hard frost. Zoysiagrass has a unique protogynous flowering mechanism: the pistil usually comes out from the glume first and then the anther 2 or 3 d later, such that crossing between different panicles of the same plant is possible.

The three *Zoysia* species used as turf or forage grass (i.e., *Z. japonica*, *Z. matrella*, and *Z. tenuifolia*) are native to eastern Asia. *Zoysia japonica* was introduced into the United States (US) in 1895 from the Manchurian Province of China (Hanson et al. 1974). In 1911, *Z. matrella* was introduced into the US from Manila by a US Department of Agriculture (USDA) botanist, C. V. Piper. Because of its origin, the grass is usually called Manila grass (Piper 1915). *Zoysia tenuifolia* was introduced into the US from the Mascarene Islands.

Z. japonica and *Z. matrella* have been utilized extensively as turf in Japan and other countries in eastern Asia (Shoji 1983; Fukuoka 1989). These species also are used on golf courses to create fairways and teeing areas, as well as for grazing livestock. *Z. japonica* (Japanese lawngrass or Korean lawngrass) has a coarser leaf texture and a greater cold tolerance than *Z. matrella*. *Z. japonica* which can be seeded, but vegetative planting is preferred because seed germination is erratic and slow. *Z. tenuifolia* (Korean velvet grass or Mascarene grass) is a very fine-textured species, but it is the least cold tolerant of the three commonly grown species.

Members of the genus *Zoysia* have a chromosome number of $2n = 40$ and are mostly cross-pollinated, tetraploid species with a small genome size (421 Mb for *Z. japonica*; Forbes 1952; Arumuganathan et al. 1999). At present, there are 74 genetic resource collections of *Zoysia* species, mostly *Z. japonica*, throughout the world. The National Plant Germplasm System of the USDA-ARS (http://www.ars-grin.gov/npgs/index.html) houses 53 *Zoysia* genetic resource collections, and there are three in the Eurisco system (http://eurisco.ecpgr.org/) in Europe and 18 in the Margot Forde Germplasm Centre (http://www.agresearch.co.nz/seeds) in New Zealand.

Most *Zoysia* species can hybridize with other members of the genus and produce fertile F_1 hybrids (Yaneshita et al. 1997). Zoysiagrass usually outcrosses, and most *Zoysia* species (with the exception of some *Z. japonica* lines) produce only a limited number of seeds. Therefore, most breeding involves clonal selections, with only minimal breeding by crossing two different lines because of the difficulties of making F_1 hybrids. The breeding objectives include winter hardiness, heat and drought tolerance, lengthy green period, weak dormancy, high forage and seed yield and resistance to plant pathogens such as rust.

Although zoysiagrass is bred only in the US, Japan, China, and Korea, more than 30 cultivars have been released in the US (e.g., "Meyer", "Emerald", "El Toro", "Empire", and "Zenith") and Japan (e.g., "Asagake", "Asamoe", and "Tanezo"). At present, however, only three *Z. japonica* cultivars ("Compadre", "Companion", and "Zenith") from the United States are included in the Organization for Economic Co-operation and Development's (2009) list of cultivars eligible for seed certification.

9.2 Diversity Analyses

9.2.1 Phenotype-based Diversity Analyses

Choi et al. (1997b) analyzed 93 native zoysiagrass lines collected from the southern and western coastal regions of South Korea. Among them, 69 lines showed normal growth in the experimental field plots. Leaf width, seed length, cotton hair and the number of seeds per spikelet were considered as main factors for the classification of native zoysiagrasses. When nine morphological traits were used for cluster analysis, the 93 collected lines were classified into four main groups, which were believed to be useful for identifying hybrid types. Group I included *Z. sinica* and *Z. macrostachya* types, which had an average seed length of 6.5 mm and an average seed width of 1.2 mm. Group II included many hybrid-type zoysiagrasses with characteristics of both *Z. japonica* and *Z. sinica* types. Group III included the typical *Z. japonica* type, and Group IV included *Z. matrella* and *Z. tenuifolia* types.

Li et al. (2002) analyzed 78 accessions of zoysiagrass from China on the basis of 18 aboveground morphological traits and found that the accessions could be divided into six groups: *Z. sinica*; *Z. japonica*; *Z. tenuifolia* and *Z. matrella*; *Z. macrostachya* and *Z. sinica* var. *nipponica*, and other two small groups. The principal component analysis revealed that spikelet number, spikelet length, spikelet density, ratio of spikelet length to spikelet width, leaf width and leaf hairs were the main indexes used to identify *Zoysia* species. In particular, leaf hairs provide a simple index to distinguish *Z. sinica* from *Z. japonica*. Shoot height, spikelet length, leaf length, flowering culm length and ratio of leaf length to leaf width were the main indexes for identifying subgroups within the species.

Hashiguchi et al. (2006) studied 37 ecotypes of *Z. matrella* and *Z. tenuifolia* collected from the southwestern islands of Japan by using six morphological characteristics and covering gain. The results of cluster analysis suggested that the 37 ecotypes and four cultivars could be classified into three major groups: *Z. matrella* except cultivars "Oujima1" and "Miyakojima" (Cluster I), *Z. tenuifolia* except "Minatogawa 2" (Cluster II), and "Tanegashima 2" (Cluster III). On the basis of the high correlations between leaf width and other morphological characteristics, the zoysiagrass ecotypes were classified as two species, *Z. matrella* and *Z. tenuifolia*. Moreover, covering gain showed high correlations with leaf width, plant height and runner length.

Yang et al. (1995) used morphological characteristics and isozyme band patterns for the identification of zoysiagrass species. They reported that quantitative characteristics such as leaf width and seed length do not provide clear standards for the identification of species. These characteristics showed wide ranges of variation within the population of each species because of environmental effects and natural hybridization. Expressed sequenced Tag (EST) band patterns showed differences among the five major species as well as intermediate types. However, intermediate types showed mixed band patterns of two or more of the species.

According to Anderson (2000), *Zoysia* species could be delineated by differences in a combination of morphological traits, both floral and vegetative. Floral traits important for distinguishing among the species were peduncle diameter; raceme length; number of spikelets per raceme; pedicel apex shape; fusion of glume to pedicel; glume texture; and spikelet length, width, and shape. Floral traits by themselves provided good resolution among most species, although boundaries were not distinct between *Z. pacifica* (= *Z. matrella* var. *pacifica*) and *Z. matrella* or between *Z. macrantha* and *Z. japonica*. Only six species could be recognized on the basis of vegetative traits alone. Important vegetative traits were the presence of blade unrolling, blade diameter and/or width, collar length, culm elongated internode diameter and the presence of elongated culm internodes.

9.2.2 Molecular Marker-based Diversity Analyses

Yamada and Fukuoka (1984) used peroxidase isozymes to analyze 24 populations of Japanese lawngrass (*Z. japonica*) collected from across Japan. The peroxidase zymogram pattern of the leaf extracts revealed about 20 bands. Among the 24 populations, those from pastures generally showed very low intrapopulation variability. Although no clear trend between the degree of relatedness and geographical location was noted in the dendrogram, three populations in Hokkaido and two populations in Miyazaki Prefecture were significantly distant from all other populations. The populations in Honshu and Shikoku were close to one another, and the similarity among the Oozasa, Sotoyama, and Morioka populations was particularly striking. Zymograms of esterase and acid phosphatase isozymes were determined by Weng (2002) using 182 *Z. matrella* and *Z. sinica* individuals collected from the coasts around Taiwan and Penghu Island. There were 26 bands and 108 patterns in the esterase zymogram, and nine bands and 12 patterns in the acid phosphatase zymogram. In the Euclidean distance analysis of the dendrogram, the clones collected from the northern part of the east coast of Taiwan (EN, consisting mainly of limestone) were significantly distant from those of the other five regions. The distances between the clones from these five regions were shorter. Moreover, the clones from the EN region showed lower diversity in both isozyme patterns and less adaptation to salinity because of the protection afforded by the calcium in the limestone.

Sixty-eight zoysiagrass lines collected from the southwestern coastal regions of Korea were evaluated by random amplified polymorphic DNA (RAPD) analysis by Choi et al. (1997a). Seventeen polymorphic bands were obtained by PCR analysis with five random primers. On the basis of the band patterns, zoysiagrass lines were classified into four major groups by cluster analysis and were subdivided into eight groups according to growth characteristics. Group I included *Z. macrostachya*, which was characterized by wide seeds (1.4 mm). Group II included *Z. sinica* and natural hybrids exhibiting *Z. sinica* characteristics, such as long, narrow seed shape; this group showed the highest ratio (5.4) of seed length to seed width among the groups. Group III included *Z. japonica*, which had hairy, wide (5.2 mm) leaves, and natural hybrids exhibiting *Z. japonica* characteristics. Group IV included *Z. matrella*, *Z. tenuifolia*, and their variants, showing natural hybrid characteristics. Distinguishing between *Z. sinica* and *Z. macrostachya* on the basis of morphological characteristics is very difficult, but the use of RAPD markers enabled this to be done. Most of the collected zoysiagrasses that showed mutual characteristics had banding patterns similar to those of *Z. japonica* or *Z. sinica*, indicating an abundance of their natural hybrids.

Weng et al. (2007) also analyzed genetic diversity by using RAPD and isozyme markers. A total of 131 clones of *Zoysia* species collected from 59 sites on the coasts of Taiwan and its neighboring islets were analyzed. In the RAPD analysis with 12 primers, 92 polymorphic bands were used to distinguish 131 genotypes. There were 19 polymorphic bands and 81 zymogram patterns for esterase and nine bands and 10 patterns for acid phosphatase. Cluster analysis by the unweighted pair-group method with arithmetic means of RAPD data indicated that clones collected from the same geographic region were clustered together. However, isozyme data showed discordant patterns. Interestingly, both RAPD data and isozyme fingerprinting revealed less correlation with the intuitive taxonomic classification of the tested clones, but stronger relationships with the plants' specific adaptation to geographic or geological aspects of their habitats.

Ten screened RAPD primers were used to assess the genetic diversity of 105 plants from seven *Z. sinica* populations collected in different regions of China (Li and Tong 2004). About 4.84, 30, and 70% of the genetic variation was detected among groups, populations, and within populations, respectively. The highest genetic diversity was found among populations from Ningguo in Anhui province and Sheyang in Jiangsu province, whereas the lowest genetic diversity was among populations from Shexian in Anhui province.

Yaneshita et al. (1997) used restriction fragment length polymorphism (RFLP) analyses of chloroplast DNA (cpDNA) and nuclear DNA to study genetic variations among 17 zoysiagrass accessions collected from natural populations in Japan. These accessions were classified into five species on the basis of morphological characteristics: *Z. japonica*, *Z. matrella*, *Z. tenuifolia*, *Z. sinica*, and *Z. macrostachya*. On the basis of eight kinds of RFLPs in cpDNAs detected across accessions, six chloroplast genome types (types A to F) were identified. Genetic relationships among the 17 accessions were also investigated by RFLP analyses of nuclear DNA with 20 genomic and gene probes. A dendrogram constructed with genetic distances calculated from the RFLP patterns indicated four major groups. Six *Z. japonica* accessions made up one group, whereas one accession of *Z. japonica* possessing the type F cpDNA was clustered with *Z. macrostachya* and *Z. sinica*. Four *Z. matrella* accessions with type A cpDNA constituted another group in the dendrogram, showing a closer relationship to the *Z. japonica* accessions than to the other two accessions of *Z. matrella*. The remaining two *Z. matrella* and *Z. tenuifolia* accessions were grouped together. These data indicate that zoysiagrasses distributed in Japan harbor high genetic variation and that interspecific hybridization has occurred in natural populations.

Hashiguchi et al. (2007) used simple sequence repeat (SSR) markers to analyze the genetic diversity in zoysiagrass indigenous to the southwestern islands of Japan. Forty-one lines of zoysiagrass, which included 19 *Z. matrella*, 18 *Z. tenuifolia*, and four cultivars (*Z. japonica* "Asagake", *Z. japonica* × *tenuifolia* 'Emerald", *Z. matrella* "Tottori-korai", and *Z. tenuifolia* "Velvet"), were used. From 12 SSR markers, 155 SSR bands were scored. The number of putative alleles ranged from 6 to 22, with an average value of 12.9. The polymorphic information content (PIC) values of *Z. matrella* and *Z. tenuifolia* were 0.79 and 0.72, respectively. *Zoysia matrella* had a significantly higher PIC than that of *Z. tenuifolia* ($p = 0.02$). Cluster analysis based on the 155 SSR bands revealed that the 41 lines were classified into seven groups.

The genetic diversity and interspecific relationships of 96 germplasms of *Zoysia* belonging to five species and one mutation were analyzed by using sequence-related amplified polymorphisms (SRAP) and SSR markers (Guo et al. 2008, 2009). According to the SSR-marker-based diversity analysis, 282 bands were detected by 29 pairs of SSR primers, and 272 were polymorphic (percentage of polymorphic bands 96.45%). Nei's gene diversity index (*H*) was 0.2203, and the Shannon diversity index was 0.3504, indicating a high level of genetic diversity. The genetic similarity among the 96 *Zoysia* germplasms ranged from 0.5922 to 0.9362, with an average of 0.7811. The genetic similarity between *Z. japonica* and *Z. sinica* and between *Z. matrella* and *Z. tenuifolia* was relatively high, whereas both *Z. macrostachya* (Z010) and *Z. sinica* var. *nipponica* (Z122) had lower genetic similarity than the other species. The ranges of genetic similarity within *Z. japonica*, *Z. sinica*, and *Z. matrella* were 0.5922 to 0.9362, 0.6879 to 0.9078, and 0.6738 to 0.8582, respectively. Genetic distances and relationships among the 96 *Zoysia* germplasms were clearly revealed in the cluster dendrogram, and the germplasms were classified into six major groups. Group I included *Z. japonica*, *Z. sinica*, and some accessions of *Z. matrella*, together accounting for 90.6% of all germplasms. Group II consisted of some accessions of *Z. matrella* and *Z. tenuifolia*, and Groups III to VI comprised of one sample each of *Z. macrostachya* (Z010), *Z. sinica* var. *nipponica* (Z122), *Z. matrella* (Z095), and *Z. sinica* (Z115). The cluster results of some secondary groups showed that most accessions of *Z. japonica* and *Z. sinica* were clustered into one group; *Z. matrella* and *Z. tenuifolia* formed another group; and some accessions of *Z. japonica*, *Z. sinica*, and *Z. matrella* comprised a third group. The SRAP-marker-based analysis showed similar results to those obtained by using SSR markers (Guo et al. 2008).

Tsuruta et al. (2008) developed cpDNA gene region markers and cpDNA SSR markers. Twenty-four cpDNA regions of *Z. japonica* "Asagake" were amplified by using primer pairs designed from the complete chloroplast genome sequences of rice (*Oryza sativa*) and maize (*Zea mays*). Of the 24 cpDNA regions, six amplified regions (*trnK* intron [*matK*], intergenic region of *trnD-psbM*, *atpA-rps*14, *petE-psaJ*, *atpB-rbcL*, and *rps*3-*rps*12) were polymorphic among seven turfgrass species and were used for the phylogenetic clustering of seven turfgrasses and two cereal crops. In addition, 21 chloroplast microsatellites were developed from 18 regions. Ten of the 21 primer pairs exhibited polymorphisms among three species of *Zoysia*. These cpDNA gene and SSR markers will be useful in the study of genetic relationships among *Zoysia* species.

9.3 Molecular Linkage Maps: Strategies, Resources and Achievements

9.3.1 Restriction Fragment Length Polymorphism Markers

Yaneshita et al. (1999) first developed restriction fragment length polymorphism (RFLP) probes from *Pst*I-digest, *Taq*I-digest, and *Sau*3AI-digest genomic libraries from *Z. japonica* line JTZJ-24, cDNA probes constructed by using mRNAs from mature leaves and rhizomes of JTZJ-24, and seven gene clones encoding photosynthetic enzymes isolated from JTZJ-24. Of the 115 DNA clones tested, 100 (87.0%), including 55 genomic clones, 38 cDNA clones, and 7 gene clones, showed allelic-RFLP banding patterns among the parental accessions and were used in RFLP linkage map construction. Cai et al. (unpubl. data) also developed and screened 127 probes showing high levels of identity with rice genome sequences from *Sau*3AI partly digested genomic libraries of *Z. japonica* "Akemidori", and we used these probes in zoysiagrass–rice synteny analyses.

Anderson (2000) tested 227 heterologous probes (Lin et al. 1995) to survey the polymorphisms among *Zoysia* species. Of the 227 probes surveyed, 100 (58% of the 172 with a signal) had low copy numbers, 38 (22%) had multiple copies (>4 bands per probe), and 34 (20%) were repetitive sequences. In total, 112 probes (65% of those with a signal) resulted in RFLPs, and 60 were not polymorphic. The remaining 55 probes yielded no signal with the *Zoysia* fragments and were considered non homologous with the *Zoysia* genome.

9.3.2 Amplified Fragment Length Polymorphism Markers

The amplified fragment length polymorphism (AFLP) technique was developed by Vos et al. (1995), and it is based on the amplification of subsets of genomic restriction fragments using the PCR. Most alleles are present as single copies in the parental lines and most AFLP markers are dominant, so the method can score markers individually on the basis of their presence or absence in the progeny. The AFLP marker system allows the rapid detection of a large number of polymorphic loci through relatively modest levels of experimental activity. Consequently, the technique has been widely used in linkage map construction and marker development (Keim et al. 1997; Simons et al. 1997; Alonso-Blanco et al. 1998; Qi et al. 1998).

9.3.3 Sequence-Related Amplified Polymorphism Markers

The sequence-related amplified polymorphism (SRAP) marker technique was developed by Li and Quiros (2001). SRAP is a PCR-based marker system with two types of primers, forward primers of 17 bases and reverse primers of 18 bases. The forward primers consist of a core sequence of 14 bases. The first 10 bases starting at the 5' end are filler sequences of no specific constitution, followed by the sequence CCGG and then by three selective nucleotides at the 3' end. Variation in these three selective nucleotides generates a set of primers sharing the same core sequence. In the reverse primers, the filler sequence is followed by AATT (instead of the CCGG) and three selective bases are added to the 3' end. Like AFLP markers, SRAP markers do not require any genome sequence information and they generate highly repeatable results and numerous polymorphic dominant markers. However, the SRAP technique is simpler than using AFLP markers because it does not require enzyme digestion, adapter ligation and two-step PCRs (cf. preselective and selective PCR in AFLP analysis). Guo et al. (2008) used SRAP markers in a *Zoysia* diversity study, as described above.

9.3.4 Simple Sequence Repeat Markers

Simple sequence repeats (SSRs) or microsatellites are tandem repeated units of two to six nucleotides that are widely distributed in eukaryotic genomes and are numerous and polymorphic. Although the development of SSR markers is expensive and time-consuming, SSR markers have some advantage over RFLP, RAPD, and AFLP markers. They are PCR-based, multi-allelic, highly reproducible and codominant.

Tsuruta et al. (2005) developed a genomic library enriched for the AG/TC motif from *Z. japonica* "Asagake" and screened it by using 5'-biotin-labeled oligonucleotides (AG_{20}). On the basis of a sequencing analysis of 162 clones, 119 clones containing eight to 35 SSR repeats were identified. Thirty-two primer pairs designed from these clones were evaluated for their ability to detect polymorphisms among six additional zoysiagrass cultivars. The average values of observed heterozygosity and PIC for the individual loci were 0.57 and 0.69, respectively. These primers were successful in generating several informative cross-amplification products in additional *Zoysia* species.

Cai et al. (2005) developed 1044 SSR markers from four SSR-enriched genomic libraries and tested the PCR amplification ability of the SSR markers in one screening panel consisting of eight *Zoysia* clones. Of the 1,044 SSR markers, 170 segregated in a mapping population, and 161 markers were mapped on existing AFLP-based linkage groups. Ma et al. (2007) also developed 30 polymorphic microsatellite markers for *Z. japonica*. The 30 markers revealed 125 alleles, with an average of 4.2 alleles per locus.

9.3.5 Molecular Linkage Map Construction and Quantitative Trait Losi (QTL) Analyses

Several genetic maps of zoysiagrass have been published. Two molecular linkage maps were constructed from interspecific hybrids of *Z. japonica* and *Z. matrella* on the basis of RFLP (Yaneshita et al. 1999) and AFLP markers (Ebina et al. 1999).

Yaneshita et al. (1999) constructed the RFLP linkage map of *Zoysia* species by using self-pollinated progenies obtained from the *Z. japonica* and *Z. matrella* interspecific hybrid. Of the 115 DNA clones tested, 100 (87.0%), including 55 genomic clones, 38 cDNA clones, and seven gene clones encoding photosynthetic enzymes, showed allelic-RFLP banding patterns among the parental accessions. Twenty-six probes detected two or more loci segregating independently in the self-pollinated progenies. The RFLP linkage map consists of 115 loci in 22 linkage groups ranging in size from 12.5 to 141.3 cM, with a total map distance of 1506 cM. Five pairs of linkage groups sharing a series of duplicated loci with approximately the same order were identified. On the basis of this finding, the authors concluded that *Zoysia* species with $2n = 40$ should be considered as allotetraploids, which may have evolved from progenitors with a basic chromosome number of 10 ($x = 10$).

Cai et al. (2004) constructed an AFLP-based molecular linkage map of *Z. japonica*. An F_2 population consisting of 78 individuals was used to analyze 471 AFLP markers derived from 126 *Pst*I/*Mse*I primer combinations. Of these markers, 364 were grouped into 26 linkage groups. The map covered a total length of 932.5 cM, with an average spacing of 2.6 cM between markers.

A consensus genetic linkage map with 447 SSR markers was constructed for *Z. japonica* by using 86 F_1 individuals from the cross "Muroran 2" × "Tawarayama Kita 1" (Li et al. 2009). The consensus map identified 22 linkage groups and had a total length of 2,009.9 cM, with an average map density of 4.8 cM. This zoysiagrass consensus map contained 35 SSR markers exhibiting strong identity with rice genomic sequences from known chromosomal locations; this allowed synteny to be identified between zoysiagrass linkage groups 2, 3, 9, and 19 and rice chromosomes 3, 12, 2 and 7, respectively.

Li ML et al. (2010) constructed a linkage map of *Z. matrella* that contained 213 loci and covered a map distance of 1,351.2 cM in 32 linkage groups. This was then integrated into a consensus linkage map by using two previously developed *Z. japonica* maps. The integrated map is composed of 507 loci at a mean interval of 4.1 cM, covering a map distance of 2066.6 cM in 22 linkage groups (Fig. 9-1). This is the first integrated SSR linkage map derived from different mapping populations of both *Z. japonica* and *Z. matrella*. Use of this integrated consensus map will allow researchers to streamline mapping

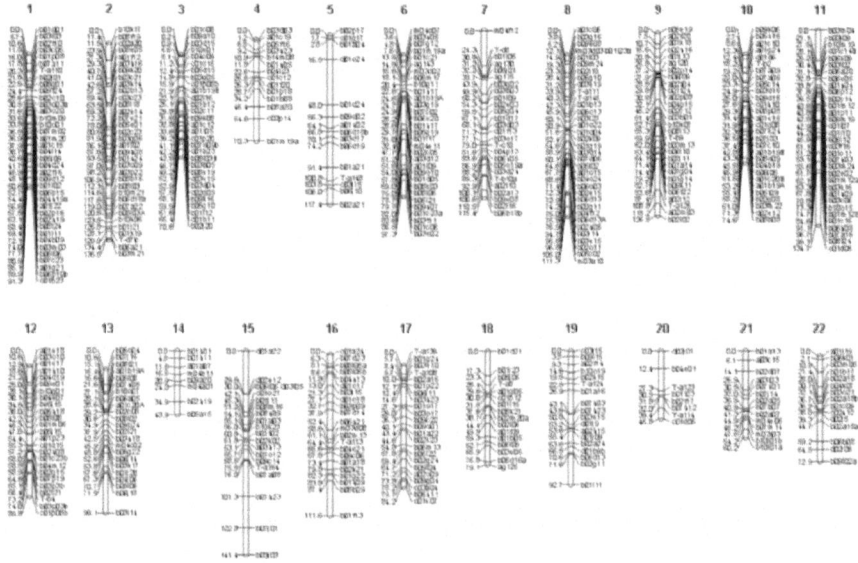

Figure 9-1 An integrated SSR-based linkage map for *Zoysia japonica* and *Z. matrella* (M.L. Li et al. 2010). The linkage groups are represented as vertical bars, with the names of markers on the right and distances (cM) between markers on the left. Markers designed from the same clone sequence are indicated by an "a" or "b" at the end of their name. Markers with multiple copies are indicated by names that end in "A" or "B."

and cloning of agriculturally or horticulturally important genes or QTLs and will promote the development of physical maps for *Zoysia* species.

There are only a few reports on QTL analyses of zoysiagrass, including two molecular linkage map-based QTL studies published in conference proceedings. Kobayashi et al. (2000) constructed an AFLP-RFLP linkage map by using an F_1 pseudo-test cross progeny derived from a cross between *Z. japonica* and *Z. matrella* to conduct a QTL analysis of morphological and physiological traits, such as leaf length and width, internode length of rhizome, leaf color retention in autumn and cold tolerance. They detected only a few clear QTLs for leaf width. Jessup et al. (2006) constructed a zoysiagrass linkage map based on 405 AFLPs, 54 buffelgrass EST-SSRs, and 29 bermudagrass EST-SSRs by using 73 F_1 individuals from a cross between two zoysiagrass cultivars, "Diamond" and "Cavalier". Mendelian mapping identified markers linked to a fall armyworm (*Spodoptera frugiperda*) resistance locus in the parent "Cavalier". The QTL analysis confirmed the Mendelian mapping results and identified another fall armyworm resistance locus in "Diamond" as well. In addition, QTL analyses of six parameters related to salinity tolerance revealed four salinity-tolerance QTLs in "Diamond" and one in "Cavalier".

Chen et al. (2009) performed a bulked segregant analysis to identify SRAP molecular markers linked to salt tolerance in zoysiagrasses by using two extremely salt-tolerant types of DNA pools. From 400 pairs of SRAP primer combinations, 111 bands were found to be polymorphic. These were further screened by using two extremely salt-tolerant samples of *Z. japonica*, and 22 pairs of polymorphism primer combinations were selected. Finally, the primer combinations with specific bands were verified by using salt-tolerant individuals, and seven SRAP molecular markers closely linked to salt tolerance were obtained.

9.4 Genetically Modified *Zoysia*

9.4.1 Embryogenic Callus Induction and Plant Regeneration

By using mature seeds, young inflorescences and stem nodes, researchers have established embryogenic callus induction and plant regeneration systems for zoysiagrass (Al-Khayri et al. 1989; Asano 1989; Yoo and Kim 1991; Noh et al. 1995; Inokuma et al. 1996).

Asano (1989) reported the induction of embryogenic callus from mature caryopsis of *Z. japonica*, and plant regeneration was achieved through precocious germination of somatic embryos. However, regeneration from protoplasts isolated from the callus was not successful. Inokuma et al. (1996) also induced the formation of embryogenic callus of *Z. japonica* from sterile mature seeds and successfully regenerated plants from protoplasts isolated from suspension cells. Al-Khayri et al. (1989) reported over 90% efficiency of a method for inducing embryogenic callus produced from embryos aseptically excised from seeds. Noh et al. (1995) developed a method for regenerating zoysiagrass through somatic embryogenesis by using immature inflorescences of various lengths and caryopses at different stages of maturity. Poeaim et al. (2005) also used immature inflorescences to obtain regenerated plants. From five cultivar of zoysiagrass, however, regenerated plants were obtained only from one cultivar and a higher frequency of callus formation was achieved at earlier developmental stages (length of inflorescences < 20 mm).

Dhandapani et al. (2008) optimized the conditions for efficient regeneration of the vegetatively propagated cultivar *Z. matrella* "Konhee". Both young inflorescences and stem nodes were used, and they displayed different responses to various combinations and concentrations of plant growth regulators in callusing, embryogenic callus formation and regeneration.

9.4.2 Gene Transformation

The first transgenic zoysiagrass plants were produced by polyethylene glycol-mediated direct gene transfer to protoplasts (Inokuma et al. 1998). In this study, plasmid pBC1 was used to deliver the hygromycin phosphotransferase (*hph*) and β-glucuronidase (*gus*) genes into protoplasts. Resistant calli and about 400 plants were generated after selection with a high concentration (400 mg/L) of hygromycin. PCR and Southern hybridization analyses confirmed that all of the plants tested contained the introduced genes. The *gus* gene regulated by the maize alcohol dehydrogenase-1 (*Adh-1*) promoter was expressed in the leaves and roots of transgenic plants.

Because protoplast transformation is a delicate system to work with and often results in multiple insertions, *Agrobacterium*-mediated transformation has been attempted. Chai and Kim (2000) reported transformation of the GUS gene by using the co-integrated vector pTOK 233, harboring *hph*, and GUS genes in *Agrobacterium* LBA 4404 in the zoysiagrass cultivar "Zenith". Transient GUS reaction was detected in callus, and its expression was promoted with sonication and vacuum treatments. GUS expression was also detected in the green shoot clusters. Toyama et al. (2003) developed a herbicide-resistant *Z. japonica* by *Agrobacterium*-mediated transformation. Maximum GUS expression was observed when a Type 3 callus was co-cultivated on 2,4-D-free co-cultivation medium for 9 d. In addition, removal of calcium and the addition of 50 to 100 mg/L acetosyringone during co-cultivation enhanced GUS expression. When this optimized protocol was applied to the transformation of the bialaphos-resistance gene (*bar*), four plants per 700 mg of infected calluses survived on the selective medium. DNA gel-blot analysis showed that two copies of the transgene had been integrated. After application of 2 g/L bialaphos for 1 wk, the transgenic plants survived herbicide spraying, whereas untransformed zoysiagrasses and invading weeds died. Zhang et al. (2007) transferred a synthetic *cryIA(b)* gene from *Bacillus thuringiensis* under the control of a maize ubiquitin promoter by *Agrobacterium*-mediated transformation. The embryogenic calli derived from dormancy-removal mature seeds were co-cultured with the disarmed strain EHA105 harboring the binary vector pKUB. Three days after co-culture with EHA105 in the dark at 21°C, the transient GUS expression frequency was 74.2%. After selection with 100 mg/L hygromycin B, more than 50 independent resistant cell clones and 25 regenerated plants were obtained. Integration and expression of the *cryIA(b)* gene into the genome was confirmed in 22 regenerated plants by GUS histochemical assay, PCR amplification, and Southern and Western blot analyses, with a transformation efficiency of 1.4%.

Ge et al. (2006) developed a straightforward and efficient transformation protocol in *Z. japonica* without callus induction and propagation. Sterilized stolon nodes were infected and co-cultivated with *Agrobacterium tumefaciens* harboring pCAMBIA vectors. The *hph* gene was used as the selectable marker, and hygromycin was used as the selection agent. Both green and albino shoots were directly regenerated from the infected stolon nodes 4 to 5 wk after hygromycin selection. Greenhouse-grown plants were obtained 10 to 12 wk after *Agrobacterium*-mediated transformation, and the transformation frequency was 6.8%. The plants' transgenic nature was confirmed by PCR, Southern blot hybridization analyses, RT-PCR analysis and GUS staining.

Lim et al. (2004) were the first to produce transgenic herbicide-resistant *Z. japonica* plants by particle bombardment–mediated transformation with the plasmid pSMABuba, which contains hygromycin resistance (*hpt*) and bialaphos resistance (*bar*) genes. The parameters of DNA delivery efficiency of the particle bombardment were partly optimized by using a transient expression assay of a chimeric β-glucuronidase (*gusA*) gene driven by the CaMV 35S promoter. The transgenic zoysiagrass plants were confirmed by PCR analysis with a *bar*-specific primer, and expression of the transgene in transformed plants was demonstrated by reverse transcriptase (RT)-PCR analysis. All the tested transgenic plants showed resistance to the herbicide BastaR at field application rates of 0.1 to 0.3%.

In all the transformation experiments described above, the transgenic plants were *Z. japonica*. Recently, however, *Agrobacterium*-mediated transformation of *Z. sinica* and *Z. tenuifolia* has been reported as well. Li et al. (2006) used a highly regenerable callus line Zh44 of *Z. sinica* for *Agrobacterium*-mediated transformation. The gene, encoding C-repeat/ dehydration-responsive element binding factor 1 (CBF1/DREB1b) from *Arabidopsis* driven by the cauliflower mosaic virus (CaMV) 35S promoter, was introduced into *Z. sinica* under selection by using the *bar* gene with glufosinate. Three independent transgenic lines were successfully obtained by using a three-step selection-and-regeneration procedure. The transgenic plants were confirmed by PCR, Southern blot and RT-PCR analyses. Two transgenic lines exhibited growth retardation and decreased tillering. The transgenic lines had a significantly stronger chilling tolerance than that of wild-type (WT) plants, measured by ion leakage, total chlorophyll content, and yellowed leaf rate under chilling conditions.

Li et al. (2010) reported *Agrobacterium*-mediated transformation of *Z. tenuifolia*. Initial calli were induced from stem nodes, and the compact calli were selected and subcultured monthly on fresh medium. For genetic transformation, calli were incubated with *A. tumefaciens* strain EHA105 harboring the binary vector pCAMBIA 1301, which contains the *hpt* gene as a selectable marker for hygromycin resistance and an intron-containing β-glucuronidase gene (*gus-int*) as a reporter gene. The transgenic nature of the transformants was demonstrated by the detection of β-glucuronidase activity in the primary transformants and by PCR and Southern hybridization analysis. About 5% of the total inoculated callus explants produced transgenic plants after approximately 5 mon.

9.4.3 Risk Evaluation of Genetically Modified Zoysia

Two recent studies have evaluated the risk of introducing genetically modified (GM) zoysiagrass into the environment. Bae et al. (2008) investigated the environmental and biodiversity risks of herbicide-tolerant GM *Zoysia* before applying to regulatory agencies for approval for commercial release. The GM and WT zoysiagrasses' substantial trait equivalence, ability to cross-pollinate and gene flow in confined and unconfined test fields were analyzed. No differences were noted between the substantial traits of GM and WT zoysiagrasses. To assess the potential for cross-pollination and gene flow, the nonselective herbicide Basta was used. No unintended cross-pollination or gene flow from GM zoysiagrass was detected in the neighboring weed species examined, but cross-pollination and gene flow were observed in WT zoysiagrass (on average, 6% at proximity, 1.2% at 0.5 m, 0.12% at 3 m, and 0% at > 3 m). Kang et al. (2009) tested the viability and longevity of pollen for both GM and WT *Zoysia* by using a newly developed germination medium containing 20% sucrose and 50 ppm H_3BO_3. After about 1000 hr, pollen grains transferred to the medium had a germination rate of > 90%. Pollen was predominantly shed at approximately 1000 hr, with viability declining to nearly 0% at 1200 hr. All germinability was lost within 150 min when pollen in the medium was stored at 25°C. No significant differences in pollen viability or longevity were found between GM and WT plants.

9.5 Future Prospects

As is the case with most forage crops, the genomic study of zoysiagrass has lagged behind that of other major crops. Only one study has characterized a zoysiagrass gene by using a PCR method. Oishi and Ebina (2005) isolated cDNAs encoding betaine aldehyde dehydrogenase (BADH, EC 1.2.1.8) from the salt-tolerant *Z. tenuifolia* by PCR. *Zoysia* betaine aldehyde dehydrogenase 1 (*ZBD1*) is 1892 bp long and codes for 507 amino acids. The deduced amino acid sequence of ZBD1 shows 88% identity with the sequence of rice BADH. As investigations of synteny between zoysiagrass and other well-studies crop species like rice, maize and sorghum are conducted, *Zoysia* gene isolation could be achieved more easily.

Although genetic and genomic studies of forage crops such as *Lolium* species and alfalfa have been conducted and there are many reports of the genetic diversity of *Zoysia* species and GM *Zoysia*, much genetic and genomic research on *Zoysia* still remains to be done. Therefore, in the future, molecular genetics research should focus on map construction, QTL mapping, gene isolation, and functional genomics of *Zoysia* species.

References

Al-Khayri JM, Huang FH, Thompson LF, King JW (1989) Plant regeneration of zoysiagrass from embryo-derived callus. Crop Sci 29: 1324–1325.

Alonso-Blanco C, Peeters AJM, Koornneef M, Lister C, Dean C, Van Den Bosch N, Pot J, Kuiper MTR (1998) Development of an AFLP based linkage map of L*er*, Col and Cvi *Arabidopsis thaliana* ecotypes and construction of a L*er*/Cvi recombinant inbred line population. Plant J 14: 259–271.

Anderson SJ (2000) Taxonomy of *Zoysia* (Poaceae) morphological and molecular variation. PhD Diss, Texas A&M University, 335 p.

Arumuganathan K, Tallury SP, Fraser ML, Bruneau AH, Qu R (1999) Nuclear DNA content of thirteen turfgrass species by flow cytometry. Crop Sci 39: 1518–1521.

Asano Y (1989) Somatic embryogenesis and protoplast culture in Japanese lawngrass (*Zoysia japonica*). Plant Cell Rep 8: 141–143.

Bae TW, Vanjildorj E, Song SY, Nishiguchi S, Yang SS, Song IJ, Chandrasekhar T, Kang TW, Kim JI, Koh YJ, Park SY, Lee J, Lee YE, Ryu KH, Riu KZ, Song PS, Lee HY (2008) Environmental risk assessment of genetically engineered herbicide-tolerant *Zoysia japonica*. J Environ Qual 37: 207–218.

Cai HW, Inoue M, Yuyama N, Nakayama S (2004) An AFLP-based linkage map of zoysiagrass (*Zoysia japonica*). Plant Breed 123: 543–548.

Cai HW, Inoue M, Yuyama N, Takahashi W, Hirata M, Sasaki T (2005) Isolation, characterization and mapping of simple sequence repeat markers in zoysiagrass (*Zoysia* spp.). Theor Appl Genet 112: 158–166.

Chai ML, Kim DH (2000) *Agrobacterium*-mediated transformation of Korean lawngrass (*Zoysia japonica*). J Kor Soc Hort Sci 41(5): 455–458.

Chen X, Guo HL, Xue DD, Zheng YQ, Liu JX (2009) Identification of SRAP molecular markers linked to salt tolerance in zoysiagrass. Acta Pratacul Sin 18(2): 66–75.

Choi JS, Ahn BJ, Yang GM (1997a) Classification of zoysiagrasses (*Zoysia* spp.) native to the southwest coastal regions of Korea using RAPDs. J Kor Soc Hort Sci 38(6): 789–795.

Choi JS, Ahn BJ, Yang GM (1997b) Distribution of native zoysiagrasses (*Zoysia* spp.) in the south and west coastal regions of Korea and classification using morphological characteristics. J Kor Soc Hort Sci 38(4): 399–407.

Dhandapani M, Hong SB, Aswath CR, Kim DH (2008) Regeneration of zoysia grass (*Zoysia matrella* L. Merr.) cv. Konhee from young inflorescences and stem nodes. *In vitro* Cell Dev Biol Plant 44: 8–13.

Dong HD, Gong LJ (2001) Ecology of *Zoysia japonica* and the exploitation and application of its resources in China. China Forestry Press, Beijing, pp 1–8 (in Chinese).

Ebina M, Kobayashi M, Kasuga S, Araya H, Nakagawa H (1999) An AFLP based genome map of zoysiagrass. In: Proceeding of the Plant and Animal Genome VII conference, San Diego, CA, USA, 17–21 January, p 278. Available at http: //www.intl-pag.org/pag/.

Forbes I Jr (1952) Chromosome numbers and hybrids in *Zoysia*. Agron J 44: 147–151.

Fu PY (1995) Clavis Plantarum Chinae Boreali-orientalis (Editio secunda). Science Press, Beijing. (In Chinese with Latin Name Index).

Fukuoka H (1989) Breeding of *Zoysia* spp. J Jpn Soc Turfgrass Sci 17: 183–190 (in Japanese).

Ge Y, Norton T, Wang ZY (2006) Transgenic zoysiagrass (*Zoysia japonica*) plants obtained by *Agrobacterium*-mediated transformation. Plant Cell Rep 25: 792–798.

Guo HL, Liu JX, Zhou ZF, Xuan JP (2008) Interspecific relationship and genetic diversity of zoysiagrass revealed by SSR markers. Acta Agrestia Sin 16(6): 552–558.

Guo HL, Zheng YQ, Chen X, Xue DD, Liu JX (2009) Genetic diversity and relationships of zoysiagrass as revealed by SRAP markers. Acta Pratacul Turae Sin 18(5): 201–210.

Hanson AA, Juska FV, Burton GW (1974) Species and varieties. In: Hanson AA, Juska FV (eds) Turfgrass science. American Society of Agronomy, Madison, WI, pp 370–409.

Hashiguchi M, Tsuruta S, Matsuo T, Akamine H, Akashi R (2006) Analysis of genetic resource in *Zoysia* spp.1. Morphological characteristics and covering gain in zoysiagrass indigenous to southwest islands of Japan. Jpn J Grassl Sci 52: 183–189.

Hashiguchi M, Tsuruta S, Matsuo T, Ebina M, Kobayashi M, Akamine H, Akashi R (2007) Analysis of genetic resource in *Zoysia* spp. 2. Evaluation of genetic diversity in zoysiagrass indigenous to southwest islands of Japan based on simple sequence repeat markers. Jpn J Grassl Sci 53: 133–137.

Inokuma C, Sugiura K, Cho C, Okawara R, Kaneko S (1996) Plant regeneration from protoplasts of Japanese lawngrass. Plant Cell Rep 15: 737–741.

Inokuma C, Sugiura K, Imaizumi N, Cho C (1998) Transgenic Japanese lawngrass (*Zoysia japonica* Steud.) plants regenerated from protoplasts. Plant Cell Rep 17: 334–338.

Jessup RW, Burson BL, Krishnaramanujam R, Engelke MC, Genovesi AD, Reinert JA, Binzel ML, Kamps TL, Shulze S, Paterson AH (2006) Genetic mapping of fall armyworm resistance and salinity tolerance in zoysiagrass [abstract]. American Society of Agronomy. Madison, WI, Paper No. 1228b.

Kang HG, Bae TW, Jeong OC, Sun HJ, Lim PO, Lee HY (2009) Evaluation of viability, shedding pattern, and longevity of pollen from genetically modified (GM) herbicide-tolerant and wild-type zoysiagrass (*Zoysia japonica* Steud.). J Plant Biol 52: 630–634.

Keim P, Schupp JM, Travis JM, Clayton SE, Zhu K, Shi T, Ferreira L, Webb A, David M (1997) A high-density soybean genetic map based on AFLP markers. Crop Sci 37: 537–543.

Kobayashi M, Tsurumi Y, Ebina M, Nakagawa H (2000) QTL analysis in zoysiagrass: morphological and physiological characteristics. Proceedings of the Plant and Animal Genome VIII Conference, San Diego, CA, January 9–12.

Li G, Quiros CF (2001) Sequence-related amplified polymorphism (SRAP), a new marker system based on a simple PCR reaction: its application to mapping and gene tagging in *Brassica*. Theor Appl Genet 103: 455–461.

Li M, Li H, Hu X, Pan X, Wu G (2010) An *Agrobacterium tumefaciens*-mediated transformation system using callus of *Zoysia tenuifolia* Willd. ex Trin. Plant Cell Tiss Org Cult DOI: 10.1007/s11240-010-9736-2.

Li ML, Yuyama N, Hirata M, Han JG, Wang YW, Cai HW (2009) Construction of a high-density SSR marker-based linkage map of zoysiagrass (*Zoysia japonica* Steud.). Euphytica 170(3): 327–338.

Li ML, Yuyama N, Hirata M, Wang YW, Han JG, Cai HW (2010) An integrated SSR based linkage map for *Zoysia matrella* L. and *Z. japonica* Steud. Mol Breed 26: 467–476.

Li RF, Wei JH, Wang HZ, He J, Sun ZY (2006) Development of highly regenerable callus lines and *Agrobacterium*-mediated transformation of Chinese lawngrass (*Zoysia sinica* Hance) with a cold inducible transcription factor, CBF1. Plant Cell Tiss Org Cult 85: 297–305.

Li Y, Tong HY (2004) Genetic differentiation in *Zoysia sinica* populations revealed by RAPD markers. Guihaia (Guangxi Zhiwu) 24(4): 345–349 (In Chinese).

Li Y, Li PP, Liu JX (2002) Morphological diversity of above-ground part of *Zoysia* spp. in China. J Plant Res Environ 11(4): 33–39.

Lim SH, Kang BC, Shin HK (2004) Herbicide resistant turfgrass (*Zoysia japonica* cv. 'Zenith') plants by particle bombardment-mediated transformation. Kor Turfgrass Sci 18(4): 211–219.

Lin YR, Schertz KF, Paterson AH (1995) Comparative analysis of QTLs affecting plant height and maturity across the Poaceae, in reference to an interspecific sorghum population. Genetics 141: 391–411.

Loch DS, Simon BK, Poulter RE (2005) Taxonomy, distribution and ecology of *Zoysia macrantha* Desv., an Australian native species with turf breeding potential. Int Turfgrass Soc Res J 10: 593–599.

Ma KH, Jang DH, Dixit A, Chung JW, Lee SY, Lee JR, Kang HK, Kim SM, Park YJ (2007) Characterization of 30 new microsatellite markers, developed from enriched genomic DNA library of zoysiagrass, *Zoysia japonica* Steud. Mol Ecol Notes 7: 1323–1325.

Noh HY, Choi JS, Ahn BJ (1995) Plant regeneration through somatic embryogenesis in zoysiagrasses (*Zoysia* spp.). J Kor Soc Hort Sci 36(4): 582–587.

Organization for Economic Co-operation and Development (2009) List of varieties eligible for seed certification. http://www.oecd.org/agr/seed. United Nations Organization for Economic Co-operation and Development, Paris.

Oishi H, Ebina M (2005) Isolation of cDNA and enzymatic properties of betaine aldehyde dehydrogenase from *Zoysia tenuifolia*. J Plant Physiol 162(10): 1077–1086.

Piper CV (1915) P.I. 34657. *Osterdamia matrella* (L.) Kuntze (*Zoysia pungens* Willd.). In: Plant inventory 33: 43. US Department of Agriculture, Washington, D.C.

Poeaim A, Matsuda Y, Murata T (2005) Plant regeneration from immature inflorescence of zoysiagrass (*Zoysia* spp.). Plant Biotechnol 22(3): 245–248.

Qi X, Stam P, Lindhout P (1998) Use of locus-specific AFLP markers to construct a high-density molecular map in barley. Theor App Genet 96: 376–384.

Shoji S (1983) Species ecology of *Zoysia* grass. J Jpn Soc Turfgrass Sci 12: 105–110 (in Japanese).

Simons G, van der Lee T, Diergarde P, van Daelen R, Groendijk J, Frijters A, Buschges R, Hollicher K, Topsch S, Schulze-Lefert P, Salamini F, Zabeau M, Vos P (1997) AFLP-based fine mapping of the *Mlo* gene to a 30-kb DNA segment of the barley genome. Genomics 44: 61–70.

Stewart A (2005) The potential for domestication and seed propagation of native New Zealand grasses for turf. In: Royal New Zealand Institute of Horticulture Conference 2003: Greening the City—Bringing Biodiversity Back into the Urban Environment. Christchurch, 21–24 October 2003. pp 277–284 (http://www.rnzih.org.nz/pages/2003ConferenceProgramme.htm).

Toyama K, Bae CH, Kang JG, Lim YP, Adachi T, Rui KZ, Song PS, Lee HY (2003) Production of herbicide-tolerant zoysiagrass by *Agrobacterium*-mediated transformation. Mol Cells 16: 19–27.

Tsuruta S, Hashiguchi M, Ebina M, Matsuo T, Yamamoto T, Kobayashi M, Takahara M, Nakagawa H, Akashi R (2005) Development and characterization of simple sequence repeat markers in *Zoysia japonica* Steud. Grassl Sci 51: 249–257.

Tsuruta S, Hosaka F, Otabara T, Hashiguchi M, Yamamoto T, Akashi R (2008) Genetic diversity of chloroplast DNA in *Zoysia* and other warm-season turfgrasses. Grassl Sci 54(3): 151–159.

Vos P, Hogers R, Bleekers M, Reijans M, van der Lee T, Hornes M, Frijters A, Pot J, Peleman J, Kuiper M, Zabeau M (1995) AFLP, A new technique for DNA fingerprinting. Nucl Acids Res 23: 4407–4414.

Weng JH (2002) Genetic variation of *Zoysia* in Taiwan as analyzed by isozyme patterns and salinity tolerance. Plant Prod Sci 5(3): 236–241.

Weng JH, Fan MJ, Lin CY, Liu YH, Huang SY (2007) Genetic variation of *Zoysia* as revealed by random amplified polymorphic DNA (RAPD) and isozyme pattern. Plant Prod Sci 10: 80–85.

Yamada T, Fukuoka H (1984) Variations in peroxidase isozyme of Japanese lawn grass (*Zoysia japonica* Steud.) populations in Japan. Jpn J Breed 34: 431–438.

Yaneshita M, Kaneko S, Engelke MC, Sasakuma T (1997) Genetic variation and interspecific hybridization among natural populations of zoysiagrasses detected by RFLP analysis of chloroplast and nuclear DNA. Gene Genet Sys 72: 173–179.

Yaneshita M, Kaneko S, Sasakuma T (1999) Allotetraploidy of *Zoysia* species with 2n=40 based on RFLP genetic map. Theor Appl Genet 98: 751–756.

Yang GM, Ahn BL, Choi LS (1995) Identification of native zoysiagrasses (*Zoysia* spp.) using morphological characteristics and esterase isozymes. J Kor Soc Hort Sci 36(2): 240–247.

Yoo YK, Kim KS (1991) Effects of plant growth regulators on callus formation and organogenesis from the shoot-tip cultures of five turfgrass species. J Kor Soc Hort Sci 32: 237–246.

Zhang L, Wu D, Zhang L, Yang C (2007) *Agrobacterium*-mediated transformation of Japanese lawngrass (*Zoysia japonica* Steud.) containing a synthetic cryIA(b) gene from *Bacillus thuringiensis*. Plant Breed 126(4): 428–432.

10

Alfalfa

Maria J. Monteros,[1,2] Yuanhong Han[1] and E. Charles Brummer[1,2,*]

ABSTRACT

Alfalfa or lucerne (*Medicago sativa* L.), one of the most productive and nutritious forage crops, is grown worldwide for hay, silage and pasture. It is an autotetraploid, outcrossing crop, characteristics that have limited its genetic gain. A substantial collection of germplasm is held in repositories throughout the world, affording breeders with materials to help improve traits including forage and seed yield; cold, drought and salinity tolerance; numerous disease and insect resistances and nutritive value. Phenotypic recurrent selection, with or without progeny testing, remains the primary breeding method, with strain crossing among various germplasm sources used as a common method to consolidate traits. Various genetic marker types have been developed in alfalfa, with thousands of SNP markers now available. The application of SNP to marker-assisted selection and genomic selection is in the early stages. However, the availability of low-cost, high-throughput assays will make the use of molecular markers a routine part of breeding programs over the next decade. Transgenes can be routinely evaluated in alfalfa, but to date, only one transgene has been commercialized in the USA. In the future, alfalfa breeders will continue to refine conventional selection methods, utilize genetic markers to select for multiple target traits and incorporate transgenes with practical and economic value in compliance with regulatory agencies.

Key words: Alfalfa, Lucerne, Breeding methods, Genetic markers, Marker-assisted breeding, Genomic selection, Transgenic crops

[1] The Samuel Roberts Noble Foundation, Ardmore, OK 73401, USA.
[2] Institute of Plant Breeding, Genetics and Genomics, The University of Georgia, Athens, GA 30602, USA.
*Corresponding author: *ecbrummer@noble.org*

10.1 Biology, Including Taxonomy, Phylogenetics and Agronomy

10.1.1 Alfalfa Taxonomy and Genetic Diversity

Alfalfa (*Medicago sativa* L.) is member of the genus *Medicago* within the Fabaceae family. Taxonomic treatment of the genus, and in particular, closely related taxa, has been comprehensively discussed by Small (2011), who provides the most straightforward organization of the genus. His monograph updates and expands several former descriptions of the genus (Sinskaya 1950; Lesins and Lesins 1979; Quiros and Bauchan 1988). Regarding alfalfa, Small (2011) states: "The taxonomic treatment of *M. sativa*, the most important species of *Medicago*, has been, in my opinion, intolerable. … [T]he classifications are based on (a) extreme taxonomic splitting … and (b) misinterpretation of geographically localized introgression and hybrid wild alfalfa populations as genuine species." The numerous taxonomic synonyms for alfalfa and its relatives are listed by Small (2011), which makes a useful cross-reference to help understand the oftentimes confusing literature.

M. sativa consists of several diploid and tetraploid interfertile taxa, as summarized in Fig. 10-1. These taxa may be discriminated from one another by ploidy (diploid, $2n = 2x = 16$ or tetraploid, $2n = 4x = 32$), flower color (yellow, purple or variegated), pod shape (coiled or falcate), and the presence or absence of glandular hairs (Small 2011). Analysis of chloroplast genome sequences clearly showed that subsp. *sativa* derived from subsp. *caerulea*, but the evolutionary history of tetraploid subsp. *falcata* is more complicated, with potential contributions from *M. prostrata* in addition to the expected diploid subsp. *falcata* (Havananda et al. 2011).

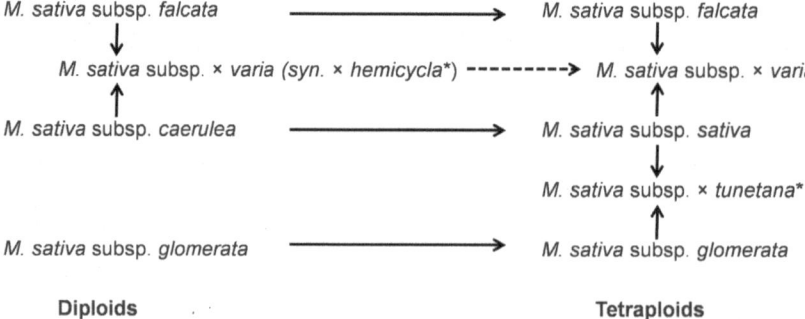

Figure 10-1 Diploid and tetrapoid taxa in the *Medicago sativa* complex as defined by Small (2011), who does not recognize the commonly used taxon names with asterisks. Distinguishing traits include yellow flowers and falcate (sickle shaped) pods for *falcata*; purple flowers and pods with multiple coils for *caerulea* and *sativa*; yellow flowers, coiled pods, and glandular hairs on pods for *glomerata*; and variegated flowers and pods with one or few coils for *varia*. Some *falcata* accessions, designated var. *viscosa*, also have glandular hairs on pods.

Alfalfa is allogamous, generally self-incompatible and pollinated by various insect species (Viands et al. 1988). Few hybridization barriers appear to exist among taxa within the complex; the ploidy barrier can be overcome through unreduced gametes to scale germplasm upward or through parthenogenesis to scale it downward (McCoy and Bingham 1988; Veronesi et al. 1986). DNA sequence analysis supports the possibility of gene flow among taxa (Havananda et al. 2011). Alfalfa can be hybridized to several other members of the genus, mainly confined to perennial species (McCoy and Bingham 1988; Small 2011). Recently, hybrids of alfalfa with *M. arborea*, a woody shrub, have been reported (Armour et al. 2008), offering a possible avenue for yield improvement of cultivated alfalfa (Irwin et al. 2010). Successful hybridization of alfalfa with tetraploidized *M. truncatula* has also been reported recently (E.T. Bingham, pers. comm.). This exciting development requires additional evaluation to determine if chromatin can indeed be transferred between these species.

The origin of alfalfa is thought to have been near the Caspian Sea, including the Caucasus Mountains and eastern Turkey, northern Iran, and western Turkmenistan (Michaud et al. 1988). A second center of origin in central Kazakhstan was proposed by Sinskaya (1950). A limited evaluation of subsp. *caerulea*, the presumed diploid progenitor of cultivated alfalfa, suggests that the region to the west of the Caspian Sea was the most likely primary center of origin (Sakiroglu and Brummer 2012). Alfalfa spread throughout much of Eurasia, the Middle East and North Africa in historical times, perhaps inadvertently disseminated as hay was transported by traders or armies for horse feed (Small 2011). Today, alfalfa is cultivated throughout dry tropical and temperate regions of the world. In the USA, nine historical germplasm sources have been identified (Barnes et al. 1977), although in reality, continual introduction and consequent mixing among germplasms has combined the USA germplasm in many cases beyond recognition of the original introductions.

Most cultivated alfalfa derives from either subsp. *sativa* or subsp. × *varia*, with some direct use of subsp. *falcata per se* or in hybrids with the other subspecies in northern regions of cultivation. Virtually all cultivated alfalfa is tetraploid, and consequently, the diploid germplasm could harbor desirable alleles that have not been previously incorporated into breeding programs (Sakiroglu et al. 2011), such as QTL for aluminum tolerance (Khu et al. 2013). Alfalfa breeders could use the methods of analytic breeding (Chase 1963) to incorporate diploid chromatin on a large scale if deemed sufficiently valuable.

The yellow and purple flowered wild diploid taxa—subspecies *falcata* and *caerulea*, respectively—are clearly differentiated genetically, based on SSR markers (Sakiroglu et al. 2010). Their natural hybrid, subsp. × *varia* (i.e., × *hemicycla*), clearly falls between the two parental taxa, with evidence that some accessions were newly formed hybrids. Limited analysis of other diploids has been done since most genetic diversity studies published to date have focused on tetraploid germplasm, with an emphasis on cultivated populations. A number of experiments have demonstrated gene flow between landraces and wild populations.

Among tetraploid germplasm, the most significant result of diversity experiments is the distinctiveness of pure subsp. *falcata* from all other germplasm (summarized in Brummer 2004 and Li and Brummer 2012). Among *sativa*-based germplasm, a rough differentiation can be made among populations or cultivars based on fall dormancy/winter hardiness status (Vandemark et al. 2006; Ariss and Vandemark 2007), which probably partially reflects the amount of *falcata* introgression that has occurred either historically or through conscious selection to increase winter hardiness.

The other primary result derived from diversity experiments in alfalfa is that virtually any population has a large reservoir of marker alleles, which presumably also means many alleles for traits of interest (Li et al. 2013). In fact, it is possible to successfully select for just about any trait in just about any population of alfalfa, attesting to the practical observation of diversity in alfalfa populations.

10.1.2 Agronomy of Alfalfa and Traits of Interest

10.1.2.1 Management

The essential keys to successful alfalfa cultivation have been addressed by Pliny (1950), whose recommendations can be summarized as follows: use a well tilled, weed free seedbed, apply lime and fertilizer as needed, plant after danger of frost, do not grow alfalfa in wet areas, harvest at early flower, control weeds and rotate crops when significant numbers of plants have died. Despite the intervening two millennia, this advice still covers the main techniques for successful alfalfa culture. A comprehensive contemporary alfalfa management guide is available elsewhere (Undersander et al. 2011), so this topic is not further elaborated here.

10.1.2.2 Traits of Importance

Three sets of traits may be generally identified as important for alfalfa cultivation: (1) yield, including forage yield stability across harvests and years, and seed yield, (2) persistence, including tolerance to cold, heat, drought, salt, grazing, and other abiotic and biotic stresses, and (3) nutritive value, including digestibility, protein content and secondary metabolite profile. Improving all three simultaneously is difficult due to negative correlations among them. Higher yield is typically accompanied by higher fiber. Improved persistence may come at the expense of autumn yield and so forth. "In economic plant breeding one frequently encounters physiologically negative correlations such as those, in alfalfa, between … yield and quality. In seeking improvement, therefore, the breeder must … search for races which violate such naturally antagonistic correlations to the greatest possible extent" (Freeman 1914). And the search goes on.

10.1.2.2.1. Forage yield is perhaps the most important trait, overall, with the caveat that plants must be able to persist in order to yield. Assuming a plant genetic package that enables the plant to survive in a location, higher yield is preferred. Yield can have several different meanings—total yearly yield, total yield across years, yield in specific years (e.g., first post-seeding year yield) or yield in particular harvests within a year. In addition to yield per se, the stability of production across harvests and across years is also of interest to the forage producer. From a plant improvement perspective, clearly understanding which yield definition is most important to the grower and thus the breeder, is the key to success.

10.1.2.2.2. Seed yield represents an important trait for any commercialized alfalfa cultivar. Seed yield is generally negatively correlated with forage yield, and consequently, conscious efforts to maintain or improve seed yield are needed. Because seed is generally produced in regions different from the target area of forage production, selection for seed yield typically needs to be done separately from forage yield selection, and may necessitate the improvement of other traits—such as additional disease or insect resistances—in order for successful seed production. As an example, the Imperial Valley in California has been designated as a non-transgenic seed production area (Mueller 2004). In the Imperial Valley, silverleaf whiteflies are a serious pest (L. Teuber, Univ. Calif.-Davis, pers. comm.), so resistance in alfalfa germplasm utilized for seed production in these areas is necessary (Jiang et al. 2003).

10.1.2.2.3. Abiotic stresses of all types affect alfalfa, as they do other crops. Frost tolerance and winter hardiness, perhaps more than any other trait, define alfalfa germplasm adaptation zones (Castonguay et al. 2006). The winter hardiness trait derives from *M. sativa* subsp. *falcata* germplasm, as observed by Westgate (1909): "Most of the hardy and drought resistant races or strains of alfalfa appear to possess a small percentage of *Medicago falcata* blood in their ancestry." While winter hardy cultivars can be grown in lower latitudes that have limited winter, they will not produce large amounts of forage during autumn and winter, due to strong autumn (fall) dormancy. Conversely, non-dormant cultivars typically winterkill when grown in regions with cold winters.

Fall dormancy begins as photoperiods shorten and temperatures fall in the late summer (McKenzie et al. 1988), is characterized based on the height of regrowth in autumn (Schwab et al. 1996; Teuber et al. 1998), and is generally related to winter hardiness (Schwab et al. 1996). However, the relationship between the traits is not absolute, and selection can decouple the traits to an extent, resulting in germplasm that produces well from late summer through autumn but still remains winter hardy (Brummer et al. 2000; Weishaar et al. 2005).

Other abiotic stresses, including tolerance to drought, heat, salt, aluminum, grazing and nutrient limitations are all important in at least some alfalfa growing regions. Grazing tolerance is a trait that provides additional plant protection when alfalfa is used as pasture (Smith et al. 2000). Although alfalfa is most commonly grown as a hay or silage crop, it can perform very well as grazed forage. The primary difficulty in using alfalfa as a grazing crop is that overgrazing tends to reduce stands, but alfalfa bred under continuous stocking conditions showed improved persistence (Smith and Bouton 1993; Bouton and Smith 1998). Alfalfa selected for grazing tolerance performed better under both continuous and rotational stocking compared to non-grazing tolerant cultivars (Brummer 2006).

10.1.2.2.4. Biotic stresses. Alfalfa is afflicted by numerous disease and insect pests (Leath et al. 1988; Manglitz and Ratcliffe 1988), and the search for resistant types has been ongoing since early in the 20th century. Resistance to many of the most serious pests (Table 10-1) has been identified within cultivated germplasm (Elgin et al. 1988; Sorensen et al. 1988), and currently commercialized cultivars typically have high levels of resistance to many of the most prevalent diseases and insect pests (e.g., see the "Winter Survival, Fall Dormancy & Pest Resistance Ratings for Alfalfa Varieties" leaflet at http://alfalfa.org [Accessed 12 Jan 2013] which summarizes resistances to commercially available cultivars in the USA). Resistance has occasionally been introduced from outside the cultivated gene pool, including potato leafhopper resistance (Elden and McCaslin 1997; McCaslin 1999). Several pests remain a serious problem, including the alfalfa weevil (*Hypera postica* (Gyllenhal)) and cotton root rot (*Phymatotrichopsis omnivora* (Duggar) Hennebert), and no resistance has been identified in either the *M. sativa* complex or in species that can be hybridized with the *Medicago* complex.

10.1.2.2.5. Nutritive value, often termed "forage quality", is another trait for further improvement (Hill et al. 1988). Nutritive value traits include the fiber and protein concentration of biomass. The fibers of most interest are those that are indigestible, as they influence the availability of nutrients within the biomass and also affect the rate of forage passage through the animal. Improving nutritive value could be achieved by lowering the amount of indigestible fiber, by limiting the rate of digestion of protein in the rumen

Table 10-1 Diseases and insect pests affecting alfalfa. Standard evaluation tests have been developed and published by the North American Alfalfa Improvement Conference (NAAIC) for many of these diseases and pests. Some of the disease tests are only conducted on seedling plants, which may not reflect adult plant resistance. Unless noted, NAAIC standard tests have been developed and are available at http://www.naaic.org.

Trait	Scientific Name	Comments on NAAIC standard test
Diseases		
Alfalfa enation virus		No test
Alfalfa mosaic virus		No test
Anthracnose	*Colletotrichum trifolii* Bain & Essary	Seedling test
Aphanomyces root rot	*Aphanomyces euteiches* Drechs. (Races 1 & 2)	Seedling test
Bacterial wilt	*Clavibacter michiganense* subsp. *insidiosum* (McCull)	Seedling and adult plant tests
Brown root rot	*Phoma sclerotioides* G. Preuss ex Sacc.	
Common leaf spot	*Pseudopeziza medicaginis* (Lib.) Sacc.	
Downy mildew	*Peronospora trifoliorum* de Bary	
Fusarium wilt	*Fusarium oxysporum* Schlecht f. sp. *medicaginis* (Weimer) Snyd. & Hans.	Young plant and adult plant test
Lepto leaf spot	*Leptosphaerulina briosiana* (Poll.) Graham and Luttrel	
Phymatotrichum (Cotton) root rot	*Phymatotrichum omnivorum* (Shear) Duggar	No test; no resistance identified
Phytophthora root rot	*Phytophthora medicaginis* (Hansen and Maxwell)	Seedlings in greenhouse or young plants in field or greenhouse
Pythium seed rot	*Pythium* spp.	
Rust	*Uromyces striatus* Schroet.	
Sclerotinia crown and stem rot	*Sclerotinia trifoliorum* Eriks.	
Spring blackstem and leaf spot	*Phoma medicaginis* (Malbr. & Roum.) var. *medicaginis* Boerema	
Stemphylium leaf spot	*Stemphylium botryosum* Wallr.	
Summer black stem and leaf spot	*Cercospora medicaginis* Ellis & Everh.	No test

Table 10-1 contd.....

Table 10-1 contd.

Trait	Scientific Name	Comments on NAAIC standard test
Verticillium wilt	*Verticillium albo-atrum* Reinke & Berth.	Young plants in greenhouse or field
Yellow leaf blotch	*Leptotrochila medicaginis* (Fckl.) Schuepp.	
Nematodes		
Columbia root knot nematode	*Meloidogyne chitwoodi* (race 2)	
Root-knot nematode	*Meloidogyne* spp.	
Root-lesion nematode	*Pratylenchus penetrans* Cobb, Filipjev and Schur-Stekhoven	
Stem nematode	*Ditylenchus dipsaci* (Kuhn) Filipjev	
Insects		
Alfalfa blotch leafminer	*Agromyza frontella* (Rondani)	No test
Alfalfa plant bug	*Adelphocoris lineolatus* (Goeze)	No test
Alfalfa weevil	*Hypera postica* (Gyllenhal)	No test; no resistance identified
Blister beetles	*Epicauta* spp.	No test
Blue alfalfa aphid	*Acyrthosiphon kondoi* Shinji	
Clover leaf weevil	*Hypera punctata* (F.)	No test
Clover root curculio	*Sitona hispidulus* (F.)	No test
Cowpea aphid	*Aphis craccivora* Koch	
Meadow spittlebug	*Philaenus spumarius* (L.)	No test
Pea aphid	*Acyrthosiphon pisum* (Harris)	
Potato leafhopper	*Empoasca fabae* (Harris)	
Silverleaf whitefly	*Bemisia argentifolii* Bellows & Perring	
Spotted alfalfa aphid	*Therioaphis maculata* (Buckton)	
Tarnished plant bug	*Lygus lineolaris* (Palisot)	No test
Variegated cutworm	*Peridroma saucia* (Hübner)	No test

(thereby limiting both bloat potential and excreted nitrogen) (Coulman et al. 2000), and by various other mechanisms, such as increasing pectin concentration.

Genetic variation exists for most nutritive value traits among alfalfa germplasms (Buxton et al. 1987; Julier et al. 2000; Jung et al. 1997) and breeding programs have been relatively effective at modifying quality (Hill et al. 1988). The primary difficulty of changing quality substantially in alfalfa is that it is highly, negatively correlated with maturity. Hence, an ill-timed rainy period can delay harvest and easily overcome any positive change made through breeding. Transgenes offer an alternative to obtain substantial changes to nutritive value regardless of management problems.

10.2 Classical Breeding Objectives and Outcomes

10.2.1 Classical Intra-population Breeding Methods

Most alfalfa cultivars are synthetic populations (Hill et al. 1988) (Fig. 10-2). These cultivars are the product of an initial polycross among selected parents followed by several generations of random intercrossing for seed increase. The number of parents in a given cultivar can range widely, up to several hundred. Due to the population-based nature of alfalfa cultivars and breeding populations, fixing traits is difficult, although the positive aspect of the residual variation is predicted stability in the face of environmental perturbations across years (Bradshaw 1965).

Selection in alfalfa breeding programs typically involves phenotypic recurrent selection on individual plants (Fig. 10-3), useful for highly heritable traits like disease and insect resistances. Phenotypic recurrent selection has been highly successful at increasing desirable allele frequencies, so that most elite cultivars on the market in the USA today have high levels of resistance to a suite of disease and insect pests. In contrast, family selection—most commonly, half-sib families developed from a polycross nursery—are used for more complex traits like yield, which are not easily evaluated on a single plant basis. The half-sib progeny test, or often, among and within half-sib family selection, are commonly used (Casler and Brummer 2008). Typical breeding programs frequently use a combination of these two methods in selecting parents for potential synthetics. In addition to intra-population breeding methods described here, strain crosses between two or more highly adapted cultivars with complementary traits is also a strategy used in alfalfa breeding programs.

Selection for any of the key traits requires a clear and reliable phenotypic screen. For many of the diseases and insects, standard tests have been developed, reviewed and approved by the alfalfa research community

Figure 10-2 Sward plot trial of alfalfa synthetic populations and cultivars.

Color image of this figure appears in the color plate section at the end of the book.

(http://naaic.org/resource/stdtests.php [Accessed 1/2/2013]). These tests provide a consistent methodology not only for selection but also for evaluation and comparison among cultivars. However, for many of the more complex traits including yield, no standard method has been developed, and this may account, in many cases, for the limited progress on yield enhancement. The alfalfa community should focus more on clearly defining complex traits, and developing standard methods to accelerate genetic gain. For instance, a highly controlled laboratory screen developed

Figure 10-3 Alfalfa spaced plant evaluation trial.

Color image of this figure appears in the color plate section at the end of the book.

for freezing tolerance led directly to improved winter hardiness in the field (Castonguay et al. 2009a; Castonguay et al. 2009b). Standard methods for aluminum tolerance have also been developed (Khu et al. 2012), and their value for field tolerance is currently being evaluated. Similar methods for other complex traits like drought are urgently needed.

10.2.2 Hybrid Alfalfa

Hybrid alfalfa has been described for nearly 100 yr in response to the success of hybrid maize. Although in theory, capturing heterosis in breeding programs should result in increased gains, the successful development of alfalfa hybrids is hampered by several factors. First, severe inbreeding depression makes the development of true inbred genotypes nearly impossible (Jones and Bingham 1995). However, some near-inbred genotypes have been developed (Bingham 1993), although their actual homozygosity has not been determined and is likely less than expected (Brouwer and Osborn 1997). However, even if inbred genotypes were developed, single-cross hybrids would not maximize heterozygosity (or complementary gene interactions) in autotetraploid alfalfa. Heterosis is progressive in autotetraploids (Bingham et al. 1994; Groose et al. 1989), which means not only that a double cross is necessary if starting with inbred genotypes, but also that homogeneous hybrids are not feasible.

Hybrid alfalfa could be developed in several ways. First, self-incompatibility could be used to ensure cross-pollination between clonal genotypes. Self-compatibility varies among genotypes but appears amenable to improvement by selection (Campbell and He 1997). Second, cytoplasmic male sterility (Davis and Greenblatt 1967), which can be unreliable in alfalfa (Viands et al. 1988), has been investigated the most (Davis 1971; Sun et al. 2002; Sun et al. 2004; Tysdal and Kiesselbach 1944), and commercial "hybrid" alfalfa cultivars are currently on the market that use cytoplasmic male sterility but which are actually narrow based populations, rather than canonical single-cross hybrids in the maize sense.

Finally, population hybrids may offer a means to partially capture heterosis (Brummer 1999). Although hybrids between *M. sativa* subsp. *falcata* and subsp. *sativa* may produce high yielding progeny (Riday and Brummer 2002a; Riday and Brummer 2005; Segovia-Lerma et al. 2004), they often suffer from less ideal characteristics, such as slower regrowth or early autumn dormancy (Riday and Brummer 2002b; Riday and Brummer 2004). Among subsp. *sativa*-based germplasm, some evidence suggests that Peruvian germplasm forms a heterotic group with other germplasms (Al Lawati et al. 2010; Madril et al. 2008; Maureira-Butler et al. 2007), and that cultivars with diverse autumn dormancies also combine well (Bhandari et al. 2007).

The search for heterosis has often focused on identifying heterotic groups –populations that are differentiated based on neutral genetic markers. In reality, developing heterotic groups *de novo* using adapted populations and then selecting with a method like reciprocal recurrent selection may be more useful. The canonical heterotic group story in maize, where two differentiated populations, Reid Yellow Dent and Lancaster Surecrop, were used as heterotic groups, is actually not true: these populations have partially common backgrounds and the differentiation occurred after cycles of selection (Tracy and Chandler 2006). Thus, the key is to start with good material, keep the populations separate while breeding for hybrid performance, and in this way, create the heterotic groups.

10.3 Structural Genomics and Molecular Marker Resources

Advances in genomics, sequencing technologies, and genotyping platforms provide resources for the implementation of marker-assisted breeding and integration of biotechnology traits to alfalfa breeding programs. These technologies can enhance breeding and efficiency of trait-integration to increase genetic gains and keep up with increasing demands of high-value forage.

10.3.1 Alfalfa Transcriptome Sequences

Using genomic approaches for alfalfa improvement has benefited from the resources available for the model legume and close relative of alfalfa, *Medicago truncatula* (Young and Udvardi 2009). These resources include a large collection of *M. truncatula* expressed sequence tags (ESTs) and bacterial artificial chromosome (BAC) sequences. Alfalfa-specific resources are now becoming widely available as well. The initial development of an EST library for alfalfa was reported for progeny grouped according to the presence or absence of glandular hairs (Hays and Skinner 2001). Alfalfa transcriptome sequences were isolated from root and shoot tissues collected during progressively developing water stress, and thousands of ESTs were generated using 454 sequencing technology from single genotypes of *M. sativa* subsp. *sativa* var. "Chilean" and *M. sativa* subsp. *falcata* var. "Wisfal" (Han et al. 2011). These alfalfa transcriptome sequences are available in the GenBank transcript assembly database (http://www.ncbi.nlm.nih. gov [Accessed Nov 2012]) under the accession numbers SRX040822 and SRX040823 for the Chilean (57.5 Mb) assembly and Wisfal (67.6 Mb) assembly, respectively. An additional set of transcriptome sequences was generated from 27 alfalfa genotypes including elite breeding genotypes relevant to existing commercial breeding programs, parents of mapping populations, and unimproved wild genotypes (Li et al. 2012). These transcriptome sequences are available in the Legume Information System (Gonzales et al. 2005) (http://medsa.comparative-legumes.org [Accessed Nov 2012]). Collectively, these transcriptome sequences represent a valuable resource for the discovery of sequence variation among and within alfalfa genotypes and for the development of molecular markers.

10.3.2 Molecular Marker Resources

Molecular markers have been used to estimate the location and importance of genomic regions underlying quantitative traits important to plant breeders (Tanksley et al. 1989). DNA markers are unaffected by the conditions in which plants are grown and can be used to determine genetic relatedness between plant accessions, identify novel sources of genetic variation, confirm the pedigree and identity of new varieties and to assess evolutionary relationships (Ganal et al. 2009). Initially, restriction fragment length polymorphism (RFLP) markers were used and their variation was based on differences at sites recognized by restriction enzymes. The development of microsatellite or simple sequence repeat (SSR), markers in alfalfa followed and was largely driven by the availability of *M. truncatula* EST sequences (Euyayl et al. 2004; Julier et al. 2003; Sledge et al. 2005). SSR markers are particularly useful because they are co-dominant meaning that heterozygotes can be distinguished from homozygotes and because multiple alleles can be detected at a single locus. Primer sets are available for SSR loci distributed over a large portion of the alfalfa genome (Khu et al. 2013; Sledge et al. 2005). SSR markers were used to characterize the genetic diversity of alfalfa genotypes from different geographical regions (Falahati-Anbaran et al. 2007; Sakiroglu et al. 2010), and of 190 perennial *Medicago* accessions that were mostly collected from the wild in the former Soviet Union area (Han et al. 2012a).

Single nucleotide polymorphisms (SNP) are the most common type of sequence variation among plants, are also co-dominant, and are often functionally important. SNP development in alfalfa has accelerated as advances in sequencing technologies facilitated transcriptome sequencing efforts (Han et al. 2011; Li et al. 2012). Nearly one million SNP have been identified in these two experiments, covering germplasm from wild diploids to elite cultivated genotypes. A subset of the identified SNPs from both studies were validated using high-resolution melting (HRM) analysis (Han et al. 2011; Han et al. 2012b), which enabled the identification of allelic dosage (TTTT, TTTC, TTCC, TCCC, CCCC) for genotyping tetraploid alfalfa (Fig. 10-4). An advantage of SNPs is that they are amenable to high-throughput genotyping assays that are faster and at a lower cost per data point than SSRs. To that end, an Illumina® iSelect Infinium array that includes approximately 10,000 biallelic SNPs was developed and used to

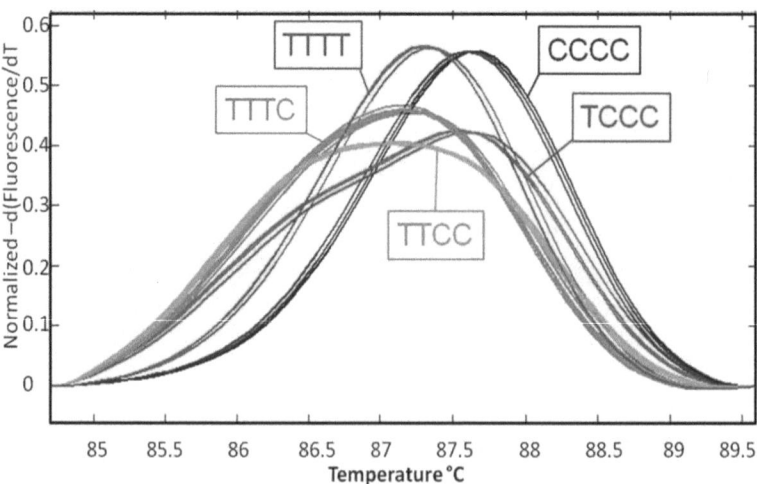

Figure 10-4 Normalized melting curves of alfalfa genotypes showing the variation in melting profile due to variation in allelic dosage at a biallelic SNP.

Color image of this figure appears in the color plate section at the end of the book.

evaluate 576 genotypes, including cultivated alfalfa, wild diploid alfalfa and other related species (Li et al. 2013). The five alternate genotypes could be identified in tetraploid alfalfa (Fig. 10-5) using various software programs developed for or adapted to tetraploid genotype calling and that can also distinguish allelic dosage (Voorrips et al. 2011).

Because SNPs are biallelic and thus have only two alternative bases per locus, they are less useful than SSR loci, which can have numerous alleles at a locus, for some applications. The limitation of uniquely tagging a genomic region with a target gene using a single SNP marker may be overcome by defining a haplotype that is represented by a set of DNA bases at several linked SNPs that span the target locus. The number of SNPs defining a haplotype will be determined by the levels of linkage disequilibrium, which is expected to be relatively small in alfalfa, compared to autogamous legumes such as soybean (*Glycine max* L.) and *M. truncatula*. Although the development and utilization of molecular markers in alfalfa has lagged behind other major crop species, significant progress to develop molecular marker resources that are publicly available to the alfalfa community has been made in the past few years.

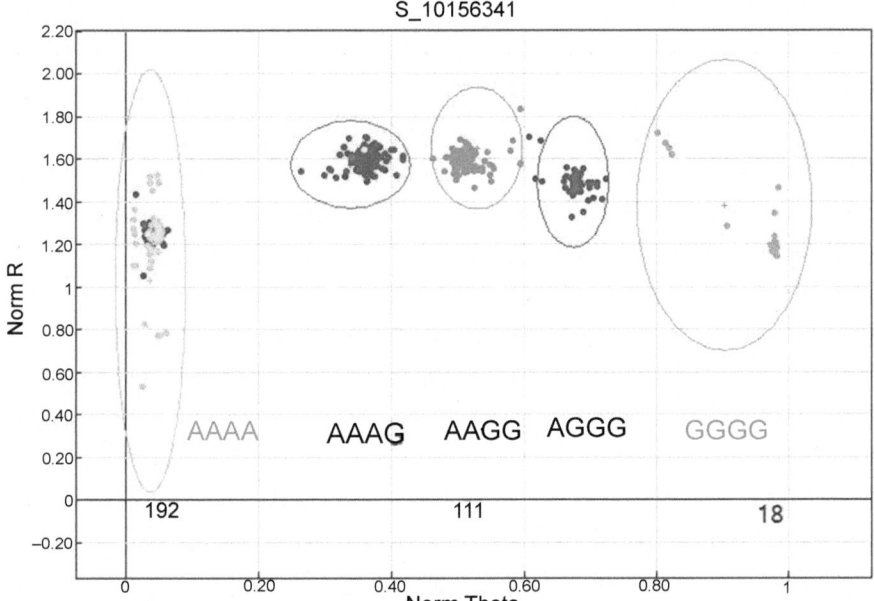

Figure 10-5 Allelic dosage variation identified in diverse alfalfa genotypes, including cultivated alfalfa, wild diploid alfalfa and other related species, using an Illumina® array.

10.3.3 Genetic Linkage Maps

Linkage map development in alfalfa was hindered by a complex autotetraploid genome and complex segregation patterns, and thus early mapping studies focused on diploid populations (Brummer et al. 1993; Echt et al. 1994; Kiss et al. 1993). These early linkage maps included a combination of RFLP, RAPD, isozyme and morphological markers (Table 10-2). Many of the markers used to generate these linkage maps had segregation distortion, often skewed toward excess heterozygosity, reflecting the endemic inbreeding depression observed in alfalfa (Brouwer and Osborn 1999). Using non-inbred F_1 populations for mapping purposes helped reduce distortion (Tavoletti et al. 1996) partly because not all genotypes are fully informative (Li et al. 2011). The availability of a large number of EST-SSR markers from *M. truncatula*, genome-wide SNP (Han et al. 2011; Li et al. 2012a), and mapping software suitable for tetraploids (Hackett et al. 2007) greatly enhanced mapping capabilities in tetraploid alfalfa (Table 10-2) (Eujayl et al. 2004; Julier et al. 2003; Khu et al. 2013; Robins et al. 2007b; Sledge et al. 2005).

   Comparisons among the existing linkage maps in alfalfa are limited given the small number of markers mapped in multiple populations. Tavoletti (1996) identified seven markers that were assigned to a different linkage group or had a different order compared to a previously generated linkage map (Echt et al. 1994), further highlighting the need for a consensus linkage map in alfalfa. The high rates of marker polymorphism and widespread distribution of SSR and SNP markers in alfalfa would make it possible to assemble consensus linkage maps using marker data from several independent populations. However, the lack of a coordinated community-wide marker development and integration attempt in alfalfa has hampered the integration of the various alfalfa linkage maps generated (Table 10-2). Incorporating a common set of markers in the various existing alfalfa maps will enable cross-referencing of marker positions and distances to facilitate comparisons of the relevant marker-trait associations and validate quantitative trait loci (QTL).

10.3.4 QTL Identification

Molecular marker-based linkage maps facilitate the identification of chromosomal regions that contain genes underlying important traits, the positional cloning of target disease resistance genes and marker-assisted selection (Mohan et al. 1997; Paterson et al. 1988; Young 1996). Initially, alfalfa diploid linkage maps were used to identify QTLs for aluminum tolerance (Narasimhamoorthy et al. 2007). Linkage maps at the tetraploid level were used to locate the genomic positions of QTL associated with

Table 10-2 Description of the main alfalfa genetic linkage maps available.

Ploidy level	Parental genotypes	Number of progeny	Number of linkage groups	Map length (cM)	Number and type of mapped marker	Reference
Diploid (F₂)	Mqk93 (ssp. quasifalcata) × Mcw2M. (ssp. coerulea)	138	8	659	89 (RFLP, RAPD, isozyme and morphological)	(Kiss et al. 1993)
Diploid (F₂)	W2xiso #3 (ssp. sativa) × 440501 #2 (ssp. coerulea)	86	10	467.5	108 (RFLP)	(Brummer et al. 1993)
Diploid (BC₁)	F2-16 (CADL*) × 74-25-20(CADL)	87	7	603	153 RFLP, RAPD	(Echt et al. 1994)
Diploid (F₁)	PG-F9 (CADL) × W2x-1 (CADL)*	57	8 × 2**	234 / 236**	50/55 (RFLP)**	(Tavoletti et al. 1996)
Diploid (F₂)	Mqk93 (ssp. quasifalcata) × Mcw2M. (ssp. coerulea)	137	8	754	868 (RFLP, RAPD, isozyme and seed protein)	(Kaló et al. 2000)
Tetraploid (BC₁)	Blazer XL 17(ssp. sativa) × Peruvian 13 (ssp. sativa)	101	7	443	82 (RAPD)	(Brouwer and Osborn 1999)
Diploid (BC₁)	AL1(ssp. coerulea) × AL2 (ssp. coerulea)	130	10	764	132 (SSR, SNP, INDEL and RFLP)	(Narasimhamoorthy et al. 2007)
Tetraploid (BC₁)	Wisfal (ssp. falcata) × Chilean (ssp. sativa)	186	8	624	286 (SSR)	(Sledge et al. 2005)
Tetraploid (F₁)	Magali2 (ssp. sativa) × Mercedes4.11 (ssp. sativa)	168	8	709	107 (SSR)	(Julier et al. 2003)
Tetraploid (F₁)	Altet-4 (ssp. sativa) × NECS-141 (ssp. sativa)	185	8	761	249 (SSR)	(Khu et al. 2013)

*CADL = cultivated alfalfa at the diploid level
**Two linkage maps were constructed

biomass production, winter hardiness, freezing injury, plant re-growth, persistence (Brower et al. 2000; Robins et al. 2007b; Robins et al. 2008b), self-fertility (Robins and Brummer 2010), biomass production under limited water availability (Han et al. 2010a), water-use efficiency (Julier et al. 2010) and aluminum tolerance (Khu et al. 2013) (Table 10-3). Most of these experiments used non-inbred F_1 populations to minimize segregation distortion that would be found in inbred populations due to uncovering deleterious recessive alleles during the inbreeding process (Osborn et al. 1997).

The value of QTL mapping for practical breeding has been limited and no improved alfalfa cultivars generated through molecular breeding are currently available. Quantitative traits often have low heritability and often many "small-effect" QTL contribute to the trait. Thus, multiple genomic regions must be integrated simultaneously to obtain a significant impact on the phenotype. Often there is a significant QTL by environment interaction compromising the stability of the relevant QTL across environments. In soybean, the designation of "cq" for "confirmed QTL" was proposed as a way to allow breeders and researchers to recognize the QTLs that were mapped and confirmed in a population derived by independent meiotic events (Monteros et al. 2008). This designation of confirmed QTL could then be used by breeders to prioritize trait-integration using molecular markers through "molecular breeding".

10.3.5 Alfalfa Nuclear Genome

Alfalfa has an estimated haploid genome size of 830–860 Mbp (Blondon et al. 1994). An international effort including researchers from the USA, Canada and France under the Alfalfa Genome Sequencing Consortium, is sequencing the alfalfa genome and generating transcriptome sequences for additional SNP discovery. The alfalfa sequence and SNP markers will facilitate the translation and application of genomics to alfalfa cultivar development. The tetraploid genotype NECS-141, which had high biomass yield in Iowa, USA and was used to generate a segregating mapping population (Khu et al. 2013), is being sequenced using Illumina sequencing and Whole Genome Profiling (van Oeveren et al. 2011). The National Center for Genome Resources (NCGR) in Santa Fe, NM is currently working on the sequence assembly, which should be completed in mid-2013, and will include Illumina reads from short- and large insert paired-end libraries (SIPE and LIPE) tied to a physical map developed by Keygene's Whole Genome Profiling (van Oeveren et al. 2011). The heterozygosity of tetraploid alfalfa poses some challenges for *de novo* genome sequence assembly. However, the synteny of alfalfa to the *M. truncatula* genome provides opportunities to assist the assembly of alfalfa scaffolds into super-scaffolds and align

Table 10-3 Target traits and quantitative trait loci (QTL) mapping studies in alfalfa.

Trait	Ploidy	Population	Number of QTLs	Relevant LGs	Reference
Aluminum tolerance	Diploid	AL1 × AL2	7	1, 2, 3	(Narasimhamoorthy et al. 2007)
Forage yield	Tetraploid	WISFAL-6 × ABI408	32*	1, 2, 3, 4, 6, 7, 8	(Robins et al. 2007a)
Plant height	Tetraploid	WISFAL-6 × ABI408	23*	3, 4, 5, 6, 7, 8	(Robins et al. 2007a)
Plant regrowth	Tetraploid	WISFAL-6 × ABI408	31*	2, 3, 4, 5, 6, 7, 8	(Robins et al. 2007a)
Persistence	Tetraploid	WISFAL-6 × ABI408	7	1, 2, 7	(Robins et al. 2008a)
Biomass production	Tetraploid	WISFAL-6 × ABI408	41*	1, 2, 3, 4, 5, 6, 7, 8	(Robins et al. 2007b)
Biomass production under limited water availability	Tetraploid	'Chilean' × 'Wisfal'	5	1, 2, 3, 5, 6	(Han et al. 2010b)
Self-fertility	Tetraploid	WISFAL-6 × ABI408	9	2, 4, 8	(Robins and Brummer 2010)
Water use efficiency	Tetraploid	Magali-A × Gabes-2355	6	2, 3, 7, 8	(Julier et al. 2010)
Colletotrichum trifolii resistance	Tetraploid	D × W126	3	4, 6, 8	(Mackie et al. 2007)
Stagonospora meliloti resistance	Tetraploid	D × W126	3	2, 6, 7	(Musial et al. 2007)
Aluminum tolerance	Tetraploid	Altet-4 × NECS-141	3	1, 4, 7	(Khu et al. 2013)

*Identified using single-marker analysis.

the existing alfalfa transcriptome sequences from various genotypes to a reference alfalfa genome and further identify sequence variation among genotypes. Anchoring of scaffolds to chromosomes with genetic markers will be complemented by sequence annotation and the development of tools to manage and visualize the available sequences.

An alfalfa genome BLAST Server that includes the alfalfa scaffolds generated at NCGR is being developed at the Noble Foundation and may be used to identify genes of interest based on target query sequences. The ultimate goal is to make these resources publicly available to the community through the www.alfalfagenome.org and Legume Information System (LIS) (Gonzales et al. 2005) (http://medsa.comparative-legumes.org) websites [Accessed 12 Jan 2013]. The genomic resources available for *Medicago* spp. have already proven valuable to identify candidate genes relevant to QTL positions associated with abiotic stress tolerance (Han et al. 2010a; Khu et al. 2013). Beyond that, sequencing the alfalfa genome has both basic and applied uses: to enable a better understanding of legume evolution, gene content and genome structure in alfalfa and to generate and apply molecular tools directly in breeding programs targeting key traits for alfalfa improvement.

10.3.6 From Models to Crops

Useful tools and resources have been developed in *M. truncatula*, including high-density genetic and physical maps (Choi et al. 2004a), genetic mutants (Tadege et al. 2008), and multiple websites and bioinformatics resources (Young and Udvardi 2009). However, in order to be maximally useful, these resources need to be applied to alfalfa and other crops to enhance the sustainability of agricultural systems. Translational genomics can leverage infrastructure in model legumes to enhance applied breeding programs. More efficient selection of desirable individuals/alleles via genomics-enhanced breeding will ultimately accelerate the rates of genetic gain.

10.3.6.1 Comparative Genomics

The goal of comparative genomics is the transfer of knowledge between model and crop species, facilitated by mapping studies using common DNA markers (or sequence) to evaluate synteny among related species and enable gene discovery. Despite the differences in the genome size, base chromosome number and ploidy level, comparative genetic mapping studies have shown that legumes sharing a close phylogenetic relationship have extensive genome conservation (Zhu et al. 2005). The use of *M. truncatula* EST and BAC sequences to develop cross-species gene-specific PCR primers designed to anneal with conserved exon sequences was previously described for alfalfa and other crop legumes (Choi et al. 2004b). A simplified consensus comparative map of eight legume species, including alfalfa, indicates the degree of synteny is correlated with phylogenetic distance (Zhu et al. 2005).

M. *truncatula* and alfalfa exhibit nearly perfect macrosynteny between the two genomes and share highly conserved nucleotide sequences (Choi et al. 2004a). Alfalfa maps can also be aligned with those of white and red clover, two other major forage legumes (Zhang et al. 2007). Recently efforts to increase the number of markers suitable for cross-species amplification by targeting transcription factors (TF) were successful (Monteros and Han unpubl.). The sequence conservation of TF extends across multiple kingdoms and their genome-wide distribution in eukaryotes (Riechmann et al. 2000) facilitates their use as markers to transfer information from models to crops (Han et al. 2010b). Amplicons from 86% of the TF-derived markers amplified in alfalfa and at least one other species. Integrating the alfalfa genome with the genome sequences of the model species M. *truncatula*, L. *japonicus*, G. *max* (Sato et al. 2008; Schmutz et al. 2010; Young et al. 2011), and other crop legumes will be facilitated by the infrastructure currently available through the Legume Information System (Available at http:// www.comparative-legumes.org [Accessed 12 Jan 2013]).

10.3.6.2 Microarrays

An Affymetrix® GeneChip containing probe sets for 51,000 M. *truncatula* genes and for genes of its symbiotic, nitrogen-fixing partner *Sinorhizobium meliloti* (Young and Udvardi 2009), was successfully used to monitor genome-wide gene expression of leaf and root tissues from alfalfa seedlings (Tesfaye et al. 2006), to compare profiles of genotypes divergently selected for cell wall composition (Yang et al. 2010), and to identify possible genes involved in biomass heterosis (Li et al. 2009). This GeneChip was the basis for the first global Gene Expression Atlas for a legume based on extensive DNA-chip hybridization (Benedito et al. 2008). The results from that study demonstrate massive differences in gene expression between organs and numerous cases of legume-specific genes that are preferentially expressed in nodules. Plans for a similar Gene Atlas in alfalfa are currently underway and may provide unique insights into the perennial nature of alfalfa and its ability to persist throughout multiple years of growth in the field.

10.3.6.3 Proteomics and Metabolomics

Proteomics complements transcriptomic data by evaluating whether changes in transcript levels lead to changes in the levels of the corresponding proteins (Young and Udvardi 2009). A proteome reference map for M. *truncatula* root proteins was created using two-dimensional gel electrophoresis and peptide mass fingerprinting to aid the dissection of nodulation and root developmental pathways (Mathesius et al. 2001). Other proteomics studies in *Medicago* evaluated the proteins produced during seed development and germination (Gallardo et al. 2007), plant-pathogen interactions (Colditz et al. 2004), and symbiosis (van Noorden et al. 2007).

Metabolomics studies separate, measure and identify cellular metabolites produced by plants. These experiments measure hundreds of metabolites and represent only a fraction of the total metabolites produced in any given plant tissue at a specific point in time. Metabolomics studies in alfalfa were used to evaluate the effect of over-expressing the *WXP1* transcription factor on wax accumulation (Zhang et al. 2005), and to identify the accumulation of osmolytes and flavonoids in roots and shoots of alfalfa genotypes with differential responses to drought stress (Kang et al. 2011).

10.4 Molecular Breeding and Transgenics

10.4.1 Marker-assisted Breeding

Although marker-assisted breeding has been the justification for investment over the past 20 yr in programs to develop molecular markers and to genetically map QTL (Bernardo 2008), attempts to apply the use of markers to alfalfa cultivar development programs have been limited. As the preceding breeding discussion has intimated, the application of markers to alfalfa cultivar development encounters several difficulties. First, the need to work with inter-mating populations rather than inbred lines limits the breeder's ability to introgress and fix new alleles/traits. Second, many of the most important diseases and insect pests of alfalfa can be evaluated and screened with relatively robust, but simple greenhouse tests that can evaluate many plants simultaneously (see http://www.naaic.org/resource/stdtests.php [Accessed 1/1/2013]), so markers may not be a better option for selection with current breeding methods (but see below). Third, phenotypic evaluation of some of the more complex traits to improve, from biomass yield to drought tolerance, is difficult and the heritability of these traits is low. Thus, marker-trait associations are hard to identify.

Despite these limitations, molecular markers offer significant potential to improve alfalfa breeding. In order to realize their potential, however, alfalfa breeding methods may need to be changed to better incorporate markers and gain the efficiencies they offer. While using markers to select for diseases may not be more efficient than large-scale greenhouse screening with the pathogens, markers would be very valuable to verify that plants selected based on yield or other traits also possess disease resistances, and further, the resistance alleles are present in multiple alleles at the relevant loci.

We can imagine using markers to identify and increase the frequencies of desirable traits (or QTL) within breeding populations. Marker-trait associations can be identified using genetic mapping or by comparing selected and unselected populations to identify allele frequency shifts during selection. Comparing selected and unselected populations has successfully identified markers and genes for winter hardiness (Castonguay et al. 2011; Castonguay et al. 2010; Castonguay et al. 2012; Rémus-Borel et al. 2010) and for cell wall composition (Castonguay, pers. comm.). Identifying markers based on frequency changes within selected populations provides immediate benefits, by enabling breeders to further increase—and ultimately fix—the desirable alleles directly in breeding populations.

Using traditional QTL mapping in biparental populations is necessary when desirable QTL alleles reside in wild or non-elite germplasm and are not present in elite breeding populations. The use of QTL mapping and subsequent allele introgression and/or concentration can be useful, particularly if methods to track allele dosage can enable the selection of higher dosage individuals (duplex, triplex or quadriplex), thereby quickly increasing allele frequencies in the population. Currently, we have identified QTL for aluminum tolerance in alfalfa (Khu et al. 2013; Reyno et al. 2013) using a whole plant assay in media and a soil-based assay in the greenhouse (Khu et al. 2012). These QTL, derived from wild diploid germplasm, were transferred via unreduced gametes to tetraploid germplasm, and these are now being introgressed into commercially elite breeding populations.

The most exciting development in marker-assisted selection is the recent ability to generate genome-wide markers through DNA sequencing (Baird et al. 2008; Elshire et al. 2011), which could enable whole genome selection (WGS) (Heffner et al. 2009). The general idea of WGS is that a population under selection—or a representative subset of that population—is genotyped at sufficient marker density to effectively cover the entire genome. The density depends on the extent of linkage disequilibrium (LD), but given sufficiently extensive LD as may be expected within a narrow-based breeding population, and the ability of genotype-by-sequencing (GBS) to generate 10,000 or more markers reliably, this may be feasible. The same individuals are also phenotyped for traits of interest, and a model that assigns breeding values to individual marker loci is developed. Assuming the model explains a significant amount of the additive genetic variation for the trait, selection based only on marker breeding values could be accomplished without phenotyping, resulting in multiple cycles of selection per year. The model would be updated periodically to account for changes in breeding values as allele frequencies change (Jannink et al. 2010).

We have discussed a possible application of GBS to alfalfa breeding previously (Li and Brummer 2012). Under this scenario, multi-year phenotyping would continue as currently done in breeding programs, for the sole reason of model refinement. Marker-only selection could be done for multiple cycles during each phenotyping cycle. The primary issue to be resolved from a commercial standpoint is whether the additional cost of genotyping—phenotyping costs will remain!—is worth the added genetic gain that genomic selection can provide. A second issue to be considered is whether a different breeding system should be adopted; one focused less on recurrent selection but more on in-depth characterization of a select set of elite genotypes, which are recombined in narrow-based synthetics or hybrids. This type of method would be very different from current breeding, but may help incorporate genome-wide selection more effectively into alfalfa improvement programs, as well as lead to the routine commercial development of cultivars that benefit from specific combining abilities of the parental genotypes.

10.4.2 Transgenic Improvement

Alfalfa can be easily regenerated, although variation among genotypes for regeneration ability is present (Bingham et al. 1988). In general, germplasm derived from subsp. *falcata* is easiest to regenerate; non-dormant germplasm is most difficult. Alfalfa can be genetically transformed using either biolistic or *Agrobacterium*-mediated methods and many genes have been successfully inserted and evaluated (Brummer et al. 2004; Samac and Austin-Phillips 2006; Tesfaye et al. 2009). Some of the (many) traits recently targeted with transgenes in alfalfa include staygreen (Zhou et al. 2011), aluminum tolerance (Barone et al. 2008), alfalfa weevil resistance (Kumar 2011), tolerance to various abiotic stresses including drought and salt (Bao et al. 2009; Jin et al. 2010; Suarez et al. 2009), phosphorus uptake (Ma et al. 2012), and cell wall composition (Dien et al. 2011; Pattathil et al. 2012; Zhao et al. 2010). Given the expansion of genomic resources over the past decade, identifying, cloning and transforming genes into alfalfa has become routine, and the evaluation and functional characterization of genes for diverse traits will likely continue to increase in the future.

Despite the pace of transgene production and characterization, to date, the only commercialized transgenic trait in alfalfa is glyphosate tolerance (RoundupReady®). Currently, a low lignin trait is in the deregulation process in the USA, with deregulation expected in 2014 (Mark McCaslin, pers. comm.). Both glyphosate tolerance and low lignin represent traits of broad value to alfalfa growers and represent value-added traits that could command a large share of the seed market across the nation. Most other transgenes would offer less broad-based appeal or would not have the strikingly obvious phenotypes of these genes. Therefore, given that other genes may have less overall penetration into the market, and given the current high cost of deregulation, the likelihood that future transgenes will be commercialized in the USA and elsewhere in the near future is low.

As a consequence, perhaps the best value of transgenes to alfalfa improvement is simply as proof-of-concept for the role of various genes in phenotype expression. Following transgenic confirmation with allele prospecting in diverse germplasm, marker-assisted selection, targeted mutagenesis, and/or other approaches could be used to incorporate particular genes (or alleles) into elite breeding populations in lieu of transgenesis. Of course, should the regulatory barriers decrease, a broad array of new traits and new variation for existing traits could potentially be incorporated into commercial cultivars; when that day will come is unclear at this point (Wang and Brummer 2012).

10.5 Prospects

Alfalfa production in the USA today faces several challenges, from agricultural policies that favor cultivation of annual row crops to genetic gains for yield that are lower than those row crops. Environmentally, alfalfa is a desirable crop, fixing atmospheric nitrogen—which it contributes to succeeding crops in rotation—breaking up pest cycles, minimizing weeds, improving soil structure and minimizing erosion. Thus, genetic improvements that make farmers more interested in cultivating alfalfa are needed. We see the development of genome-wide markers and high-throughput genetic marker systems, a toolbox of genomic technologies, and transgenic plants expressing new traits as all promising avenues to make alfalfa a crop of choice. While sociological and political constraints on alfalfa cultivation may remain, the advances in genomics, coupled with a rethinking of current alfalfa breeding methods to maximize gain from genomics, will lead to better yielding and persisting alfalfa cultivars, which require less external inputs and are more nutritious. The rapid changes in genomics technologies now make realization of long-promised efficiency gains likely.

References

Al Lawati A, Pierce C, Murray L, Ray I (2010) Combining ability and heterosis for forage yield among elite alfalfa core collection accessions with different fall dormancy responses. Crop Sci 50: 150–158.

Ariss JJ, Vandemark GJ (2007) Assessment of genetic diversity among nondormant and semidormant alfalfa populations using sequence-related amplified polymorphisms. Crop Sci 47: 2274–2284.

Armour D, Mackie J, Musial J, Irwin JAG (2008) Transfer of anthracnose resistance and pod coiling traits from *Medicago arborea* to *M. sativa* by sexual reproduction. Theor Appl Genet 117: 149–156.

Baird NA, Etter PD, Atwood TS, Currey MC, Shiver AL, Lewis ZA, Selker EU, Cresko WA, Johnson EA (2008) Rapid SNP discovery and genetic mapping using sequenced RAD markers. PLoS One 3: e3376.

Bao AK, Wang SM, Wu GQ, Xi JJ, Zhang JL, Wang CM (2009) Overexpression of the Arabidopsis H⁺-PPase enhanced resistance to salt and drought stress in transgenic alfalfa (*Medicago sativa* L.). Plant Sci 176: 232–240.

Barnes DK, Bingham ET, Murphy RP, Hunt OJ, Beard DF, Skrdla WH, Teuber LR (1977) Alfalfa germplasm in the United States: Genetic vulnerability, use, improvement, and maintenance. USDA-ARS Tech Bull 1571, US Gov Print Office, Washington, DC.

Benedito VA, Torres-Jerez I, Murray JD, Andriankaja A, Allen S, Kakar K, Wandrey M, Verdier J, Zuber H, Ott T et al. (2008) A gene expression atlas of the model legume *Medicago truncatula*. Plant J 55: 504–513.

Bernardo R (2008) Molecular markers and selection for complex traits in plants: learning from the last 20 years. Crop Sci 48: 1649–1664.

Bhandari H, Pierce C, Murray L, Ray I (2007) Combining abilities and heterosis for forage yield among high-yielding accessions of the alfalfa core collection. Crop Sci 47: 665–671.

Bingham E, Groose R, Woodfield D, Kidwell K (1994) Complementary gene interactions in alfalfa are greater in autotetraploids than diploids. Crop Sci 34: 823–829.

Bingham ET (1993) Registration of alfalfa inbred parental line MAG7. Crop Sci 33: 1427–1427.

Bingham ET, McCoy TJ, Walker KA (1988) Alfalfa tissue culture. In: Hanson AA et al. (eds) Alfalfa and alfalfa improvement. Agron Monogr 29, ASA, CSSA, and SSSA, Madison, WI, pp 903–929.

Blondon F, Marie D, Brown S, Kondorosi A (1994) Genome size and base composition in *Medicago sativa* and *M. truncatula* species. Genome 37: 264–270.

Bouton J, Smith SR Jr (1998) Standard test to characterize alfalfa cultivar tolerance to intensive grazing with continuous stocking. Standard tests to characterize alfalfa cultivars. 3rd edn North Amer Alfalfa Improve Conf, Beltsville, MD (Available online at http://www.naaic.org) (Verified 13 Jan 2013).

Bradshaw A (1965) Evolutionary significance of phenotypic plasticity in plants. Adv Genet 13: 115–155.

Brouwer DJ, Osborn TC (1997) Molecular marker analysis of the approach to homozygosity by selfing in diploid alfalfa. Crop Sci 37: 1326–1330.

Brouwer DJ, Osborn TC (1999) A molecular marker linkage map of tetraploid alfalfa (*Medicago sativa* L.). Theor Appl Genet 99: 1194–1200.

Brouwer DJ, Duke SH, Osborn TC (2000) Mapping genetic factors associated with winter hardiness, fall growth, and freezing injury in autotetraploid alfalfa. Crop Sci 40: 1387–1396.

Brummer EC (2006) Grazing-tolerant alfalfa cultivars have superior persistence under continuous and rotational stocking. Forage and Grazinglands doi:101094/FG-2006-0825-01-RS.

Brummer EC, Bouton JH, Kochert G (1993) Development of an RFLP map in diploid alfalfa. Theor Appl Genet 86: 329–332.

Brummer, EC, Shah MM, Luth D (2000) Reexamining the relationship between fall dormancy and winter hardiness in alfalfa. Crop Sci 40: 971–977.

Brummer EC (2004) Genomics research in alfalfa, *Medicago sativa* L. In: Wilson RF et al. (eds) Legume crop genomics. AOCS Press, Champaign, IL, pp 110–142.

Buxton D, Marten G, Hornstein J (1987) Genetic variation for forage quality of alfalfa stems. Can J Plant Sci 67: 1057–1067.

Campbell T, He Y (1997) Factorial analysis of self-incompatibility in alfalfa. Can J Plant Sci 77: 69–73.

Casler MD, Brummer EC (2008) Theoretical expected genetic gains for among-and-within-family selection methods in perennial forage crops. Crop Sci 48: 890–902.

Castonguay Y, Laberge S, Brummer EC, Volenec JJ (2006) Alfalfa winter hardiness: A research retrospective and integrated perspective. Adv Agron 90: 203–265.

Castonguay Y, Michaud R, Nadeau P, Bertrand A (2009a) An indoor screening method for improvement of freezing tolerance in alfalfa. Crop Sci 49: 809–818.

Castonguay Y, Bertrand A, Michaud R, Laberge S (2011) Cold-induced biochemical and molecular changes in alfalfa populations selectively improved for freezing tolerance. Crop Sci 51: 2132–2144.

Castonguay Y, Cloutier J, Michaud R, Bertrand A, Laberge S (2009b) Development of marker-assisted selection for the improvement of freezing tolerance in alfalfa. In Yamada T, Spangenberg G (eds) Molecular Breeding of Forage and Turf. Springer, New York, pp 221–228.

Castonguay Y, Cloutier J, Bertrand A, Michaud R, Laberge S (2010) SRAP polymorphisms associated with superior freezing tolerance in alfalfa (*Medicago sativa* spp. *sativa*). Theor Appl Genet 120: 1611–1619.

Castonguay Y, Dubé MP, Cloutier J, Michaud R, Bertrand A, Laberge S (2012) Intron-length polymorphism identifies a Y_2K_4 dehydrin variant linked to superior freezing tolerance in alfalfa. Theor Appl Genet 124: 809–819.

Chase SS (1963) Analytic breeding in *Solanum tuberosum* L.—a scheme utilizing parthenotes and other diploid stocks. Can J Genet Cytol 5: 359–363.

Choi HK, Mun JH, Kim DJ, Zhu H, Baek JM, Mudge J, Roe B, Ellis N, Doyle J, Kiss GB, Young ND (2004a) Estimating genome conservation between crop and model legume species. Proc Natl Acad Sci USA 101: 15289–15294.

Choi HK, Kim D, Uhm T, Limpens E, Lim H, Mun JH, Kalo P, Penmetsa RV, Seres A, Kulikova O, Roe BA, Bisseling T, Kiss GB, Cook DR (2004b) A sequence-based genetic map of *Medicago truncatula* and comparison of marker colinearity with *M. sativa*. Genetics 166: 1463–1502.

Colditz F, Nyamsuren O, Niehaus K, Eubel H, Braun HP, Krajinski F (2004) Proteomic approach: identification of *Medicago truncatula* proteins induced in roots after infection with the pathogenic oomycete *Aphanomyces euteiches*. Plant Mol Biol 55: 109–120.

Coulman B, Goplen B, Majak W, McAllister T, Cheng KJ, Berg B, Hall J, McCartney D, Acharya S (2000) A review of the development of a bloat-reduced alfalfa cultivar. Can J Plant Sci 80: 487–491.

Davis WH (1971) Hybrid alfalfa production. US Patent No 3,570,181 Issued March 16, 1971.

Davis WH, Greenblatt IM (1967) Cytoplasmic male sterility in alfalfa. J Hered 58: 301–305.

Dien BS, Miller DJ, Hector RE, Dixon RA, Chen F, McCaslin M, Reisen P, Sarath G, Cotta MA (2011) Enhancing alfalfa conversion efficiencies for sugar recovery and ethanol production by altering lignin composition. Bioresour Technol 102: 6479–6486.

Echt CS, Kidwell KK, Knapp SJ, Osborn TC, McCoy TJ (1994) Linkage mapping in diploid alfalfa (*Medicago sativa*). Genome 37: 61–71.

Elden T, McCaslin M (1997) Potato leafhopper (Homoptera: Cicadellidae) resistance in perennial glandular-haired alfalfa clones. J Econ Entomol 90: 842–847.

Elgin JH, Welty RE, Gilchrist DB (1988) Breeding for disease and nematode resistance. In: Hanson AA et al. (eds) Alfalfa and Alfalfa Improvement. Agron Monogr 29, ASA, CSSA, and SSSA, Madison, WI, pp 827–858.

Elshire RJ, Glaubitz JC, Sun Q, Poland JA, Kawamoto K, Buckler ES, Mitchell SE (2011) A robust, simple genotyping-by-sequencing (GBS) approach for high diversity species. PLoS One 6: e19379.

Eujayl I, Sledge MK, Wang L, May GD, Chekhovskiy K, Zwonitzer JC, Mian MAR (2004) *Medicago truncatula* EST-SSRs reveal cross-species genetic markers for *Medicago* spp. Theor Appl Genet 108: 414–421.

Falahati-Anbaran M, Habashi AA, Esfahany M, Mohammadi SA, Ghareyazie B (2007) Population genetic structure based on SSR markers in alfalfa (*Medicago sativa* L.) from various regions contiguous to the centres of origin of the species. J Genetics 86: 59–63.

Freeman GF (1914) Physiological correlations and climatic reactions in alfalfa breeding. Am Nat 48: 356–368.

Gallardo K, Firnhaber C, Zuber H, Hericher D, Belghazi M, Henry C, Kuster H, Thompson R (2007) A combined proteome and transcriptome analysis of developing *Medicago truncatula* seeds. Mol Cell Proteom 6: 2165–2179.

Ganal MW, Altmann T, Roder MS (2009) SNP identification in crop plants. Curr Opin Plant Biol 12: 211–217.

Gonzales MD, Archuleta E, Farmer A, Gajendran K, Grant D, Shoemaker R, Beavis WD, Waugh ME (2005) The Legume Information System (LIS): an integrated information resource for comparative legume biology. Nucl Acids Res 33: D660–D665.

Groose R, Talbert L, Kojis W, Bingham ET (1989) Progressive heterosis in autotetraploid alfalfa: Studies using two types of inbreds. Crop Sci 29: 1173–1177.

Hackett CA, Milne I, Bradshaw JE, Luo Z (2007) TetraploidMap for Windows: linkage map construction and QTL mapping in autotetraploid species. J Hered 98: 727–729.

Han Y, Alarcon Y, Greene SL, Kisha T, Monteros MJ (2012a) Characterization of perennial *Medicago* germplasm diversity using molecular markers. North America Alfalfa Impr Conf, Ithaca, NY, 10–13 July 2012.

Han Y, Khu DM, Monteros MJ (2012b) High resolution melting analysis for SNP genotyping and mapping in tetraploid alfalfa (*Medicago sativa* L.). Mol Breeding 29: 489–501.

Han Y, Ray IM, Sledge MK, Bouton JH, Monteros MJ (2010a) Molecular mapping of QTLs associated with drought tolerance in alfalfa. 6th Int Symp Mol Breeding Forage and Turf, Buenos Aires, Argentina, 15–19 March 2010.

Han Y, Khu DM, Torres-Jerez I, Udvardi M, Monteros MJ (2010b) Plant transcription factors as novel molecular markers for legumes. In: Huyghe C (ed) Sustainable use of genetic diversity in forage and turf breeding. Springer, The Netherlands, pp 421–425.

Han Y, Kang Y, Torres-Jerez I, Cheung F, Town C, Zhao P, Udvardi M, Monteros MJ (2011) Genome-wide SNP discovery in tetraploid alfalfa using 454 sequencing and high resolution melting analysis. BMC Genomics 12: 350.

Havananda T, Brummer EC, Doyle JJ (2011) Complex patterns of autopolyploid evolution in alfalfa and allies (*Medicago sativa*: Leguminosae). Am J Bot 98: 1633–1646.

Hays DB, Skinner DZ (2001) Development of an expressed sequence tag (EST) library for *Medicago sativa*. Plant Sci 161: 517–526.

Heffner EL, Sorrells ME, Jannink JL (2009) Genomic selection for crop improvement. Crop Sci 49: 1–12.

Hill RR, Shenk JS, Barnes RF (1988) Breeding for yield and quality. In: Hanson AA et al. (eds) Alfalfa and Alfalfa Improvement. Agron Monogr 29, ASA, CSSA, and SSSA, Madison, WI, pp 809–825.

Irwin JAG, Armour D, Pepper P, Lowe K (2010) Heterosis in lucerne testcrosses with *Medicago arborea* introgressions and Omani landraces and their performance in synthetics. Crop Pasture Sci 61: 450–463.

Jannink JL, Lorenz AJ, Iwata H (2010) Genomic selection in plant breeding: from theory to practice. Brief Funct Genom 9: 166–177.

Jiang Y, Zareh N, Walker G, Teuber L (2003) Characterization of alfalfa germplasm expressing resistance to silverleaf whitefly, *Bemisia argentifolii*. J Appl Entomol 127: 447–457.

Jin T, Chang Q, Li W, Yin D, Li Z, Wang D, Liu B, Liu L (2010) Stress-inducible expression of GmDREB1 conferred salt tolerance in transgenic alfalfa. Plant cell Tiss Org Cult 100: 219–227.

Jones J, Bingham ET (1995) Inbreeding depression in alfalfa and cross-pollinated crops. Plant Breed Rev 13: 209–234.

Julier B, Huyghe C, Ecalle C (2000) Within-and among-cultivar genetic variation in alfalfa: forage quality, morphology, and yield. Crop Sci 40: 365–369.

Julier B, Bernard K, Gibelin C, Huguet T, Lelièvre F (2010) QTL for water use efficiency in alfalfa. In: Huyghe C (ed) Sustainable use of genetic diversity in forage and turf breeding. Springer, The Netherlands, pp 433–436.

Julier B, Flajoulot S, Barre P, Cardinet G, Santoni S, Huguet T, Huyghe C (2003) Construction of two genetic linkage maps in cultivated tetraploid alfalfa (*Medicago sativa*) using microsatellite and AFLP markers. BMC Plant Biol 3: 9.

Jung H, Sheaffer C, Barnes D, Halgerson J (1997) Forage quality variation in the US alfalfa core collection. Crop Sci 37: 1361–1366.

Kaló P, Endre G, Zimányi L, Csanádi G, Kiss GB (2000) Construction of an improved linkage map of diploid alfalfa (*Medicago sativa*). Theor Appl Genet 100: 641–657.

Kang H, Han Y, Torres-Jerez I, Sinharoy S, Tang Y, Monteros MJ, Udvardi MK (2011) System responses to long-term drought and re-watering of two contrasting alfalfa varieties. Plant J 68: 871–889.

Khu DM, Reyno R, Brummer EC, Monteros MJ (2012) Screening methods for aluminum tolerance in alfalfa. Crop Sci 52: 161–167.

Khu DM, Reyno R, Han Y, Zhao PX, Bouton JH, Brummer EC, Monteros MJ (2013) Identification of aluminum tolerance QTLs in tetraploid alfalfa. Crop Sci 53: 1–16.

Kiss GB, Csanadi G, Kalman K, Kalo P, Okresz L (1993) Construction of a basic genetic map for alfalfa using RFLP, RAPD, isozyme and morphological markers. Mol Gen Genet 238: 129–137.

Kumar S (2011) Biotechnological advancements in alfalfa improvement. J Appl Genet 52: 111–124.

Leath KT, Erwin DC, Griffin GD (1988) Diseases and nematodes. In: Hanson AA et al. (eds) Alfalfa and Alfalfa Improvement. Agron Monogr 29, ASA, CSSA, and SSSA, Madison, WI, pp 621–670.

Lesins KA, Lesins I (1979) Genus *Medicago* (Leguminosae): A taxogenetic study. Dr W Junk bv, The Hague, The Netherlands.

Li X, Han Y, Wei Y, Acharya AR, Farmer A, Ho J, Monteros M, Brummer EC (2013) Development and use of a high density SNP array to evaluate genetic diversity in alfalfa. Int Plant Animal Genome Conf XXI, San Diego, CA, 12–16 Jan 2013.https://pag.confex.com/pag/xxi/webprogram/Paper7784.html (Accessed 9 August 2013).

Li X, Brummer EC (2012) Applied genetics and genomics in alfalfa breeding. Agronomy 2: 40–61.

Li X, Wei Y, Nettleton D, Brummer EC (2009) Comparative gene expression profiles between heterotic and non-heterotic hybrids of tetraploid *Medicago sativa*. BMC Plant Biol 9: 107.

Li X, Wang X, Wei Y, Brummer EC (2011) Prevalence of segregation distortion in diploid alfalfa and its implications for genetics and breeding applications. Theor Appl Genet 123: 667–679.

Li X, Acharya A, Farmer AD, Crow JA, Bharti AK, Kramer RS, Wei Y, Han Y, Gou J, May GD (2012) Prevalence of single nucleotide polymorphism among 27 diverse alfalfa genotypes as assessed by transcriptome sequencing. BMC Genomics 13: 568.

Ma XF, Tudor S, Butler T, Ge Y, Xi Y, Bouton J, Harrison M, Wang ZY (2012) Transgenic expression of phytase and acid phosphatase genes in alfalfa (*Medicago sativa*) leads to improved phosphate uptake in natural soils. Mol Breed 30: 377–391.

Mackie JM, Musial JM, Armour DJ, Phan HTT, Ellwood SE, Aitken KS, Irwin JAG (2007) Identification of QTL for reaction to three races of *Colletotrichum trifolii* and further analysis of inheritance of resistance in autotetraploid lucerne. Theor Appl Genet 114: 1417–1426.

Madril C, Pierce C, Ray IM (2008) Heterosis among hybrids derived from genetically improved and unimproved alfalfa germplasm. Crop Sci 48: 1787–1792.

Manglitz GR, Ratcliffe RH (1988) Insects and mites. In: Hanson AA et al. (eds) Alfalfa and Alfalfa Improvement. Agron Monogr 29, ASA, CSSA, and SSSA, Madison, WI, pp 671–704.

Mathesius U, Keijzers G, Natera SHA, Weinman JJ et al. (2001) Establishment of a root proteome reference map for the model legume *Medicago truncatula* using the expressed sequence tag database for peptide mass fingerprinting. Proteomics 1: 1424–1440.

Maureira-Butler I, Udall J, Osborn TC (2007) Analyses of a multi-parent population derived from two diverse alfalfa germplasms: testcross evaluations and phenotype-DNA associations. Theor Appl Genet 115: 859–867.

McCaslin MH (1999) Potato leafhopper resistant alfalfa. US Patent No 5,908,974 Issued June 1, 1999.

McCoy TJ, Bingham ET (1988) Cytology and cytogenetics of alfalfa. In: Hanson AA et al. (eds) Alfalfa and Alfalfa Improvement. Agron Monogr 29, ASA, CSSA, and SSSA, Madison, WI, pp 737–776.

McKenzie JS, Paquin R, Duke SH (1988) Cold and heat tolerance. In: Hanson AA et al. (eds) Alfalfa and Alfalfa Improvement. Agron Monogr 29, ASA, CSSA, and SSSA, Madison, WI, pp 259–302.

Michaud RA, Lehman WF, Rumbaugh MD (1988) World distribution and historical development. In: Hanson AA et al. (eds) Alfalfa and Alfalfa Improvement. Agron Monogr 29, ASA, CSSA, and SSSA, Madison, WI, pp 25–91.

Mohan M, Nair S, Bhagwat A, Krishna TG, Yano M, Bhatia CR, Sasaki T (1997) Genome mapping, molecular markers and marker-assisted selection in crop plants. Mol Breed 3: 87–103.

Monteros MJ, Burton JW, Boerma HR (2008) Molecular mapping and confirmation of QTLs associated with oleic acid content in N00-3350 soybean. Crop Sci 48: 2223–2234.

Mueller S (2004) Seed production issues for genetically enhanced alfalfa. Proc National Alfalfa Symposium, San Diego, CA. http: //alfalfa.ucdavis.edu/+symposium/proceedings/indexaspx?yr=2004. Accessed 16 Jan 2013.

Narasimhamoorthy B, Bouton JH, Olsen KM, Sledge MK (2007) Quantitative trait loci and candidate gene mapping of aluminum tolerance in diploid alfalfa. Theor Appl Genet 114: 901–913.

Osborn TC, Brouwer D, McCoy TJ (1997) Molecular marker analysis in alfalfa. In: McKersie BD, Brown DCW (eds) Biotechnology and the Improvement of Forage Legumes. CAB International, Guelph, Canada, pp 91–109.

Paterson A, Lander E, Hewwitt JD, Peterson S, Lincoln SS, Tanksley SD (1988) Resolution of quantitative traits into Mendelian factors by using a complete linkage map of restriction fragment length polymorphisms. Nature 335: 721–726.

Pattathil S, Saffold T, Gallego-Giraldo L, O'Neill M, York WS, Dixon RA, Hahn MG (2012) Changes in cell wall carbohydrate extractability are correlated with reduced recalcitrance of HCT downregulated alfalfa biomass. Indust Biotechnol 8: 217–221.

Pliny (1950) Natural History, Vol V: Books 17–19. Harvard, Cambridge, MA.

Quiros CR, Bauchan GR (1988) The Genus *Medicago* and the origin of the *Medicago sativa* complex. In: Hanson AA et al. (eds) Alfalfa and Alfalfa Improvement. Agron Monogr 29, ASA, CSSA, and SSSA, Madison, WI, pp 93–124.

Rémus-Borel W, Castonguay Y, Cloutier J, Michaud R, Bertrand A, Desgagnés R, Laberge S (2010) Dehydrin variants associated with superior freezing tolerance in alfalfa (*Medicago sativa* L.). Theor Appl Genet 120: 1163–1174.

Reyno R, Khu DM, Monteros MJ, Parrott W, Brummer EC (2013) Evaluation of two transgenes for aluminum tolerance in alfalfa. Crop Sci 53: 1581–1588.

Riday H, Brummer EC (2002a) Heterosis of agronomic traits in alfalfa. Crop Sci 42: 1081–1087.

Riday H, Brummer EC (2002b) Forage yield heterosis in alfalfa. Crop Sci 42: 716–723.

Riday H, Brummer EC (2004) Morphological variation of *Medicago sativa* subsp. *falcata* genotypes and their hybrid progeny. Euphytica 138: 1–12.

Riday H, Brummer EC (2005) Heterosis in a broad range of alfalfa germplasm. Crop Sci 45: 8–17.

Riechmann JL, Heard J, Martin G, Reuber L, Jiang C, Keddie J, Adam L, Pineda O, Ratcliffe OJ, Samaha RR, Creelman R, Pilgrim M, Broun P, Zhang JZ, Ghandehari D, Sherman BK, Yu G (2000) Arabidopsis transcription factors: genome-wide comparative analysis among eukaryotes. Science 290: 2105–10.

Robins JG, Brummer EC (2010) QTL underlying self-fertility in tetraploid alfalfa. Crop Sci 50: 143–149.

Robins JG, Bauchan GR, Brummer EC (2007a) Genetic mapping forage yield, plant height, and regrowth at multiple harvests in tetraploid alfalfa (*Medicago sativa* L.). Crop Sci 47: 11–18.

Robins JG, Viands DR, Brummer EC (2008a) Genetic mapping of persistence in tetraploid alfalfa. Crop Sci 48: 1780–1786.

Robins JG, Hansen JL, Viands DR, Brummer EC (2008b) Genetic mapping of persistence in tetraploid alfalfa, Crop Sci 48: 1780–1786.

Robins JG, Luth D, Campbell TA, Bauchan GR, He C, Viands DR, Hansen JL, Brummer EC (2007b) Genetic mapping of biomass production in tetraploid alfalfa. Crop Sci 47: 1–10.

Sakiroglu M, Brummer EC (2012) Presence of phylogeographic structure among wild diploid alfalfa accessions (*Medicago sativa* L. subsp. *microcarpa* Urb.) with evidence of the center of origin, Genet Resour Crop Evol 60: 23–31.

Sakiroglu M, Doyle JJ, Brummer EC (2010) Inferring population structure and genetic diversity of broad range of wild diploid alfalfa (*Medicago sativa* L.) accessions using SSR markers. Theor Appl Genet 121: 403–415.

Sakiroglu M, Moore KJ, Brummer EC (2011) Variation in biomass yield, cell wall components, and agronomic traits in a broad range of diploid alfalfa accessions. Crop Sci 51: 1956–1964.

Samac DA, Austin-Phillips S (2006) Alfalfa (*Medicago sativa* L.). Meth Mol Biol 343: 301–311.

Sato S, Nakamura Y, Kaneko T, Asamizu E, Kato T, Nakao M, Sasamoto S et al. (2008) Genome structure of the legume *Lotus japonicas*. DNA Res 15: 227–239.

Schmutz J, Cannon SB, Schlueter J, Ma J, Mitros T, Nelson W, Hyten DL et al. (2010) Genome sequence of the palaeopolyploid soybean. Nature 463: 178–183.

Schwab P, Barnes DK, Sheaffer C (1996) The relationship between field winter injury and fall growth score for 251 alfalfa cultivars. Crop Sci 36: 418–426.

Segovia-Lerma A, Murray L, Townsend M, Ray IM (2004) Population-based diallel analyses among nine historically recognized alfalfa germplasms. Theor Appl Genet 109: 1568–1575.

Sinskaya EN (1950) Flora of cultivated plants of the USSR: XIII perennial leguminous plants, Part 1 Medic, Sweetclover, Fenugreek. Translated from Russian in 1961 by the Israel Program for Scientific Translations, Jerusalem.

Sledge M, Ray IM, Jiang G (2005a) An expressed sequence tag SSR map of tetraploid alfalfa (*Medicago sativa* L.). Theor Appl Genet 111: 980–992.

Small E (2011) Alfalfa and Relatives: Evolution and classification of *Medicago*. CAB International, Wallingford, Oxfordshire.

Smith SR, Bouton JH, Singh A, McCaughey W (2000) Development and evaluation of grazing-tolerant alfalfa cultivars: A review. Can J Plant Sci 80: 503–512.

Smith SR, Bouton JH (1993) Selection within alfalfa cultivars for persistence under continuous stocking. Crop Sci 33: 1321–1328.

Sorensen EL, Byers RA, Horber EK (1988) Breeding for insect resistance. In: Hanson AA et al. (eds) Alfalfa and Alfalfa Improvement. Agron Monogr 29, ASA, CSSA, and SSSA, Madison, WI, pp 859–902.

Suarez R, Calderon C, Iturriaga G (2009) Enhanced tolerance to multiple abiotic stresses in transgenic alfalfa accumulating trehalose. Crop Sci 49: 1791–1799.

Sun P, Velde M, Gardner D (2002) Alfalfa hybrids having at least 75% hybridity. US Patent 6,774,280, Issued Aug 10, 2004.

Tadege M, Wen J, He J, Tu H, Kwak Y, Eschstruth A, Cayrel A, Endre G, Zhao PX, Chabaud M, Ratet P, Mysore KS (2008) Large-scale insertional mutagenesis using the Tnt1 retrotransposon in the model legume *Medicago truncatula*. Plant J 54: 335–347.

Tanksley SD, Young ND, Paterson AH, Bonierbale MW (1989) RFLP mapping in plant breeding: New tools for an old science. BioTechnology 7: 257–264.

Tavoletti S, Veronesi F, Osborn T (1996) RFLP linkage map of an alfalfa meiotic mutant based on an F1 population. J Hered 87: 167–170.

Teuber LR, Taggard KL, Gibbs LK, McCaslin MH, Peterson MA, Barnes DK (1998) Fall dormancy. In: Standard tests to characterize alfalfa cultivars. North American Alfalfa Improve Conf. http://www.naaic.org/stdtests/Dormancy2html. Accessed 16 Jan 2013.

Tracy W, Chandler M (2006) The historical and biological basis of the concept of heterotic patterns in corn belt dent maize. In KR Lamkey, Lee M (eds) Plant breeding: the Arnel R Hallauer international symposium. Blackwell, Ames, IA, pp 219–233.

Tysdal H, Kiesselbach T (1944) Hybrid alfalfa. Agron J 36: 649–667.

Undersander D, Cosgrove D, Cullen E, Grau C, Rice ME, Renz M, Sheaffer C, Shewmaker G, Sulc M (2011) Alfalfa Management Guide. ASA, CSSA, SSSA, Madison, WI.

van Noorden GE, Kerim T, Goffard N, Wiblin R, Pellerone FI, Rolfe BG, Mathesius U (2007) Overlap of proteome changes in *Medicago truncatula* in response to auxin and *Sinorhizobium meliloti*. Plant Physiol 144: 1115–1131.

van Oeveren J, de Ruiter M, Jesse T, van der Poel H, Tang J, Yalcin F, Janssen A, Volpin H, Stormo KE, Bogden R, van Eijk MJ, Prins M (2011) Sequence-based physical mapping of complex genomes by whole genome profiling. Genome Res 21: 618–625.

Vandemark G, Ariss J, Bauchan G, Larsen R, Hughes T (2006) Estimating genetic relationships among historical sources of alfalfa germplasm and selected cultivars with sequence related amplified polymorphisms. Euphytica 152: 9–16.

Veronesi F, Mariani A, Bingham ET (1986) Unreduced gametes in diploid *Medicago* and their importance in alfalfa breeding. Theor Appl Genet 72: 37–41.

Viands DR, Sun P, Barnes DK 1(988) Pollination control: mechanical and sterility. In: Hanson AA et al. (eds) Alfalfa and Alfalfa Improvement. Agron Monogr 29, ASA, CSSA, and SSSA, Madison, WI, pp 931–960.

Voorrips RE, Gort G, Vosman B (2011) Genotype calling in tetraploid species from bi-allelic marker data using mixture models. BMC Bioinformatics 12: 172.

Wang ZY, Brummer EC (2012) Is genetic engineering ever going to take off in forage, turf and bioenergy crop breeding? Ann Bot 110: 1317–1325. doi: 101093/aob/mcs027.

Weishaar MA, Brummer EC, Volenec JJ, Moore KJ, Cunningham S (2005) Improving winter hardiness in nondormant alfalfa germplasm. Crop Sci 45: 60–65.

Westgate JM (1909) Methods of breeding alfalfa by selection. J Hered 5: 144–147.

Yang SS, Xu WW, Tesfaye M, Lamb JAFS, Jung HJG, VandenBosch KA, Vance CP, Gronwald JW (2010) Transcript profiling of two alfalfa genotypes with contrasting cell wall composition in stems using a cross-species platform: optimizing analysis by masking biased probes. BMC Genomics 11: 323.

Young ND (1996) QTL mapping and quantitative disease resistance in plants. Annu Rev Phytopath 34: 479–501.

Young ND, Udvardi M (2009) Translating *Medicago truncatula* genomics to crop legumes. Curr Opin Plant Biol 12: 193–201.

Young ND, Debelle F, Oldroyd GED, Geurts R, Cannon SB, Udvardi MK, Benedito VA et al. (2011) The *Medicago* genome provides insight into the evolution of rhizobial symbioses. Nature 480: 520–524.

Zhang J-Y, Broeckling CD, Blancaflor EB, Sledge MK, Sumner LW, Wang Z-Y (2005) Overexpression of WXP1, a putative *Medicago truncatula* AP2 domain-containing transcription factor gene, increases cuticular wax accumulation and enhances drought tolerance in transgenic alfalfa (*Medicago sativa*). Plant J 42: 689–707.

Zhang Y, Sledge MK, Bouton JH (2007) Genome mapping of white clover (*Trifolium repens* L.) and comparative analysis within the Trifolieae using cross-species SSR markers. Theor Appl Genet 114: 1367–1378.

Zhao Q, Gallego-Giraldo L, Wang H, Zeng Y, Ding SY, Chen F, Dixon RA (2010) An NAC transcription factor orchestrates multiple features of cell wall development in *Medicago truncatula*. Plant J 63: 100–114.

Zhou C, Han L, Pislariu C, Nakashima J, Fu C, Jiang Q, Quan L, Blancaflor EB, Tang Y, Bouton JH (2011) From model to crop: functional analysis of a STAY-GREEN gene in the model legume *Medicago truncatula* and effective use of the gene for alfalfa improvement. Plant Physiol 157: 1483–1496.

Zhu H, Choi HK, Cook DR, Shoemaker RC (2005) Bridging model and crop legumes through comparative genomics. Plant Physiol 137: 1189–1196.

Red Clover

Sachiko Isobe,[1,]* Roland Kölliker,[2] Beat Boller[2] and Heathcliffe Riday[3]

ABSTRACT

Red clover (*Trifolium pratense* L.) is an important forage legume grown on approximately four million hectares worldwide. Around the world red clover is grown in pure stands or mixed with grasses for hay, haylage, silage or grazing. It is known for rapid establishment, shade tolerance, and tolerance to low pH, low fertility and poorly drained soils. A high and reliable dry matter yield is a key objective of traditional breeding. However, realized yield potential of red clover is most often limited by insufficient persistence, which in turn is affected by various diseases. Therefore, breeding red clover for yield per se is rarely carried out without simultaneously paying attention to resistance against biotic and abiotic stresses affecting persistence. More than a century of breeding has increased expected stand longevity of red clover from two or three to four years. Seed yield is also an important character for the market success of red clover cultivars. A recent QTL analysis of seed yield and persistence in red clover dispelled the suspicion of negative correlation between the two traits. Morphological, isozyme and DNA marker polymorphisms suggested a large amount of genetic diversity

[1]Kazusa DNA Research Institute, 2-6-7 Kazusa-Kamatari, Kisarazu, Chiba, 292-0818, Japan.
[2]Agroscope Reckenholz-Taenikon Research Station ART, Reckenholzstr. 191, 8046 Zurich, Switzerland.
[3]US Dairy Forage Research Center, 1925 Linden Drive West Madison, WI 53706, USA.
*Corresponding author: *sisobe@kazusa.or.jp*

among red clover accessions. Most of the variation resided within populations and amount of variation among populations was higher for wild populations when compared to cultivars or landraces. The nuclear DNA content of red clover was estimated as 0.89–0.97 pg/2C and 440 Mb. Linkage maps were constructed based on DNA markers, such as RFLP, AFLP, and microsatellite markers. Based on the maps, QTLs were identified for persistence and seed yield. The tissue culture and gene transfection protocols for red clover have been established.

Key words: Red clover, History, Characteristics, Breeding, Structural genomics

11.1 Introduction

11.1.1 Brief History

Red clover (*Trifolium pratense* L.) is an important forage legume grown on approximately four million hectares worldwide. It has a long and varied history in agriculture (Westgate and Hillman 1911; Fergus and Hollowell 1960; Smith et al. 1985; Undersander et al. 1990; Lacefield and Ball 1999). It is uncertain when red clover was first cultivated. European macrofossil pollen records show red clover appearing with the clearing of forests and introduction of pastures during the late Bronze Age (Hodgson et al. 1999; Bradshaw 2005). It is probable that some of these early pastures were used for making hay by the time of the Iron Age (Hodgson et al. 1999). Archeological evidence from ancient Egypt indicates that berseem clover (*Trifolium alexandrinum*) was cultivated and fed as fodder to cattle (Cagle 2001). The earliest written records of red clover are from ancient herbalist manuscripts. However, determining which clover species was discussed in early texts is not always obvious (Seven 1991). Early (405–1050 AD) Anglo-Saxon herbal manuscripts mentions clover (*clæfre* or *χίρσιον*); these manuscripts themselves are based on earlier Roman and Greek texts of "*Herbarium Apuleius*" (Cockayne 1961). Furthermore, an old Anglo-Saxon "Leechbook" (925) mentions red clover (*reád clæfre*) as an ingredient in a salve to cure "cancer" (Cockayne 1961). The first recorded mention of clover in an agricultural setting was by Albert Magnus (1193–1280) in "*De Vegetabilibus*" book VII who observed that clover (*Trifolium*) was good for the soil and as an animal feed (Jessen 1982).

Other early agricultural descriptions of clover come from Brabant, the Netherlands (Dodoens 1566), and Brescia, Italy (Gallo 1565) (Blomeyer 1889; Merkenschlager 1934a, 1934b; Zeven 1991). In Brabant, Rembert Dodoens describes the cultivation of pure stands of clover (*gemeyne claveren*) (Zeven 1991). Dodoens describes clover cultivation as an established practice indicating it had already existed for some time before the mid 1500s

(Blomeyer 1889). It is likely that the seed for cultivating red clover in Brabant originally came from Spain since cultivated red clover was called "Spanish clover". This Spanish clover in turn probably originated from the Near East and was brought to Spain around 1300 by the Moors ruling Spain in the Middle Age (Zeven and de Wet 1982). According to Freudenthaler et al. (1998), Dodoens (1566) mentions that red clover was cultivated on arable fields in Brabant "grows much vigorously and taller than the red clover on the meadows". In the late medieval period, Flanders was known for its advanced agricultural system and, although earlier descriptions of red clover exist, it was in Flanders that deliberate improvement/cultivation of pastures became prevalent (Lane 1980; Hopcroft 2003). Evidence of intensification started appearing in the 1300s in Flanders as records of complex agricultural rotations appear (Hopcroft 2003). From Flanders, clover cultivation spread to England and Central Europe in the mid-1600s (Michell 1974; Lane 1980), although cultivation of clover in Europe was mentioned in English herbal manuscripts as early as the late 1500s (Gerard 1597). The arrival of red clover in the New World may have occurred as early as 1625 when ships from Holland landed in North America. Clover (*clover-grass*) cultivation is described in the mid-1600s in North American records (Fergus and Hallowell 1960). In the 1780s J.C. Schubart popularized red clover cultivation in Germany, for which he was elevated to nobility by the emperor (Merkenschlager 1934a and 1934b). Russian farms started using red clover in the 1790s (Merkenschlager 1934a; Semerikov et al. 2002). By the 1800s almost all temperate agricultural regions in the world used red clover as an integral part of cropping rotations to improve soil and provide fodder.

The demise of horse-powered agriculture and the rise of widespread synthetic fertilizer use during the 20th century contributed to a steep decline in red clover acreage. This decline is evident in the United States from 20th century red clover seed production data (Fig. 11-1). A similar worldwide decline has been observed in other parts of the world. Around 1990 red clover acreage appears to have stabilized in the United States (Fig. 11-1).

11.1.2 Economic Importance

In most parts of the world red clover acreage is not recorded, making it difficult to determine exact acreage and usage. However, red clover seed production is recorded in many countries. An estimated 2.8 million kg of red clover seed per year was produced worldwide in 2005–2007 (Table 11-1). This amount of seed would be enough to maintain approximately 4 million hectares of red clover per year. Around the world red clover is grown in pure stands or mixed with grasses for hay, haylage, silage or

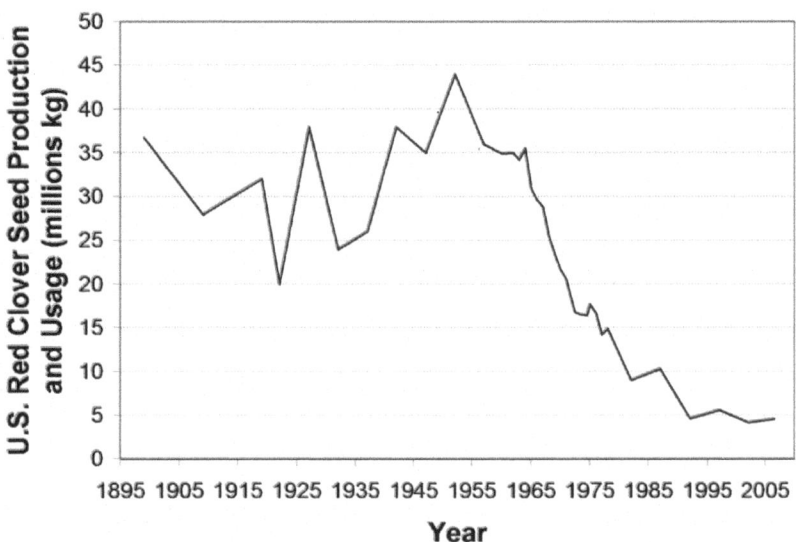

Figure 11-1 United States red clover (*Trifolium pratense* L.) seed usage or production 1899–2007 (Taylor and Qusenberry, 1996; NASS, 1899–2007).

Table 11-1 Red clover (*Trifolium pratense* L.) seed production amounts.

Country or Region	Kg (Million)
Argentina	0.07[a]
Canada	1.23[a]
European Union-25	3.70[b]
New Zealand	0.16[a]
Norway	0.10[a]
Switzerland	0.05[a]
United States of America	4.57[a]
Total	9.88

[a]2006/07 growing season estimates (unpubl.)
[b]2005 growing season (EU 2007)

grazing (Undersander et al. 1990; Lacefield and Ball 1999). Red clover cut for hay, haylage or silage is usually harvested at first bloom (35 to 40 d harvest intervals) resulting in two to four harvests per year, usually yielding approximately 9 Mg/ha dry matter per year in North America (Taylor et al. 1997). Typical annual DM yields in temperate Europe are between 12 and 16 Mg/ha (e.g., Boller et al. 2004). Red clover is known for rapid establishment, shade tolerance, tolerance to low pH, low fertility and poorly drained soils (Fergus and Hollowell 1960; Smith and Taylor 1985; Taylor et al. 1997; Undersander et al. 2000). Traditionally red clover is considered a short-lived perennial, even biannual (Westgate and Hilleman 1911). However, more than a century of breeding has increased expected stand longevity of red clover to four years (Smith 2000).

Red clover utilization in agricultural systems remains an economic way of providing animal feed protein and soil nitrogen, especially in rotational grazing systems (Burdine et al. 2005). Red clover is also a good cover crop and green manure, often utilized in organic agriculture (Knorek and Staton 1996; Sullivan 2003). In cropping rotations red clover is often established under small grains or other nurse crops. In warmer climates red clover can be used as a winter annual (Quesenberry and Blount 2006). Another specialized use of red clover includes growing it in rotations for use as a trap crop for pathogens and pests (Chen et al. 2006).

11.1.3 Morphological Characteristics and Uses

Establishing red clover is accomplished with drills, rollers and broadcast seed equipment both into firm seed beds or surface seeded (Taylor et al. 1997; Lacefield and Ball 1999; Iepema et al. 2006). Stands are usually established in spring or autumn depending on climate and desired stand usage (Wheaton 1993; Quesenberry and Blount 2006). To maintain red clover in permanent pasture, pasture renovation techniques are frequently used (Leep 1989; Wheaton and Roberts 1993; Lacefield et al. 1997).

Morphologically, red clover is an herbaceous plant consisting of a crown with a tap root that is usually about 1 meter in length (Bowley et al. 1984). Hollow stems grow from the crown, with four to nine stems observed under sward conditions (Williams 1927). Leaves are trifoliate with a delta shaped leaf mark observed in most populations. North American red clover tends to be thickly pubescent on leaves and stems with European and Chilean germplasm being much less so (Williams 1927). Fully developed stems produce seven to 15 flowering heads (Bowley et al. 1984). Each flowering head consists of 30 to 70 florets, with each floret normally producing one seed. It has long been observed that non-North American germplasm is not persistent in North America (Pieters and Hollowell 1937; Taylor and Queensberry 1996), while in turn North American germplasm persists poorly in Europe. It has been hypothesized that the pubescence of North American red clover prevents potato leafhopper (*Empoasca fabae* Harris) damage (Hollowell 1937).

Ensiling and especially haying red clover is more difficult than other forage legumes due to higher moisture content, slower drying of stems than of leaves and stem pubescence (Albrecht and Beauchemin 2003).

However, under less frequent cutting, red clover hay and silages better maintain forage quality as compared to alfalfa (*Medicago sativa* L.) (Smith 1965). Also, compared with alfalfa, red clover has lower animal intake; however, red clover is more energy dense (Broderick et al. 2000; Broderick 2001; Hoffman and Broderick 2001). Ensiling red clover does increase its phytoestrogenic activity initially (prior to the 180th day of fermentation). On the other hand, making red clover hay reduces phytoestrogen activity by 50% (Bush et al. 2007). Phytoestrogens in red clover are known to cause reproductive problems when fed to sheep (Bush et al. 2007). However, red clover phytoestrogens (mainly formononetin) are of interest to the human nutraceutical industry for their possible human medicinal benefits (Dixon 2004). One advantage of red clover silage is reduced silage protein loss compared to alfalfa silages (Albrecht and Beauchemin 2003). The reduced proteolytic activity of red clover silage better maintains silage forage nutritive value and reduces silage effluent pollution severity (Savoie and Jofriet 2003; Sullivan and Hatfield 2006). Polyphenol oxidases responsible for proteolytic inhibition in red clover give red clover hay its characteristic brown color compared to alfalfa hay.

Red clover is a perennial diploid (2n = 14). Almost all red clover varieties utilized in North America are diploid, while chromosome-doubled tetraploid red clover varieties are often bred and utilized in Europe in addition to diploid varieties (Taylor and Queensberry 1996). Red clover has an estimated genome size of 435Mb (Sato et al. 2005). Genomic work and infrastructure developed in other forage and model legumes species (i.e., *Lotus japanicus*, *Medicago sativa*, *Medicago truncatula*, and *Trifolium repens*) frequently are applicable to red clover genomics (Choi et al. 2004). Taxonomically, red clover is a member of the Leguminosae in the genus *Trifolium* L. subgenus *Trifolium* sect. *Trifolium* (Zohary and Heller 1984; Ellison et al. 2006). The center of origin of red clover is believed to be in Eurasia along the Mediterranean Sea (Fergus and Holowell 1960). Phylogenetic analysis using molecular markers group red clover with *T. pallidum*, *T. diffusum*, and *T. andricum* (Ellison et al. 2006) (Fig. 11-2). Several interspecific hybrids have been attempted (Taylor and Quesenberry 1996; Abertton 2007; Fig. 11-2). Interspecific hybridization has proven to be difficult, one reason being that most closely-related species of red clover have a base chromosome number of *2n* = 16. Interspecific hybrids between red clover and *T. medium* have been studied the most (Abberton 2007), which has expanded the red clover germplasm base.

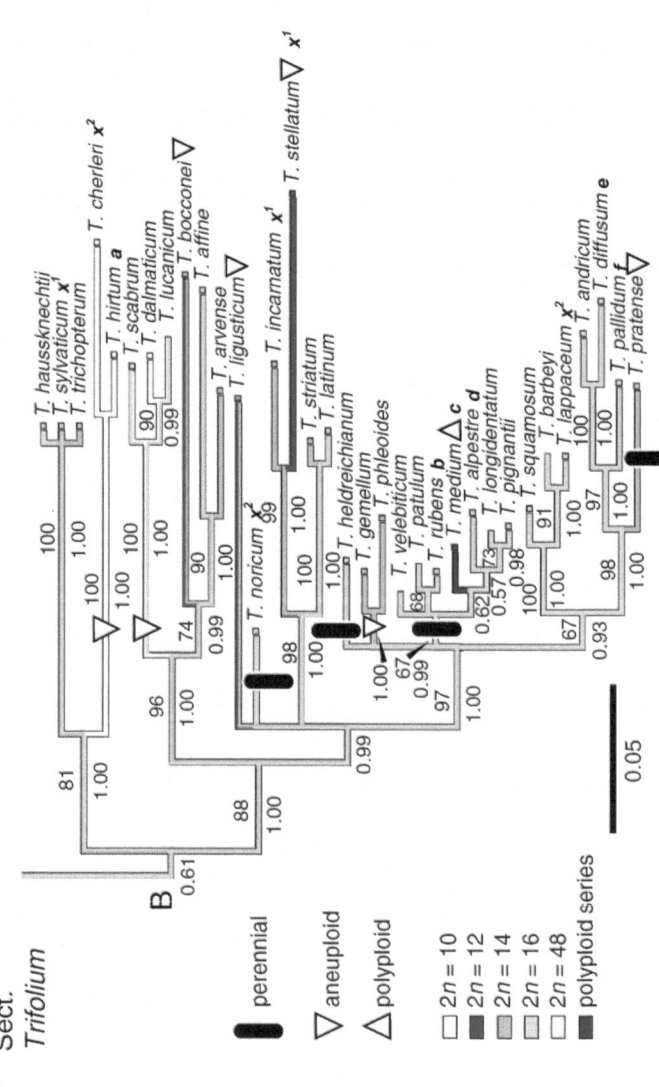

Figure 11-2 Genus *Trifolium* subsec. *Trifolium* group B species and attempted interspecific hybrids with red clover (Adapted from Ellison et al. 2006). (A) *T. pratense* x *T. hirtum*, shriveled seed produced (Taylor et al. 1963); (B) 2x *T. pratense* x *T. rubens*, hybrid embryo produced (Evans 1962); (c) successful fertile hybrids produced (Phillips et al. 1982; Collins et al. 1983; Taylor et al. 1983; Merker 1984; Sawai and Ueda 1987; Sawai et al. 1990; Sawai et al. 1995; and Isobe et al. 2002); (d) hybrids produced (Evans 1962; Merker 1988; and Phillips et al. 1992); (e) hybrid produced (Taylor et al. 1963; and Schwer and Cleveland 1972); and (f) hybrids produced (Armstrong and Cleveland 1970). Failed attempts at hybridization: (x¹) Evans and Denward 1955; and Taylor et al. 1963; and (x²) Quesenberry 1975.

11.2 Classical Genetics and Traditional Breeding

11.2.1 Breeding Objectives

11.2.1.1 Forage Yield and Persistence

Red clover is a high yielding forage legume suitable for cutting to produce fresh or conserved fodder but does not tolerate intensive grazing. A high and reliable dry matter yield, resulting from about four cuts per full harvest year, is therefore a key objective of traditional breeding, like for most forage species. However, realized yield potential of red clover is most often limited by insufficient persistence, which in turn is affected by various diseases. Therefore, breeding red clover for yield per se is rarely carried out without simultaneously paying attention to resistance against biotic and abiotic stresses affecting persistence.

Key fungal pathogens threatening the desired survival of red clover depend on the target region of cultivation. In cooler areas with strong winters and long snow cover, *Sclerotinia trifoliorum* Eriks causes crown rot and is a major disease deserving attention in breeding programs. Techniques to improve *Sclerotinia* resistance by inoculation with mycelium suspensions were developed by Frandsen (1946) and refined by Dixon and Doodson (1974). Marum et al. (1994) suggested the use of ascospores for more predictable inoculation. These various techniques are widely used by breeders, and *Sclerotinia* resistance is systematically assessed in official variety testing. In warmer climates, southern anthracnose, caused by *Colletotrichum trifolii,* can lead to loss of individual plants and significant sward damage during summer. In the southern part of the clover belt of the United States, already in the 1950s the disease was controlled by the use of resistant cultivars (Taylor 2008). However, as these regions are less important for clover cultivation, breeders have paid more attention to northern anthracnose and its causal agent, *Kabatiella caulivora* (Taylor and Smith 1990). More recently, *Colletotrichum trifolii* appeared to benefit from rising temperatures in many areas and, by causing leaf and stem symptoms but more importantly crown decay, the disease has become a limiting factor for yield and persistence in previously less affected regions (Boller et al. 1998). Consequently, inoculation techniques were developed and applied to improve resistance (Schubiger et al. 2003). *Fusarium* spp. are often encountered in association with root rot but their role as pathogens is not quite clear. Often, an increased abundance of *Fusarium* spp. is observed as a side-effect of root breakdown due to various other stresses such as summer drought or southern anthracnose. However, some breeding efforts are devoted to resistance against *Fusarium* with the aim of improving persistence. Rufelt (1985) developed a dipping technique to inoculate roots with conidia of *Fusarium* spp. but progress in resistance using recurrent selection based on this technique was very slow (Venuto et al. 1999). Nedelnik (1992) concluded that mortality of red clover plants in the field was caused by a complex of biotic and abiotic factors and was not related to *Fusarium* resistance observed in greenhouse conditions.

Red clover persistence is also negatively affected by insect and nematode pests. Although root-parasitizing insects like the root borer (*Hylastinus obscurus* Marsham) and the clover root curculio (*Sitona hispidula* F.) have been advocated as being involved strongly in lowering stand persistence of red clover (Leath and Byers 1973), no targeted selection for resistance has been carried out. In contrast, nematode resistance is a major concern in red clover breeding. In temperate Europe, the stem eelworm *Ditylenchus dipsaci* often affects red clover grown in narrow crop rotations and therefore, systematic selection for nematode resistance including artificial inoculation has been carried out almost since the establishment of modern breeding programs. Related techniques were first described by Akerberg et al. (1946) and evaluated in detail by Bingefors (1951). These techniques are used systematically by red clover breeders to improve persistence. In the United States, successful selection for resistance against root-knot nematodes (*Meloidogyne* spp.) has been carried out (Quesenberry et al. 1989).

11.2.1.2 Resistance to Foliar Diseases

Among the various foliar diseases occurring on red clover, powdery mildew (causal agent *Erysiphe polygoni*) has got the most attention in plant improvement programs. This may be due to its prominent appearance, often completely covering the foliage with its white mycelium in late summer. However, it usually does not affect yield of the affected growth cycle much and disappears without weakening regrowth of the plants after cutting. Nevertheless, resistance to powdery mildew is assessed and taken into account in official cultivar testing and progress in resistance is sought by plant breeders. Greenhouse conditions favor the disease and offer easy opportunities for screening. *Stemphylium sarciniforme* (Cav.) Wiltshire occurs widespread, it causes target spots and can decrease yield of affected growth cycles quite considerably. Important differences in cultivar susceptibility have been observed (Berg et al. 1996).

11.2.1.3 Quality Characters

Red clover is rich in protein, but contains little soluble carbohydrates and tends to be less digestible than forage grasses. However, this drawback is easily coped with by growing red clover in mixed stands with highly digestible grasses like Italian ryegrass. Red clover is an ideal complement to such grasses because of its high protein content. Therefore, nutritive value in general and digestibility in particular is a less important breeding objective than with grasses, and current attempts to improve forage quality of red clover largely focus on secondary plant metabolites. Red clover contains high concentrations of estrogenic compounds which are the basis of medications against hormonal disorders with women (Coon et al. 2007). However, when fed to ewes before mating, it can hinder conception severely. Breeding red clover for a low content of formononetine is the objective of specific breeding programs which has resulted in the release of cultivars with reduced contents, both in New Zealand and Europe (Rumball et al. 2005; Boller et al. 1994). More recently, interest has been raised for the enzyme polyphenol oxidase (PPO) contained in red clover and made responsible for the slower breakdown of red clover protein during ruminant digestion. Red clover with high levels of PPO might contribute to limit nitrogen losses from ruminant husbandry systems.

11.2.1.4 Seed Yield

Seed yield is a very important character for the market success of red clover cultivars. Intensive selection for forage characteristics seems to have hindered progress in seed yield and some cultivars with excellent agronomic features in terms of forage production are of little market importance because of their limited seed yield. This is particularly true of tetraploid cultivars which yield 20 to 50% less seed than diploid ones. Selection for seed yield is often carried out only at a late stage of the breeding process. Seed yield has also long been suspected to be negatively correlated to other important traits such as dry matter yield or persistence. However, a recent quantitative trait loci (QTL) analysis of seed yield and persistence in red clover failed to confirm this negative correlation and the results indicated that improvement of seed yield in red clover should be possible without adverse effects on persistence (Herrmann et al. 2008).

11.2.2 Breeding Achievements

Red clover was one of the first forage plants dealt with at the dawn of modern plant breeding around the turn from the 19th to the 20th century. For example, the impact of self-incompatibility on breeding techniques was discovered in red clover by one of the pioneers of a science-based plant breeding (Martinet 1903). The early importance of red clover in plant breeding is probably related to its prominent role in the historical development of farming systems in Europe, recognizing the potential of red clover to contribute to soil fertility long before the process of symbiotic nitrogen fixation had been understood. Nowadays, red clover is still the second most important forage legume in terms of cultivar breeding after lucerne, although its importance relative to white clover is decreasing. The current OECD list of cultivars eligible for seed certification (OECD 2007) contains 252 cultivars of red clover, compared to 129 in 1980. The respective figures for white clover are 182 in 2008 and 59 in 1980.

Considerable improvement of persistence and hence, yielding capacity over several years has been obtained by breeding programs in different parts of the world. The most prominent example is the development of highly persistent red clover cultivars at the Wisconsin Agricultural Experiment Station, resulting in an improvement of reliable stand duration from an initial two to four seasons after four decades of dedicated selection (Smith 2001). Similarly in Japan, targeted selection for over 20 yr, based on a combination of maternal line and individual plant selection resulted in cultivars with considerably improved persistence (Isobe et al. 2002). Great progress in persistence of early flowering, multiple-cut types of red clover was obtained by the use of Swiss Mattenklee (Boller 2000). This genetic resource, originating from local cultivars of cultivated clover in some regions of Switzerland, has been successfully used to create cultivars of outstanding longevity under temperate European conditions. This program is also a good example for the success of plant breeding in coping with new diseases. Resistance to southern anthracnose, which was clearly insufficient in existing Mattenklee cultivars, was markedly improved and new cultivars of this type are again as persistent as the old ones were before the disease became so prominent (Boller et al. 2004).

Substantial improvement in forage yield of red clover was obtained by inducing polyploidy. The first successful induction of polyploidy in red clover was reported during World War II (Levan 1940), and reproducible techniques became available in the 1950s (Brewbaker 1952). After colchicine treatment of young seedlings, stable tetraploids can be obtained after a few generations. Sexual polyploidization is also possible (Simioni et al. 2006) using unreduced pollen of diploid plants to fertilize plants of a previously colchicine-induced tetraploid variety. However, colchicine treatment of young seedlings remains the most frequently used method. Compared to their diploid ancestors, tetraploid cultivars yield significantly more forage dry matter. Disease resistance and persistence are also improved (Boller et al. 2003). However, seed yield of tetraploids is markedly lower. This is partly due to insufficient pollination because the larger flowers of tetraploids are less easily accessible to insects than those of diploids, but fertility of successfully pollinated tetraploids is also reduced. Nevertheless, tetraploid cultivars are often preferred by farmers due to their higher yielding ability and a higher seed price is accepted.

11.2.3 Limitations of Traditional Breeding and Rationale for Molecular Breeding

No studies on genetic gain in yielding potential of red clover have been published. Recent reviews of breeding progress in Europe (Abberton and Marshall 2005) and the United States (Taylor 2008) failed to identify appropriate references. However, the long lifetime of red clover cultivars supports the observation that breeding progress in red clover in terms of forage yield potential is rather slow. For example, the cultivar Mt. Calme, developed in the 1920s from an old landrace (Boller 2000) and tested successfully in independent, official tests for the first time between 1928 and 1932 (Neuweiler 1932), still persists on the very restrictive list of recommended cultivars for Switzerland and outyields newer recommended cultivars of short-lived field clover (Suter et al. 2006). Similarly, the cultivar Gumpensteiner, also an improved landrace, first released shortly after World War II and gradually improved until the early 1970s, is still the most used and best recommended red clover cultivar in Austria (Krautzer 2003). In the United States, the cultivar Arlington, registered in 1973 (Smith et al. 1973) still serves as check variety in Wisconsin and usually yields just marginally less than new releases in the first two years of stand (e.g., Smith 2001). Of the 129 red clover cultivars listed in the OECD list of 1981, 55 are still to be found on the edition of 2007/8.

The limited progress in improvement of yield potential is probably due to the multitude of stress factors breeders have to account for, preventing a more targeted selection for yield. Most conventional breeding efforts must be invested into maintenance of adequate persistence in response to newly developing strains of pathogens or changing abiotic pressures. Moreover, it seems difficult to maintain seed yield potential when selecting for these agronomic objectives. These limitations offer interesting opportunities for molecular breeding. In the first place, molecular tools may help improving resistance against diseases threatening plant survival. Markers to securely identify individuals which are homozygous for important resistance genes would greatly facilitate subsequent phenotypic selection for yield related, polygenic factors within populations with a genetically fixed resistance. Also, a targeted combination of several resistance genes in individuals or populations could be envisaged. Furthermore, markers for important traits conferring longevity per se would be very useful, because phenotypic selection for this characteristic is particularly time-consuming. Improving seed yield by conventional breeding is additionally hindered by the fact that phenotypic selection is only possible after seed set, making it impossible to select pre-flowering like with most forage production characteristics.

Targeted use of genetic resources for widening the genetic basis of breeding programs may also be aided by molecular breeding tools. Reliable information about the breeding potential of genetic resources existing in gene banks is scarce. Data on agronomic performance or specific traits are usually not available and cumbersome to obtain because seed has to be increased before proper field trials can be carried out. Therefore, molecular markers might be helpful to choose accessions with a large genetic distance to the breeding material in order to maximize the chances of adding new useful genetic variation to the breeding pool.

11.3 Genetic Diversity

11.3.1 Detecting Genetic Diversity in Red Clover

Genetic diversity is the prerequisite for any selection, either natural or human mediated. Therefore, plant breeders have long sought efficient means to characterize genetic diversity in red clover by means of morphophysiological traits as well as phenotypic, biochemical and molecular genetic markers. A large amount of genetic diversity among red clover accessions has been detected for many agronomically important traits such as dry matter yield, forage quality and resistance to diseases and pests (reviewed by Taylor and Quesenberry 1996). Naturally, phenotypic evaluations of genetic diversity in red clover largely focus on agronomically relevant traits such as dry matter yield (Dias et al. 2007) or resistance to diseases such as stemphylium leaf spot or *Fusarium oxysporum* (Berg and Leath 1996; Venuto et al. 1995). In addition, specific compounds such as isoflavones (Papadopoulos et al. 2006), morphological such as leaf dimensions or growth habit (Dias et al. 2007) or the absence of the characteristic leaf marks (Bortnem and Boe 2002) have been successfully used to describe genetic variability. Morphophysiological characteristics may allow estimation of genetic diversity based on target traits, but they are strongly influenced by the environment. Moreover, estimates of diversity are based on a relatively small number of traits and detection is often too cumbersome to allow for large scale investigations. Biochemical and molecular genetic markers on the other hand allow for detecting genetic diversity at a large number of loci, largely independent of the environment.

Isozyme polymorphism based on enzymes such as triose phosphate isomerase (TPI), alcohol dehydrogenase (ADH), glucose-6-phosphate dehydrogenase (GDH), peroxidase (PER), malate dehydrogenase (MDH), glucosephosphate isomerase (GPI), diaphorase (DIA), 6-phosphogluconate dehydrogenase (6- PGDH), malic enzyme (ME), b-galactosidase (b-GAL), leucine aminopeptidase (LAP), and fructose biphosphatase (FBP) have been extensively used to detect genetic variation and population structure in red clover (e.g., Hagen and Hamrick 1998; Mosjidis et al. 2004). Their detection does not require sophisticated equipment and they may be particularly useful to compare different taxa or populations since the detected loci are assumed to be largely homologous across species (Klaas 1998). Using five different enzyme systems, Malaviya et al. (2005) were able to show that intraspecific variability within the highly cultivated *T. pratense* was considerably lower, but still substantial, when compared to wild species such as *T. alpestre*, *T. hybridum* or *T. incarnatum*. Despite their advantages, isozyme polymorphisms are limited to a relatively small number of loci and the diversity detected may therefore not be representative for a large part of the red clover genome. With the development of novel molecular genetic markers, these tools have been continuously evaluated, adapted and applied by scientists working with red clover. While most studies focused on genomic DNA, Milligan (1991) reported unusually high levels of genetic variation in restriction digests of *T. pratense* chloroplast DNA. This investigation suggested that the plastid genome of this species was not as invariant as believed and instead showed high levels of diversity at the population levels. For analysis of the nuclear genome, hybridization based methods like restriction fragment length polymorphism (RFLP) analysis have never been widely used to characterize genetic diversity of red clover, probably due to the laborious method of detection associated with these methods. One exception in this respect is the study of Nelke et al. (1993) who used commercially available probes based on minisatellites (Jeffrey's probes) to successfully detect variation in a regenerative red clover somaclonal variant. On the other hand, PCR based randomly amplified polymorphic DNA (RAPD) analysis has been widely used to detect genetic diversity in various red clover germplasm collections (Campos-De-Quiroz and Ortega-Klose 2001; Kongkiatngam et al. 1995; Ulloa et al. 2003). In contrast to RAPD, amplified fragment length polymorphism (AFLP) suffer less from reproducibility problems and may therefore allow results to be transferred from one laboratory to another (Powell et al. 1996). Although technically more demanding, the technique is suitable to partial automation and is therefore highly suitable to analyze a large number of individuals from a large number of populations (Herrmann et al. 2005). With the development of a large number of genomic and gene associated simple sequence repeat markers (Kölliker et al. 2006; Sato et al. 2005) a highly efficient and informative marker system has become available. Dias et al. (2007) demonstrated the usefulness of SSR markers to characterize red clover germplasm and to assign individuals to their respective populations.

In addition to the choice of phenotypic traits or molecular genetic methods used, the number of individuals analyzed per cultivar, accession or population may also influence estimates of genetic diversity and structure. Outbreeding species such as red clover are characterized by large genetic diversity within populations. To account for genotypes or alleles which occur at a frequency of at least 10%, 40 plants have to be sampled, while 100 plants are needed to detect alleles occurring at a frequency of at least 5% (Crossa 1989). However, practical limitations often limit extensive sampling of populations and one to 48 individuals have in the past been used to characterize genetic diversity in red clover populations (Dias et al. 2007; Hagen and Hamrick 1998). Analysis of bulked leaf material offers an efficient way to genotype a larger number of individuals per population with reduced effort (Greene et al. 2004; Kölliker et al. 2001), but a loss of detection of rare alleles has to be taken into account. For AFLP analysis, markers present at frequencies of at least 20% are still detectable in bulked samples consisting of DNA from 20 plants. By analyzing two bulked samples per population (i.e., 40 individuals), Herrmann et al. (2005) were able to investigate genetic diversity in a large red clover germplasm collection.

11.3.2 Core Collections and Breeding Germplasm

In appreciation of the importance to preserve genetic diversity for plant improvement programs, germplasm collections have been established worldwide for many plant species. Since utilization of gene bank material is often hindered by the large size and heterogenous structure of many collections, the need for the establishment of core collections which represent the genetic diversity of a crop species was long realized (Frankel 1984). For red clover one of the most significant red clover germplasm collections was established through the US National Plant Germplasm System. In an attempt to characterize this collection, Kouamé and Quesenberry (1993) used cluster analysis and 15 morphophysiological plant descriptors of 463 accessions. They identified 10 major groups which grouped accessions according to different characteristics such as disease resistance or persistence and did not necessarily correspond to the country of origin. Based on their results, they suggested a core collection consisting of at least one accession from every country of origin and at least two accessions from each cluster group. This core collection consisting of 81 accessions from 41 different countries is characterized by a large amount of genetic diversity as detected by isozyme characterization (Mosjidis and Klingler 2006). As expected for an outbreeding species, most of the variation resided within populations and the amount of variation among populations was higher for wild populations (10.6%) when compared to cultivars (6.7%) or landraces (5.7 %). Although most accessions were clearly distinct, two pairs of identical accessions were identified and one accession was considered to be from another species (Mosjidis and Klingler 2006). In an attempt to further characterize the US core collection, Dias et al. (2007) analyzed 57 accessions using 21 morphological traits and seven SSR markers. Again, they discovered a large phenotypic and genetic diversity within populations. Based on phenotypic data, the accessions were separated into five distinct clusters mainly characterized by flowering date, persistency, dry matter production and growth habit. Surprisingly, although only one individual per accession was used for SSR analysis, some concordances between phenotypic and genotypic characterization were observed (Dias et al. 2007). This is a further indication that molecular markers can provide valuable additional information to phenotypic characterization.

Besides a well characterized core collection, detailed knowledge on genetic variability available in a particular breeding germplasm may be of equal importance. In a comparison of 12 Chilean advanced breeding populations with eight cultivars from Chile, Argentina, Uruguay and Switzerland, Ulloa et al. (2003) showed that the breeding populations formed a distinct group, while the Uruguayan and the Swiss cultivars were clearly separated. The reduced genetic diversity observed among some of the breeding populations suggested including genetically more divergent parents in future breeding efforts in order to broaden the breeding germplasm pool.

11.3.3 Landraces and Wild Populations

Landraces or farmers' varieties, as well as "wild" red clover populations occurring spontaneously in permanent grassland not only provide valuable resources for plant breeding and the conservation of genetic diversity, they may also allow to investigate the origin of cultivars. The latter could be of particular importance for plant breeders when identifying novel sources of genetic diversity. While it is generally accepted that the early center of domestication for red clover is located in Flanders and Brabant (northern Belgium, Taylor and Quesenberry 1996), the origin of wild red clover populations and their relationship to present day cultivars is largely unknown. In Switzerland, a particular form of red clover, characterized through improved persistence and known as Mattenklee, was developed through a decade long "on farm" seed production. Based on AFLP analysis, these Mattenklee landraces were shown to represent a distinct and highly diverse genetic resource of red clover (Kölliker et al. 2003). In addition, a comparison of 89 Mattenklee landraces with Swiss Mattenklee cultivars, Swiss wild clover populations, field clover cultivars from different countries and Dutch wild clover populations and landraces clearly showed that the ancestry of Swiss red clover landraces is primarily found in introduced cultivars rather than in natural wild clover populations (Herrmann et al. 2005). However, AFLP diversity among Swiss wild clover populations was significantly influenced by the distance to the closest Mattenklee landrace collection site, indicating gene flow between these two red clover groups which are often found in adjacent pastures or meadows. A similar relationship of local cultivars to wild populations was also observed in Russia, where seven populations from the Urals were clearly separated from seven Russian and two US cultivars based on isozyme analysis (Semerikov et al. 2002). The authors concluded that wild populations were autochthonous and not naturalized cultivars and all cultivars were of the same origin, probably western Europe. Due to their distinctness from cultivated accessions and their high genetic diversity (Herrmann et al. 2005; Semerikov et al. 2002), wild red clover populations may form a particularly interesting resource to broaden gene pools in red clover breeding programs especially since wild clover populations have been shown to be highly adapted to specific environments (Mosjidis et al. 2004).

11.3.4 Efficient Utilization of Genetic Diversity in Red Clover Breeding

The information available on variation in morphophysiological traits as well as biochemical and molecular genetic markers certainly assist plant breeders in their efforts to exploit genetic diversity for further improvement of red clover. However, estimates of genetic diversity based on anonymous genetic markers are often poorly correlated to the diversity of morphophysiological traits (Dias et al. 2007; Mosjidis et al. 2004). Therefore, a novel generation of molecular genetic markers directly linked to functional characteristics is needed to enable efficient utilization of genetic diversity. A valuable resource for the development of such markers is now also available for red clover (Sato et al. 2005). In addition, novel high throughput fingerprinting methods such as diversity array technology (Wenzl et al. 2004) will facilitate genotyping in the future.

11.4 Structural Genomics

11.4.1 Cytological Investigation

The nuclear DNA content of red clover was estimated as 0.97 pg/2C using chicken red blood cells as a standard (Arumuganathan and Earle 1991), and as 0.89–0.91 pg/2C using *Arabidopsis thaliana* as a standard (Sato et al. 2005). Based on the presumed nuclear DNA content, the deduced genome size was estimated to be approximately 440 Mb (Sato et al. 2005).

Cytological investigation of red clover began in the 1970s with the main aim of developing interspecific hybrids and polyploids (Taylor 2008). The karyotype of red clover was initially determined based on metaphase chromosomes, and it was determined that the largest chromosome featured a large satellite (Taylor and Chen 1988). Based on prometaphase chromosomes, a chromosome map was developed using DAPI staining and FISH analysis using 28S rDNA, 5S rDNA and BAC clone sequences as probes. All seven chromosomes could be distinguished individually (Fig. 11-3; Sato et al. 2005). Chr1 (LG5), featuring a large satellite, was

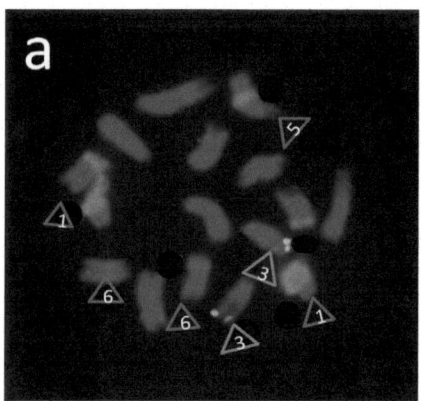

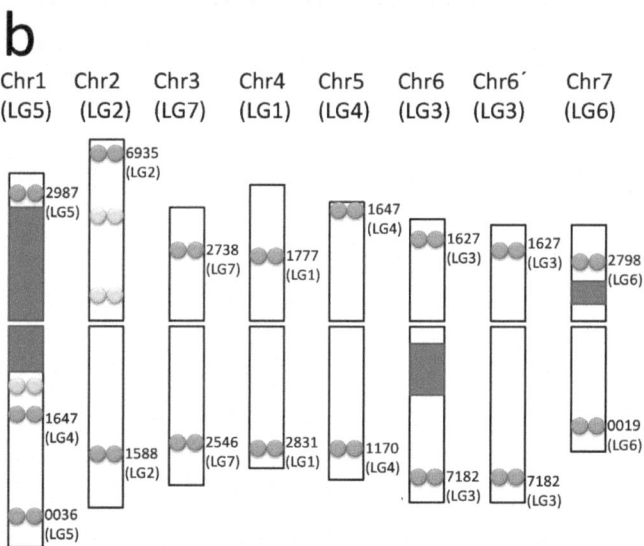

Figure 11-3 FISH analysis of red clover chromosomes. (A) FISH analysis with 28S rDNA (red signals indicated by red triangle) and a red clover BAC (RCS2546, green signals indicated by green triangle). Numbers on triangles indicate chromosome number. (B) Chromosome map of red clover. Green circles, loci of BACs corresponding to LG-specific markers; red boxes, 28S rDNA loci; yellow circles, 5S rDNA loci. Numbers represent BAC clones containing microsatellite sequences. The original figure and data were provided by Kataoka and Ohmido (Kobe University).

Color image of this figure appears in the color plate section at the end of the book.

hybridized with 28S rDNA, and produced the most intense FISH signal. The identified 28S rDNA region might correspond to a short genetic distance due to a lower degree of genetic recombination between tandem repeats (Ohmido et al. 2007).

11.4.2 Sequence Features and Structural Genomics

Based on 1920 sequence files from 960 plasmid clones from a random genomic library of red clover, a total of 244 (12.7%) sequences showed a high degree of similarity (E≤ 10^{-50}) to the chloroplast genome of *A. thaliana* (Sato et al. 2005). Only 14 (0.7%) sequences showed an equally high similarity to the *A. thaliana* mitochondrial genome. The average GC content of the genomes of nucleus, chloroplast and mitochondrion were 34.2, 33.8 and 42.9%, respectively. A total of 9339 non-redundant sequences originating from 26,356 ESTs were compared against the UniRef100 database, in an attempt to assign them putative functions. Seventy eight percent showed significant similarity (E <10^{-10}) to known genes, and most of these were derived from *A. thaliana* and rice.

A project to construct a robust physical map of red clover was recently funded under the EU ERANET Plant Genomic initiative (Skot et al. 2008). In this project, fingerprinting and BAC-end sequencing will be used to obtain 2000 BAC contigs, and to anchor these to the model legume species *Medicago truncatula* and *Lotus japonicus*. The information resource will be published along with the alignment to *M. truncatula*, and integrated with the alfalfa resource.

11.5 Genetic Linkage Mapping

11.5.1 Brief History of the Mapping Effort

A genetic linkage map is the most important fundamental genetic resource for molecular breeding. At present, three genetic linkage maps for red clover are available (Table 11-2). The populations used to create these maps were all derived from pseudo-test crosses between heterozygous parents. The first genetic linkage map for red clover was produced in 2003 (Isobe et al. 2003) using restriction fragment length polymorphism (RFLP) markers. Later, in 2005–2006, linkage maps consisting primarily of microsatellite markers (Sato et al. 2005) and amplified fragment length polymorphism (AFLP) markers (Herrmann et al. 2006) were developed.

Table 11-2 Summary of genetic linkage maps constructed in red clover.

Map parents	Population type and size	Marker type(s)	Number of loci	Number of likage groups	Total length (cm)	Mean distance (cM)	Publication
272 × WF1680	BC_1F_1 167	RFLP Morphology Total	157 1 158	7	535.7	3.4	Isobe et al. 2003
HR × R130	F_1 188	Microsatellite RFLP Total	1305 167 1472	7	868.7	0.6	Sato et al. 2005
nV × nC	F_1 254	AFLP Micro-satellit Total	216 42 258	7	444.2	1.7	Herrmann et al.

11.5.2 RFLP-based Map

Systems employing RFLP markers require more laborious methods of detection than other types of DNA markers. However, they can be used to identify co-dominant segregation, and the initial development costs are less than for other co-dominant markers. Taking these factors into consideration, RFLP markers were used for the construction of the first linkage map in red clover (Isobe et al. 2003). Markers were developed from red clover cDNA fragments, with the aim of developing gene-associated markers. The mapping population consisted of 167 progeny derived from "272" and "WF1680". "WF1680" originating from a variety cultivated in central Russia, which was characterized by very late flowering and a white flower color. "272" was a single F_1 plant derived from "1588" and "WF1680". "1588" originating from a wild accession collected in the Arhangelsk region (approximately N65°, E40°), and was characterized by early flowering and a bright red flower color.

A combined genetic linkage map containing 157 RFLP loci and one morphological locus (flower color) was obtained with seven linkage groups, a total map distance of 535.7 cM and an average distance between two loci of 3.4 cM. The donor parent ("272") specific markers were represented more frequently than the recurrent parent ("WF1680") specific markers. Distorted segregation was observed for 37% of the markers on the map.

11.5.3 Microsatellite-based Map

A high-density linkage map was developed by Sato et al. (2005) using microsatellite markers. To facilitate identification of simple sequence repeats (SSRs) in the red clover genome, four types of libraries were constructed; a SSR-enriched genomic library, a SSR-enriched cDNA library, a methylfiltration genomic library and a normalized cDNA library. A total of 83,172 clones were sequenced, then 7,244 primer pairs were designed to amplify SSRs in the sequences. Of the 7,244 resulting clones, 5,970 showed appreciable similarity to known genes. To integrate these results with the RFLP maps developed by Isobe et al. (2003), 121 cDNA probes were examined for the presence of RFLPs.

The genetic linkage map was constructed using a mapping population consisting of 188 F_1 progeny of "HR" and "R130". "HR" is an elite germplasm selected from the progeny of the Japanese cultivar "Hokuseki" and the Swiss cultivar "Renova". Characteristics of "HR" include early flowering, erect shape, red flower color and middle-sized leaves. "R130" is one of the progeny of a mapping population used in the development of the RFLP map described above (Isobe et al. 2003), which exhibits late flowering, semi-prostrate growth habit, white flower color and small leaves. The linkage analysis was performed using a combination of the color map method (Kiss et al. 1988) and JoinMap v. 3.0 (Van Ooijen and Voorrips 2001). The color-coded genotypes were displayed in a matrix for each parent and classified into seven groups representing individual chromosomes.

A total of 1434 loci (1286 microsatellite loci and 148 RFLP loci) were mapped on seven linkage groups. The total map distance was 868.7 cM and the average distance between two loci was 0.61 cM. A total of 105 loci (28%) were bi-parental, whereas 611 loci (43%) and 418 loci (29%) were HR and R130 specific, respectively. Highly distorted loci were mainly observed on LG1 and LG2. All marker information on this map is available at CloverGarden (http://clovergarden.jp). RFLP markers were used to determine the correspondence of each linkage group to the RFLP-based map, and FISH analysis using microsatellite markers as probes was used to determine the corresponding chromosome of each linkage group (Fig. 11-3B).

11.5.4 AFLP-based Map

An AFLP-based map in red clover was developed by Herrmann et al. (2007), with the aim of performing QTL analysis of seed yield components. This map also contained microsatellite markers developed by Kölliker et al. (2005) and Sato et al. (2005).

The AFLP-based genetic linkage map was constructed using a mapping population consisting of 254 F_1 progeny of a two-way pseudo-testcross between "pV" and "pC". "pV" was a genotype selected from the Belgian cultivar "Violetta", which is characterized by a high seed yield and low persistence. "pC" was a genotype selected from the Swiss Mattenklee cultivar "Corvus", which is characterized by low seed yield and long persistence.

The nine (*Eco* + 3/*Mse* + 3) primer combination yielded 5–19 polymorphic loci with a total of 117 loci, while 12 (*Pst* + 3/*Mse* + 2) primer combinations yielded 4-13 loci with a total of 99 loci. Of the 107 microsatellite loci tested, 42 exhibited polymorphisms. These 42 microsatellite loci together with 216 AFLPs were mapped on seven linkage groups, resulting in a total length of 444.2 cM and an average distance between two loci of 1.7 cM. The correspondence of each linkage group to the other two maps was determined, and a total of 58 loci (22.5%) were found to be bi-parental, while 200 loci (77.5%) were mono-parental. Distorted loci were frequently observed on LG6 and LG7.

11.5.5 Transferability of Markers among Red Clover Germplasms

The extent to which RFLP and microsatellite markers are transferable to other red clover germplasms, is an important consideration. Isobe et al. (2003) demonstrated the transferability of RFLP markers among 12 red clover individuals derived from three varieties bred in different countries. Of the cDNA probes tested, 87% detected polymorphic bands corresponding to those of the mapping parents.

Kölliker et al. (2005) analyzed the allelic variability of 27 microsatellite markers derived from genomic libraries with 24 red clover genotypes. The number of alleles per locus ranged from 2 to 25, and the mean value of the observed heterozygosity was 0.71. Similarly, Sato et al. (2005) analyzed the allelic variability of 268 microsatellite markers derived from 88 red clover genotypes (originating from 11 cultivars) and found the number of alleles per locus to be between two and 19. These results indicate that both RFLP and microsatellite markers are transferable among different red clover germplasms.

11.5.6 Comparative Mapping

Ellison et al. (2006) proposed a new infragenic classification of the genus *Trifolium* based on internal transcribed spacer sequences and chloroplast *trn*L intron sequences from nuclear ribosomal DNA. According to their results, the *Trifolium* genus can be classified into two subgenera, *Trifolium* and *Chronosemium*, with subgenus *Trifolium* containing eight sections. Although the two major forage species, red clover and white clover, were categorized into different sections, five of 27 or 181 of 599 microsatellite markers derived from red clover were successfully amplified using DNAs of white clover (*T. repense* L.) (Kölliker et al. 2005; Zhang et al. 2007). These results suggest that microsatellite markers are suitable for comparative mapping across *Trifolium* species.

Comparative mapping between red clover and white clover was reported by Zhang et al. (2006) and revealed putative macro-colinearity (for details, see chapter on white clover). The red clover sequences adjacent to mapped microsatellite markers were subjected to BLASTN searches against sequences of the two model legumes, *L. japonicus* and *M. truncatula* by Sato et al. (2005). The alignment between red clover (rc) LG1, *L. japonicus* (Lj) chr5 and *M. truncatula* (Mt) chr1 appears relatively simple. Moreover, Lj chr1 appeared to align to rc LG6-Mt chr7 and rc LG7-Mt chr3. However, the relationship of other red clover linkage groups to those of *L. japonicus* and *M. truncatula* seemed to be more complex at a larger level (macro synteny).

11.6 QTL Mapping and Molecular Breeding

11.6.1 QTL Mapping

Persistence is one of the most important breeding objectives for red clover because of its association with general adaptability and yield (Taylor et al. 2008). However, improving persistence is considered to be a difficult task, because it involves many factors, such as plant vigor, flowering characteristics, seed production, and biotic and abiotic stress tolerance. QTL mapping is expected to simplify persistence and connect to marker-assisted selection.

QTL analysis for persistence and seed yield was performed by Kölliker and colleagues (Herrmann et al. 2006; Herrmann et al. 2008) with the aim of elucidating any correlations between these components. A weighted average of vigor scores assessed over two winters and three growing season was identified as the optimal method to phenotype persistence. For this index, one QTL explaining 12.2% of the total phenotypic variation was identified. The same QTL was consistently detected for other persistence indices observed over varying time periods. For seed yield components, a total of 38 QTLs were identified which individually explained up to 33.4% of the phenotypic variation. Co-location of QTLs and correlation of phenotypic data did not indicate any negative correlation between seed yield and persistence. Thus, improvement of persistence should be possible without adverse effects on seed yield.

11.6.2 Gene Cloning and Transformation

The tissue culture protocols for red clover were initially developed by Phillips and Collins (1979) and Beach and Smith (1979). However, these methods yielded only low regeneration frequency and were genotype dependent. Quesenberry and Smith (1993) succeeded in increasing the regeneration efficiency of red clover tissue cultures using five cycles of recurrent phenotypic selection. Quesenberry et al. (1996) subsequently succeeded to generate transgenic red clover via *Agrobacterium tumefaciens*-mediated transformation. Despite this progress, no varieties of transgenic red clover have yet been developed. Sullivan et al. (2004) cloned three red clover polyphenol oxidase (PPO) genes, PPO1, PPO2 and PPO3, from a leaf cDNA library, and estimated expression of the active protein in *Escherichia coli* and transgenic alfalfa. The results demonstrated that transgenic alfalfa would be a suitable model system for characterization of cloned red clover genes.

References

Abberton MT, Marshall AH (2005) Progress in breeding perennial clovers for temperate agriculture. J Agri Sci 143: 117–135.

Abberton MT (2007) Interspecific hybridization in the genus *Trifolium*. Plant Breed 126: 337–342.

Akerberg E, Bingefors S, Lesins K (1947) Nagra aktuella problem inom förädlingen med rödklöver och lusern för Mellansverige. Sveriges Utsädesföreningen Tidskrift 57: 200–229.

Albrecht KA, Beauchemin KA (2003) Alfalfa and other perennial legume silage. In: Buxton DR et al. (eds) Silage Science and Technology. ASA-CSSA-SSSA, Madison, WI, pp 633–644.

Armstrong KC, Cleveland RW (1970) Hybrids of *Trifolium pratense* × *Trifolium pallidum*. Crop Sci 10: 354–357.

Arumuganathan K, Earle ED (1991) Nuclear DNA content of some important plant species. Plant Mol Biol Rep 9: 208–218.

Beach KH, Smith RR (1979) Plant regeneration from callus of red and crimson clover. Plant Sci Lett 15: 231–237.

Berg CC, Leath KT (1996) Responses of red clover cultivars to stemphylium leaf spot. Crop Sci 36: 71–73.

Bingefors S (1951) Studies on breeding red clover for resistance to stem nematodes. In: Osvald H (ed) Plant Husbandry. Almqvist & Wiksells Boktryckeri AB, Uppsala, Sweden Volume 8 of Växtodling.

Blomeyer A (1889) Die Cultur der landwirthschaftlichen Nutzpflanzen. Vol 1, Winter: Leipzig, Germany.

Boller B. (1994) Breeding red clover for a reduced content of formononetin. In: Reheul D, Ghesquiere A (eds) Breeding for Quality. Proceedings of the 19th Fodder Crops Section Meeting of EUCARPIA, RVP, Merelbeke, pp 187–191.

Boller B, Bigler P, Bucanovic I, Bänziger I (1998) Southern anthracnose—a new threat for red clover persistence in cooler regions. In: Boller B, Stadelmann FJ (eds) Breeding for a multifunctional agriculture. Proceedings of the 21st Meeting of the Fodder Crops and Amenity Grasses Section of EUCARPIA. FAL Reckenholz, Zürich, pp 195–197.

Boller B (2000) Altes und Neues vom schweizerischen Mattenklee, einer ausdauernden Form des Kultur-Rotklees. Vierteljahresschrift der Naturforschenden Gesellschaft in Zürich 145: 143–151.

Boller B, Schubiger F, Tanner P (2003) Kann der Biolandbau auf tetraploide Sorten von Rotklee und Raygräsern verzichten. In: Ruckenbauer P, Raab F, Kern R, Buchgraber K, Schaumberger A (eds) Bericht über die Arbeitstagung 2002 der Vereinigung der Pflanzenzüchter und Saatgutkaufleute Österreichs. BAL Gumpenstein, Irdning, Austria, pp 71–74.

Boller B, Tanner P, Schubiger F (2004) Merula und Pavo: neue, ausdauernde Mattenkleesorten Agrarforschung 11: 156–161.

Bortnem R, Boe A (2002) Frequency of the no mark leaflet allele in red clover. Crop Sci 42: 634–636.

Bowley SR, Taylor NL, Daugherty CT (1984) Physiology and morphology of red clover. Adv Agron 37: 317–347.

Bradshaw EG, Rasmussen P, Nielsen H, Anderson NJ (2005) Mid- to late-Holocene land-use change and lake development at Dallund Sø, Denmark: vegetation and land-use history inferred from pollen data. Holocene 15: 1116–1129.

Brewbaker JL (1952) Colchicine induction of tetraploids in Trifolium species. Agron J 44: 592–594.

Broderick GA, Walgenbach RP, Sterrenburg E (2000) Performance of lactating dairy cows fed alfalfa or red clover silage as the sole forage. J Dairy Sci 83: 1543–1551.

Broderick GA, Walgenbach RP, Maignan S (2001) Production of lactating dairy cows fed alfalfa or red clover silage at equal dry matter or crude protein contents in the diet. J Dairy Sci 84: 1728–1737.

Burdine KH, Eldridge RW, Trimble R (2005) The economics of renovating pastures with clover. Univ. of KY, Agricultural Economics Extension, Lexington, KY. No. 2005–04.

Bush L, Roberts CA, Schultz C (2007) Plant chemistry and antiquality components in forages. In: Barnes RF, Nelson CJ, Moore KJ, Collins M (ed) Forages: The Science of Grassland Agriculture. Vol. II, 6 ed, Balckwell Publ, Ames, IA, pp 509–528.

Cagle AJ (2001) The spatial structure of Kom el-Hisn: An old kingdom town in the Western Nile Delta, Egypt. PhD Diss, Univ of Washington, Seattle.

Campos-De-Quiroz H, Ortega-Klose F (2001) Genetic variability among elite red clover (*Trifolium pratense* L.) parents used in Chile as revealed by RAPD markers. Euphytica 122: 61–67.

Chen S, Wyse DL, Johnson GA, Porter PM, Stetina SR, Miller DR, Betts KJ, Klossner LD, Haar MJ (2006) Effect of cover crops alfalfa, red clover, and perennial ryegrass on soybean cyst nematode population and soybean and corn yields in Minnesota. Crop Sci 46: 1890–1897.

Choi HK, Mun JH, Kim DJ, Zhu H, Baek JM, Mudge J, Roe B, Ellis N, Doyle J, Kiss GB, Young ND, Cook DR (2004) Estimating genome conservation between crop and model legume species. Proc Natl Acad Sci USA 101: 15289–15294.

Cockayne TO (1961) Leechdoms, wortcunning and starcraft of early England Vol. I, II, and III. Holland Press, London.

Collins GB, Taylor NL, Phillips GC (1983) Successful hybridization of red clover with perennial *Trifolium* species via embryo rescue. Proc Int Grass Congr 14: 168–170.

Coon JT, Pittler MH, Ernst E (2007) *Trifolium pratense* isoflavones in the treatment of menopausal hot flushes: a systematic review and meta-analysis. Phytomedicine 14: 153–159.

Crossa J (1989) Methodologies for estimating the sample-size required for genetic conservation of outbreeding crops. Theor Appl Genet 77: 153–161.

Dias PMB, Julier B, Sampoux JP, Barre P, Dall'Agnol M (2007) Genetic diversity in red clover (*Trifolium pratense* L.) revealed by morphological and microsatellite (SSR) markers. Euphytica 160: 189–205.

Dixon GR, Doodson JK (1974) Techniques for Testing Resistance of Red-Clover Cultivars to *Sclerotinia-trifoliorum* Erikss (Clover Rot). Euphytica 23: 671–679.

Dixon RA (2004) Phytoestrogens. Annu Rev Plant Biol 55: 225–261.

Ellison NW, Liston A, Steiner JJ, Williams WM, Taylor NL (2006) Molecular phylogenetics of the clover genus (*Trifolium-Leguminosae*). Mol Phylogenet Evol 39: 688–705.

EU (2007) Agriculture in the European Union—statistical and economic information 2006. European Union.

Evans AM (1962) Species hybridization in Trifolium I. methods of overcoming species incompatibility. Euphytica 11: 256–262.

Evans AM, Denward T (1955) Grafting and hybridization experiment in the genus *Trifolium*. Nature 175: 687–688.

Fergus EN, Hollowell EA (1960) Red Clover. Adv Agron 12: 365–436.

Frandsen KJ (1946) Studier over *Sclerotinia trifoliorum* Eriksson. Danske Forlag, Kopenhagen, Denmark, 220 p.

Frankel OH (1984) Genetic perspectives of germplasm conservation. In: Arber WK et al. (eds) Genetic manipulation: Impact on man and society. Cambridge University Press, Cambridge, pp 161–170.

Freudenthaler P, Kainz W, Schantl S, Dachler M, Hackl G, Holaus K, Pelzmann H, Koller B, Scherenzel P (1998) Index seminum austriae—Kulturpflanzenevolution und erhaltung pflanzengenetischer ressourcen in der Landwirtschaft. Wien: AV-Druck pp 80.

Gerard J (1597) The herbal or Generall historie of plantes. Gathered by Iohn Gerarde of London Master in Chirurgerie. Imprinted at London.

Greene SL, Gritsenko M, Vandemark G (2004) Relating morphologic and RAPD marker variation to collection site environment in wild populations of red clover (*Trifolium pratense* L.). Genet Resour Crop Evol 51: 643–653.

Hagen MJ, Hamrick JL (1998) Genetic variation and population genetic structure in *Trifolium pratense*. J Hered 89: 178–181.

Herrmann D, Boller B, Widmer F, Kölliker R (2005) Optimization of bulked AFLP analysis and its application for exploring diversity of natural and cultivated populations of red clover. Genome 48: 474–486.

Herrmann D, Boller B, Widmer F, Kölliker R (2006) QTL analysis of seed yield components in red clover (*Trifolium pratense* L.) Theor Appl Genet 112: 536–545.

Herrmann D, Boller B, Studer B, Widmer F, Kölliker R (2008) Improving persistence in red clover: Insights from otl analysis and comparative phenotypic evaluation. Crop Sci 48: 269–277.

Hodgson JG, Halstead P, Wilson PJ, Davis S (1999) Functional interpretation of archaeobotanical data: making hay in the archaeological record. Veg Hist Archaeobot 8: 261–271.

Hoffman PC, Broderick GA (2001) Red clover forages for lactating dairy cows. Focus on Forage 3 (11) Wisc. Team Forage, Univ. of Wisc. Ext., Madison, WI.

Hopcroft RL (2003) Local institutions and rural development in European history. Soc Sci Hist 27: 25–74.

Iepema G, Van Eekeren N, Van Dongen M (2006) Production and persistency of red clover (*Trifolium pratense*) varieties when grown in mixtures. In: Lloveras J et al. (eds) Grassland Sci Europe 11: 388–390.

Isobe S, Sawai A, Yamaguchi H, Gau M, Uchiyama K (2002) Breeding potential of the backcross progenies of a hybrid between *Trifolium medium* × *T. pratense* to *T. pratense*. Can J Plant Sci 82: 395–399.

Isobe S, Gau M, Yamaguchi H, Uchiyama K, Maki Y, Matsu-ura M, Ueda S, Sawai A, Tsutsumi M, Takeda Y, Nakashima K (2002) Breeding of red clover 'Natsuyu' and its characteristics. National Agricultural Research Center for Hokkaido Region Research report 177: 1–13.

Isobe S, Klimenko I, Ivashuta S, Gau M, Kozlov NN (2003) First RFLP linkage map of red clover (*Trifolium pratense* L.) based on cDNA probes and its transferability to other red clover germplasms. Theor Apple Genet 108: 105–112.

Jessen C (1982) Alberti Magni ex ordine praedicatorum. De Vegetabilius Libri VII, historiae naturalis pars XVIII. Editionem crticam ab Ernesto Meyero coeptam. Minerva GMBH, Frankfurt/Main.

Kiss GB, Kereszt A, Kiss P, Endre G (1998) Colormapping: a non-mathematical procedure for genetic mapping. Acta Biol Hung 49: 125–142.

Klaas M (1998) Applications and impact of molecular markers on evolutionary and diversity studies in Allium. Plant Breed 117: 297–305.

Knorek J, Staton M (1996) Red clover. Online at The Cover Crops Program KBS/MSU website. Michigan State Univ- Ext, East Lansing, MI.

Kölliker R, Jones ES, Jahufer MZ, Forster JW (2001) Bulked AFLP analysis for the assessment of genetic diversity in white clover (*Trifolium repens* L.). Euphytica 121: 305–315.

Kölliker R, Herrmann D, Boller B, Widmer F (2003) Swiss Mattenklee landraces, a distinct and diverse genetic resource of red clover (*Trifolium pratense* L.). Theor Appl Genet 107: 306–315.

Kölliker R, Enkerli J, Widmer F (2005) Characterization of novel microsatellite loci for red clover (*Trifolium pratense* L.) from enriched genomic libraries. Mol Ecol Notes 6: 50–53.

Kongkiatngam P, Waterway MJ, Fortin MG, Coulman BE (1995) Genetic variation within and between two cultivars of red clover (*Trifolium pratense* L.)—Comparisons of morphological, isozyme, and RAPD markers. Euphytica 84: 237–246.

Kouamé CN, Quesenberry KH (1993) Cluster analysis of a world collection of red clover germplasm. Genet Resour Crop Evol 40: 39–47.

Krautzer B (2003) Entwicklung und erhaltung standortgerechter gräser und leguminosen für die grünlandwirtschaft und den landschaftsbau im Alpenraum. Abschlussbericht Projektnummer 2923, BAL Gumpenstein, A-8952 Irdning, pp 25.

Lacefield G, Henning J and Rasnake M (1997) Renovating Hay and Pasture Fields. AG-26, Univ of KY-Ext., Lexington, KY.

Lacefield G, Ball D (1999) Red Clover. Circular 99-1. Oregon Clover Commission, Salem, OR, USA.

Lane C (1980) The Development of pastures and meadows during the sixteenth and seventeenth centuries. Agri Hist Rev 28: 18–30.

Leath KT, Byers RA (1973) Attractiveness of diseased red-clover roots to clover root borer. Phytopathology 63: 428–431.

Leep RH (1989) Improving pastures in Michigan by frost seeding. Ext Bul, Univ of Michigan-Ext, East Lansing, MI, E-2185.

Levan A (1940) Framställning av tetraploid rödklöver. Sveriges Utsädesföreningen Tidskrift 50: 115–124.

Malaviya DR, Kumar B, Roy AK, Kaushal P, Tiwari A (2005) Estimation of variability of five enzyme systems among wild and cultivated species of *Trifolium*. Genet Resour Crop Evol 52: 967–976.

Martinet G (1903) Etudes et essais de plantes fourragères. Annuaire agricole de la Suisse 4: 161–169.

Marum P, Smith RR, Grau CR (1994) Development of procedures to identify red-clover resistant to *Sclerotinia-Trifoliorum*. Euphytica 77: 257–261.

Merkenschlager F (1934a) Die konstitution des rotklees. Die ernährung der pflanze 30: 81–89.

Merkenschlager F (1934b) Migration and distribution of red clover in Europe. Herb Rev 2: 88–92.

Merker A (1984) Hybrids between *Trifolium medium* and *Trifolium pratense*. Hereditas 101: 267–268.

Merker A (1988) Amphidiploids between *Trifolium alpestre* and *Trifolium pratense*. Hereditas 108: 267.

Michell AR (1974) Sir Richard Weston and the spread of clover cultivation. Agri Hist Rev 22: 160–161.

Milligan BG (1991) Chloroplast DNA diversity within and among populations of *Trifolium pratense*. Curr Genet 19: 411–416.

Mosjidis JA, Klingler KA (2006) Genetic diversity in the core subset of the US red clover germplasm. Crop Sci 46: 758–762.

Mosjidis JA, Greene SL, Klinger KA, Afonin A (2004) Isozyme diversity in wild red clover populations from the caucasus. Crop Sci 44: 665–670.

NASS (1889–2007) USDA-National Agricultural Statistic Service. Washington DC.

Nedelnik J (1992) Comparison of greenhouse resistance of *Trifolium pratense* to fungi of the genus Fusarium Link Ex Fr with persistence in field conditions. Rostlinna Vyroba 38: 395–398.

Nelke M, Nowak J, Wright JM, McLean NL (1993) DNA fingerprinting of red clover (*Trifolium pratense* L.) with Jeffrey's probes—detection of somaclonal variation and other applications. Plant Cell Rep 13: 72–78.

Neuweiler E (1932) Anbauversuche mit Rotklee. Landwirtschaftliches Jahrbuch der Schweiz 35: 50–65.

OECD (2007) OECD List of varieties eligible for certification 2007/2008. Ohmido N, Sato S, Tabata S, Fukui K (2007) Chromosome maps of legumes. Chromosome Res 15: 97–103.

Papadopoulos YA, Tsao R, McRae, KB, Mellish AE, Fillmore SAE (2006) Genetic variability of principal isoflavones in red clover. Can J Plant Sci 86: 1345–1347.

Philips GC, Collins GB (1979) *In vitro* tissue culture of selected legumes and plant regeneration from callus culture of red clover. Crop Sci 19: 59–64.

Phillips GC, Collins GB, Taylor NL (1982) Interspecific hybridization of red clover (*Trifolium pratense* L.) with T. sarosiense Hazsl. using *in vitro* embryo rescue. Theor Appl Genet 62: 17–24.

Phillips GC, Grosser JW, Berger S, Taylor NL, Collins GB (1992) Interspecific hybridization between red clover and *Trifolium alpestre* using *in vitro* embryo rescue. Crop Sci 32: 1113–1115.

Pieters AJ, Hollowell EA (1937) Clover improvement. In: Yearbook of Agriculture. USDA, Washington DC, pp 1190–1214.

Powell W, Machray GC, Provan J (1996) Polymorphism revealed by simple sequence repeats. Trends Plant Sci 1: 215–22.

Quesenberry KH (1975) Interspecific Hybridization of Perennial *Trifolium* Species related to Red Clover. PhD Diss, Univ of Kent, Lexington, KY.

Quesenberry, KH, Baltensperger DD, Dunn RA, Wilcox CJ, Hardy SR (1989) Selection for tolerance to root knot nematodes in red clover. Crop Sci 29: 62–65.

Quesenberry KH, Smith RR (1993) Recurrent selection for plant regeneration from red clover tissue culture. Crop Sci 33: 585–589.

Quesenberry KH, Wofford DS, Smith RL, Krottje PA, Tcacenco F (1996) Production of red clover transgenic for neomycin phosphotransferase II using *Agrobacterium*. Crop Sci 36: 1045–1048.

Quesenberry KH, Blount AR (2006) Southern Belle and Cherokee Red Clover in Florida. SS-AGR-40, Univ of Florida IFAS Ex., Gainesville, FL.

Rufelt S (1985) Selection for Fusarium root rot resistance in red clover. Annals of Appl Biol 107: 529–534.

Rumball W, Keogh RG, Sparks GA (2005) 'Grasslands Hf1' red clover (*Trifolium pratense* L.)—A cultivar bred for isoflavone content. New Zealand J Agri Res 48: 345–347.

Sato S, Isobe S, Asamizu E, Ohmido N, Kataoka R, Nakamura Y, Kaneko T, Sakurai N, Okumura K, Klimenko I, Sasamoto S, Wada T, Watanabe A, Kothari M, Fujishiro T, Tabata S (2005) Comprehensive structural analysis of the genome of red clover (*Trifolium pratense* L.). DNA Res 12: 301–364.

Savoie P, Jofriet JC (2003) Silage storage. In: Buxton DR, Muck RE, Harrison JH (ed) Silage Science and Technology. ASA-CSSA-SSSA, Madison, WI, pp 405–467.

Sawai A, Ueda S (1987) Embryo development of the hybrid of *Trifolium medium* L. × 4x *Trifolium pratense* L. J Japan Soc Grassl Sci 33: 157–162.

Sawai A, Ueda S, Gau M, Uchiyama K (1990) Interspecific hybrids of *Trifolium medium* L. × 4x *Trifolium pratense* L. obtained through embryo culture. J Japan Soc Grassl Sci 35: 267–272.

Sawai A, Yamaguchi H, Uchiyama K (1995) Fertility and morphology of the chromosome doubled hybrid *Trifolium medium* ×*T. pratense* (red clover) and backcross progeny. Grassl Sci 41: 122–127.

Schubiger, FX, Streckeisen P, Boller B (2003) Resistance to southern anthracnose (*Colletotrichum trifolii*) in cultivars of red clover (*Trifolium pratense*). Czech J Genet Plant Breed 39: 309–312.

Schwer JF, Cleveland RW (1972) Tetraploid and triploid interspecific hybrids of *Trifolium pratense* L., *T. diffusum* Ehrh., and some related species. Crop Sci 12: 419–422.

Semerikov VL, Belyaev AY, Lascoux M (2002) The origin of Russian cultivars of red clover (*Trifolium pratense* L.) and their genetic relationships to wild populations in the Urals. Theor Appl Genet 106: 127–132.

Simioni C, Schifino-Wittmann MT, Dall'agnol M (2006) Sexual polyploidization in red clover. Sci Agricola 63: 26–31.

Skot L, Abberton M, Donnison I, Oldroyd G, Geurts R, Mayer K, Kudrna D (2008) Using translational genomics to underoub germplasm improvement for complex traits in crop legumes. Proc Plant Anim Genome XVI Conf, San Diego, CA, 375 p.

Smith D (1965) Forage production of red clover and alfalfa under differential cutting. Agron J 57: 463–465.

Smith RR, Maxwell DP, Hanson EW, Smith WK (1973) Registration of Arlington Red Clover. Crop Sci 13: 771.

Smith RR, Taylor NL, Bowley SR (1985) Red clover. In: Clover Science and Technology. Monograph No. 25, ASA-CSSA-SSSA, Madison, WI.

Smith RR (2000) Red clover in the "twenty-first" century. In: Proc. of the 24th Forage Prod. and Use Symp. Wisc. forage council.

Smith RR (2001) Breeding for abiotic and biotic stress in perennial *Trifolium*. In: Monjardino P et al. (eds) Breeding for Stress Tolerance in Fodder Crops and Amenity Grasses. 23rd Meeting of the Fodder Crops and Amenity Grasses Section of EUCARPIA, pp 13–19.

Stam P, Van Ooijen JW (1995) JoinMap Version 2.0 Software for the calculation of genetic linkage maps. CPRO-DLO, Wageningen.

Sullivan P (2003) Overview of cover crops and green manures. Appropriate Technology Transfer for Rural Areas, National Center for Appropriate Technology, Fayetteville, AR, USA, pp 16.

Sullivan ML, Hatfield RD, Thoma SL, Samac DA (2004) Cloning and characterization of red clover polyphenol oxidase cDNAs and expression of active protein in *Echerichia coli* and transgenic alfalfa. Plant Physiol 136: 3234–3244.

Sullivan ML, Hatfield RD (2006) Polyphenol oxidase and o-diphenols inhibit postharvest proteolysis in red clover and alfalfa. Crop Sci 46: 662–670.

Suter D, Briner H, Jeangros B, Mosimann E, Bertossa M (2006) Liste der empfohlenen Sorten von Futterpflanzen 2007–2008. Agrarforschung 13, I–XVI.

Taylor NL, Stroube WH, Collins GB, Kendall WA (1963) Interspecific hybridization of red clover (*Trifolium pratense* L.). Crop Sci 3: 549–552.

Taylor NL, Collins GB, Cornelius PL, Pitcock J. (1983) Differential interspecific compatibilies among genotypes of *Trifolium sarosiense* and *T. pratense*. Proc Int Grassl Congr 14: 165–168.

Taylor NL, Smith RR, Anderson JA (1990) Selection in red clover for resistance to Northern Anthracnose. Crop Sci 30: 390–393.

Taylor NL, Quesenberry KH (1996) Red Clover Science. Kluwer Academic Publishers, Dordrecht.

Taylor NL, Henning JC, Lacefield GD (1997) Growing red clover in Kentucky. AGR-33, Univ. of KY-Ext., Lexington, KY.

Taylor NL (2008) A century of clover breeding developments in the United States. Crop Sci 48: 1–13.

Taylor, NL, Chen K (1988) Isolation of trisomics from crosses of diploid, triploid, and tetraploid red clover. Crop Sci 28: 209–213.

Ulloa O, Ortega F, Campos H (2003) Analysis of genetic diversity in red clover (*Trifolium pratense* L.) breeding populations as revealed by RAPD genetic markers. Genome 46: 529–535.

Undersander DJ, Smith RR, Kelling K, Doll J, Worf G, Wedberg J, Peters J, Hoffman P, Shaver R (1990) Red clover: Establishment, management, and utilization. A3492, Univ of Wisc-Ext, Madison, WI.

Van Ooijen JW, Voorrips RE (2001) JoinMap 3.0 software for the calculation of genetic linkage maps. Kyazma B.V, Wageningen.

Venuto BC, Smith RR, Grau CR (1995) Virulence, legume host-specificity, and genetic relatedness of isolates of *Fusarium oxysporum* from red clover. Plant Dis 79: 406–410.

Venuto BC, Smith RR, Grau CR (1999) Selection for resistance to Fusarium Wilt in red clover. Can J Plant Sci 79: 351–356.

Wenzl P, Carling J, Kudrna D, Jaccoud D, Huttner E, Kleinhofs A, Kilian A (2004) Diversity Arrays Technology (DArT) for whole-genome profiling of barley. Proc Natl Acad Sci USA 101: 9915–9920.

Westgate JM, Hillman FH (1911) Red clover. USDA Farmers Bull 455: 1–48.

Wheaton HN (1993) Red clover. Univ of Missouri Ext. Columbia, MO, G4638.

Wheaton HN, Roberts CA (1993) Renovating grass sods with legumes. Univ. of Misso.-Ext. Columbia, MO, USA, G4651.

Williams RD (1927) Red clover investigations, 1919–1926. Welsh Plant Breed. Stn Series H 7: 1–137.

Zeven AC (1991) Four hundred years of cultivation of Dutch white clover landraces. Euphytica 54: 93–99.

Zeven AC, De Wet JMT (1982) Dictionary of cultivated plants and their regions of diversity. International Book Distributors, Wageningen, pp 259.

Zhang Y, Sledge MK, Bouton JH (2007) Genome mapping of white clover (*Trifolium repens* L.) and comparative analysis within the Trifolieae using cross-species SSR markers. Theor Appl Genet 114: 1367–1378.

Zohary M, Heller D (1984) The Genus *Trifolium*. The Israel Academy of Science and Humanities, Jerusalem.

12

White Clover

John W. Forster,[1,3,4,]* *Noel O.I. Cogan*[1,4] *and Michael T. Abberton*[2]

ABSTRACT

White clover is a valuable legume component of temperate pasture-based agricultural systems supporting grazing industries for milk, meat and fibre production, generally through co-cultivation with a companion grass. The agronomic value of this species is limited by factors such as seed yield, and also a number of critical biotic and abiotic stresses. Efficient breeding systems based on exploitation of population diversity for elite cultivar development have been established, but will be highly augmented through molecular breeding. Biotechnology applications for white clover improvement are dependent on development of significant species-specific DNA sequence resources, which has been achieved by recent genomics activities. Comparative genomics with model legume species has also provided valuable information for gene identification and selection. First-generation genetic map construction and trait-dissection analysis has identified targets for marker-assisted selection, while also elucidating the sub-genome structure of this allotetraploid species. Provision of a large number of candidate genes will also assist

[1]Department of Environment and Primary Industries, Biosciences Research Division, AgriBio, the Centre for AgriBioscience, 5 Ring, Bundoora, Victoria 3083, Australia.
[2]International Institute of Tropical Agriculture, PMB 5320, Ibadan, Oyo State, Nigeria.
e-mail: *michael.abberton@cgiar.org*
[3]La Trobe University, Bundoora, Victoria 3086, Australia.
[4]Molecular Plant Breeding and Dairy Futures Cooperative Research Centres, Australia.
*Corresponding author: *john.forster@depi.vic.gov.au*

functional analysis. The current status of white clover genomics and molecular breeding is reviewed, in the context of relevant biological factors.

Keywords: Legume, Allotetraploid, Genomics, Transcriptomics, Genetic marker, Molecular breeding

12.1 Biology of White Clover

12.1.1 Taxonomy

White clover, *Trifolium repens* L., is a member of the Trifolieae tribe of the cool-season Galegoid clade in the Papilinoideae sub-family of the legume family Fabaceae (Doyle and Luckow 2003). The most closely related genus is *Melilotus* (sweet clovers), and the genus *Medicago*, including alfalfa (*Medicago sativa* L.) is also part of the Trifolieae. The clover genus *Trifolium* includes more than 250 species, of which 10 are of considerable agricultural importance (Zohary and Heller 1984). White clover (*T. repens* L.) is located in section Lotoidea of the *Trifolium* genus, along with the other perennial species *T. hybridum* L. (alsike clover) and *T. ambiguum* M. Bieb (kura clover or Caucasian clover). Centres of diversity for clovers have been identified in the eastern Mediterranean, East Africa and South America (Zohary and Heller 1984). Recent work supports a Mediterranean origin of the genus in the Early Miocene period (Ellison et al. 2006).

White clover is an allotetraploid taxon ($2n = 4x = 32$) with a fundamental chromosome number of 8. Molecular cytogenetic studies based on fluorescence *in situ* hybridization (FISH) have provided partial characterization of the white clover karyotype (Ansari et al. 1999). Molecular phylogenetics analysis based on cpDNA and nuclear ribosomal DNA internal transcribed spacer (nrDNA ITS) data identified the diploid species *T. occidentale* D.E. Coombe (Western clover) and *T. pallescens* Schreber as the contemporary taxa most closely related to the putative diploid progenitor species of white clover (Ellison et al. 2006). The taxonomy and biosystematics of white clover were reviewed by Williams (1987).

12.1.2 Economic Significance

White clover is the most widely grown temperate forage legume, and the most common in pastures grazed by sheep or cattle (Laidlaw and Teuber 2001). Estimates have been made of annual global white clover sowings totalling 3–4 Mha (Mather et al. 1996). The largest levels of usage in Europe are in more northerly and westerly parts of the continent, but it is difficult to obtain reliable wide-scale estimates of clover content for either recently reseeded or for more established pastures. A broad range of aspects relating to the use of white clover in New Zealand (NZ) were reviewed in Woodfield (1996). In this volume, Caradus et al. (1996) estimated the total financial contribution of white clover to the NZ economy in the mid-1990s, comprised of indirect (nitrogen [N] fixation, increased forage yield) and direct (seed, honey) components, to be New Zealand (NZ) $3 billion, largely due to global marketing of cultivars (Mather et al. 1996). This value provides an estimate of equivalent value to other isoclimatic regions.

12.1.3 Vegetative Morphogenesis

The agronomy of white clover and performance in mixtures with grasses has been previously reviewed (Frame and Newbould 1986; Soegaard 1994; Pederson 1995a; Frame et al. 1998; Laidlaw and Teuber 2001). The major distinguishing feature of this species is a stoloniferous habit (spreading by means of stolons, or horizontal stems, and hence providing many active growing points). Immediately following germination, white clover development is based on a large tap root and stolon establishment. After approximately 18 mon, the tap root is lost and the plant then transitions to a network of stolons, adventitious roots being formed from stolon nodes. Much of the persistence of the plant, and consequent tolerance to defoliation and other stresses, is linked to effectiveness of the stolon network as a storage reserve, growth mediator and an anchor to the soil surface. Survival of an effective network and maintenance of stolon carbohydrate reserves are crucial for competition with the companion grass in early spring. Development of a strong stolon network is a prerequisite for persistence, and has hence been a major focus of breeding efforts (e.g., Caradus and Chapman 1996; Collins et al. 1997). For development of new white clover varieties, likely on-farm use is an important determinant of the management regime used in selection, and also to some extent of the program objectives. These requirements are in turn highly dependent on leaf size, which provides a classification system for white clover morphotypes, such that small-leaved types are considered suitable for continuous hard sheep grazing, medium-leaved types are used under rotational grazing, and large or very large-leaved cultivars are mainly used for lax cattle grazing or conservation. Productivity is greater for larger-leaved varieties, but in general this trait is negatively correlated with persistency. For instance, large-leaved Ladino white clovers are usually considered to be less persistent, although considerable variation for a range of traits exists within this morphotype (Annicchiarico 1993). Breaking of the negative correlation between productivity and persistency has been a major goal for breeders, and has been achieved for some notable varieties (Woodfield and Caradus 1994).

The effect of grazing management on persistence is of considerable importance, and numerous studies have described performance differences between cultivars, ecotypes and breeding lines (e.g., Brock and Hay 1996; Mayne et al. 2000). Numerous comparisons of performance under different cutting and grazing regimes have been made (Swift et al. 1992), and persistence has also been a routine target for direct selection (e.g., Evans et al. 1996). Long term trials (over 10 or more years) have demonstrated that many modern varieties fail to show marked reduction of clover content in mixed swards, irrespective of the level of applied N (Williams et al. 2003). Such cultivars are capable of significant contributions to highly productive systems with relatively low outputs, as compared to solely low input/ output systems (Williams et al. 2000). A general review of white clover management in mixed swards has been performed by Frame and Laidlaw (1998).

Differences in the responses of grasses and clovers to temperature and light competition play major roles (Robin et al. 1994) but below-ground interactions are also likely to be important (Collins and Rhodes 1994). Some modelling-based approaches to grass/clover interactions have emphasized the effects of N build-up and consequent feedback-inhibition of N fixation on cycling over time of relative grass and clover yields. In practice, a largely empirical selection approach has been implemented based on performance over a number of years, sometimes with a range of different companion varieties, in mixed plots under cutting or a management regime more closely simulating on-farm usage (Evans and Williams 1987; Evans et al. 1995).

12.1.4 Biological Nitrogen Fixation

Several independent estimates of the amount of nitrogen fixed by white clover have been obtained. For example, Ledgard et al. (1999) reported values of 99–231 kg N/ha/yr from dairy farmlets in NZ. Considerable dependence on fixed N can be observed for white clover in mixed swards, even in the presence of high mineral levels, although the relative contribution of fixed N may be reduced under grazing, possibly due to mineralized N contributions from animal returns. The major variable appears to be the sward clover content rather than fixation-related processes per se. This observation may partially explain the reason for relatively limited, up to date, attempts to breed for improved fixation. One contributory factor is the difficulty of implementing an efficient screening system for large numbers of plants. The transfer of fixed nitrogen from clover to the companion grass has been shown to be an important factor for maintenance of total sward performance. McNeill and Wood (1996) estimated the annual N fixation by white clover in the United Kingdom (UK) to be 155 kgN/ha, of which c. 28% was transferred to the ryegrass companion in mixed swards (in turn representing c. 29% of the total grass N content). Transfer of fixed N can occur through decomposition of subterranean structures, nutrient cycling mediated by root herbivory (Ayres et al. 2007), and through animal excreta. Little is known of the former process, which is difficult to quantify. Hence, although varietal differences have been noted (Laidlaw et al. 1996), the decomposition trait has proved difficult to define in terms of selection criteria, and so has not been significantly incorporated into plant improvement programs. Greater mechanistic knowledge of the transfer process and underlying plant traits is required (Hogh-Jensen 2006). When high N fertilizer levels are typically applied, use of grass-clover mixtures can be beneficial in terms of reducing nitrate leaching into water-courses (Saarijarvi et al. 2007), but increased nitrogen use efficiency in the mixed sward represents an important objective for breeding and management. There is growing evidence that white clover cultivation can have a beneficial effect on soil quality and that, relative to perennial ryegrass, a white clover sward results in a more finely textured and less-compacted soil (Mytton et al.1993; Holtham et al. 2007).

12.1.5 *Forage Nutritional Content*

Incorporation of a substantial forage legume component, both grazed and ensiled, in the diet can enhance meat and milk quality. Lee et al. (2003) demonstrated complex interactions between composition of ingested feed and rumen activity with respect to the transformation of polyunsaturated fatty acids (PUFAS). Nonetheless, the difference between a forage (and particularly clover)-rich diet and concentrates in terms of enhanced beneficial PUFAS levels in meat and milk is now well established. Many of the important quality and health effects of forage legumes involve pathways of secondary metabolism (reviewed by Dixon and Sumner 2003).

12.1.6 **Biotic Stresses**

The major pests of white clover in UK pastures are slugs, *Sitona* weevil and stem nematode (*Ditylenchus dipsaci*). Slugs can cause considerable damage, particularly following re-seeding or over-sowing, and are generally controlled in such circumstances by pellets. Little research has been directed at breeding for slug or weevil resistance in Europe. In contrast, considerable efforts have been made to improve resistance to stem nematode (*Ditylenchus dipsaci*) (Williams et al. 2007). Although definitive evidence is not available for the effects of this pest on reduced yield or persistence, a number of field-based and experimental observations suggest that nematode infestation can significantly reduce performance, and may occasionally cause more pronounced plant number decline. In NZ and Australia, attention has focused on clover cyst nematode, and resistant varieties have been produced (Mercer et al. 2000a). However, *Sitona* weevil has emerged as the major threat to white clover in NZ pastures in recent years (Gerard et al. 2007).

White clover mosaic virus (WCMV) infection is a major cause of yield reduction in NZ pastures, while AMV is the most important viral pathogen in Australia, but the importance of viral diseases in other regions is not so well-defined. Barnett and Gibson (1975) found 37% of pastures sampled in the south-eastern USA to be infected with peanut stunt virus (PSV) and 74% to be infected with AMV, while 25% were also infected with the soybean dwarf virus (SbDV) (Damsteegt et al. 1999). Of these viruses, PSV is thought to be the most responsible for stand declines, frequency of infected plants increasing in linear fashion with pasture age (McLaughlin et al. 1992).

With respect to fungal pathogens, *Fusarium* root rot has been the focus of greenhouse- and field-based evaluation, particularly in Canada (Coulman and Lambert 1995). In Europe, *Sclerotinia trifoliorum* is probably the main fungal disease threat and resistance screens have been developed.

12.1.7 Abiotic Stresses

Lack of winterhardiness, encompassing the direct effects of low temperature, snow cover and dehydration due to desiccating winds, can contribute to poor white clover performance. This trait has been a major focus of white clover breeding programs in the UK (Helgadottir 1997). White clover breeding for European markets in NZ has also emphasized the winter-hardiness trait (Caradus and Christie 1998). Selection for cold tolerance has employed an *in situ* approach, in which survivors from field populations subjected to reliable cold stress have been used in crossing programs, or artificial selection in freezing tanks or a combination of both methods (Collins et al. 2002). Biochemical characterization of metabolic components associated with cold tolerance has been carried out for white clover and other forage species. Vegetative storage proteins (Goulas et al. 2001) and proline have been implicated in improved winter hardiness. Survivor genotypes from white clover populations grown at cold European sites were found to have higher levels of unsaturated fatty acids in stolon tissue than in the founder populations, suggesting adaptive significance in cold climates (Collins et al. 2002). Low temperature interacts with other stresses affecting plant survival such as grazing pressure, and successful efforts to produce material with enhanced on-farm winterhardiness have incorporated this complexity into their selection regimes (Rhodes et al. 1994). The adaptive behaviour of white clover, including over-wintering and spring growth, can be successfully modelled (Wachendorf et al. 2001).

On acid soils, productivity of forage legumes may be inhibited by limited aluminium (Al) tolerance. Selections for trait enhancement have been made in the field (Caradus et al. 2001) and using artificial tests (Voigt and Staley 2004), with limited success. Phosphorus (P) requirements for good clover performance are relatively high, and selection has been performed, particularly in NZ, for improved yield at lower levels of P fertilization (Caradus 1994). P pollution from farmland is widely regarded as a growing environmental problem, which is likely to be addressed by regulation in many parts of the world (e.g., the EU Water Framework Directive—http://www.euwfd.com/index.html). Selection for growth at low P will probably increase in importance, and several approaches in addition to direct selection may prove viable. Extent of inorganic phosphorus incorporation into organic compounds shows genetic variation (Caradus et al. 1998) and may be an important selection criterion for future studies. The symbiosis between white clover and arbuscular mycorrhizal fungi (AMF), or vesicular arbuscular mycorrhizae (VAM) has been investigated in the context of increased P uptake efficiency (e.g., Crush 1995; Rogers et al. 2000) although the role and importance of this association on fertilized agricultural soils is unclear.

Although the genetic heterogeneity characteristic of natural white clover populations provides considerable variation for improvement of many traits, there are some desirable attributes where this is not the case. For example, only limited variability is present for drought tolerance, a trait which has long been a major breeding objective in more arid areas (such as regions of Australia and NZ) and is of increasing importance elsewhere due to global climate change. Barbour et al. (1996) demonstrated differences between 10 cultivars with respect to their response to water stress, while Brink and Pederson (1998) detected little variation in response to a water gradient between six populations (three cultivars and three germplasm accessions). Field studies have also shown that drought, in combination with other stresses and influenced by management, can have marked effects on plant survival, and these effects may differentiate between plant populations (Jahufer et al. 1995). Direct selection for drought tolerance has been carried out in the field, and indirect methods have also been used, but success has been limited.

White clover is regarded as a salt-stress sensitive species, but variation does exist within and between cultivars and natural populations (Rogers et al. 1993). Progress has been made in breeding for saline tolerance through selection of divergent lines from the Israeli cultivar Haifa which have high and low shoot concentrations of chloride (Cl⁻) (Rogers et al. 1997).

Clover plants are also relatively sensitive to the effects of ozone (O_3) (Heagle et al. 1989), but selection for resistant genotypes has been achieved (Heagle et al. 1991).

12.1.8 Reproductive Biology

In white clover, the numbers of ripe inflorescences is the main component of, and closely correlated with, seed yield (SY) (Jahufer and Gawler 2000) and significant genotypic variation for reproductive traits has been detected (Finne et al. 2000; Jahufer and Gawler 2000). Inflorescences are produced at nodes on the developing stolon, but not all such nodes are reproductive, and a balance between reproductive nodes and nodes that develop secondary stolons is important to maintain the persistency of a variety. Annicchiarico et al. (1999) reported that persistence (as predicted by stolon density) was negatively correlated with SY and dry matter (DM) yield. Selection of traits which improve SY without impairing agronomic performance has been explored. Marshall (1995) demonstrated that selection for peduncle (flower stalk) strength improved inflorescence survival and significantly increased SY.

12.2 Genetic Diversity and Breeding Systems

12.2.1 Genetic Diversity

White clover is also an obligate outbreeding species, with a gametophytic self-incompatibility (SI) system controlled by a series of alleles at a single locus (*S*) (Attwood 1940; Attwood 1941; Attwood 1942a). Rare instances of self-compatibility have been reported (Attwood 1942b; Yamada et al. 1989), presumably due to the presence of self-fertile (S_f) alleles at the SI locus. The SI system ensures a high degree of heterogeneity within and between both natural and synthetic populations of white clover.

Genetic diversity at the population level has been analyzed using a diversity of polymorphism detection assays. Random amplification of polymorphic DNA (RAPD) marker systems have been used to analyze variation within and between white clover populations of permanent pastures in north-eastern USA (Gustine and Huff 1999). A high degree of genetic variation was detected within populations, with each of 18 populations showing significant differences but no consistent differences between the three states of origin. Similar studies were performed using collections from multiple geographical locations and several varieties, revealing a high level of similarity and a probable common European origin (Gustine et al. 2002). RAPDs have also been used for characterization of

Chinese collections (Zhang et al. 2010) and to evaluate genetic distance between inbred lines (Joyce et al. 1997).

Amplified fragment length polymorphism (AFLP) profiling (Kölliker et al. 2001b; van Treuren et al. 2005) has also been used to determine levels of genetic variability within and between white clover populations. As observed for other allogamous pasture species such as perennial ryegrass (Guthridge et al. 2001), the majority of genetic variation was detected within rather than between populations, and divergent varieties were largely discriminated on the basis of AFLP profile (Kölliker et al. 2001b). Bulking at the genotypic level followed by AFLP analysis was used to determine the level of congruence between morpho-physiological and genotypic variation in white clover.

Simple sequence repeat (SSR) markers have provided the highest degree of resolution to date in genetic diversity studies. A study of 16 elite varieties using genomic DNA-derived SSR markers (Kölliker et al. 2001a) revealed very high levels of intrapopulation diversity and overlap between populations, despite small but significant interpopulation differentiation (George et al. 2006). Complex relationships were also obtained from a study of 24 globally distributed cultivars (Jin et al. 2010), although a level of clustering on the basis of geographical and morphological similarity was observed.

12.2.2 Breeding Systems

Targets and progress in white clover breeding have been reviewed for different regions of the world (Lane et al. 1997; Quesenberry and Casler 2001; Jahufer et al. 2002; Abberton and Marshall 2005; Williams et al. 2007; Taylor 2008). Known genetic traits were reviewed by Quesenberry et al. (1991), and subsequent studies have demonstrated the genetic basis for traits such as red leaf, which is controlled by a single dominant allele (Pederson 1995b), and pink flower colour, which is controlled by homozygous alleles at two gene loci (Pederson and McLaughlin 1995). However, the majority of target traits in white clover, as for other forage species, show continuous variation, with potential for genotype x environment interaction, and are hence likely to be controlled by multiple quantitative trait loci (QTLs) at disparate chromosomal locations.

Methodology for pasture plant breeding, especially in commercial settings, has been based on variants of a common process, which initiates with mass selection (as spaced plants in the field) through evaluation of large-scale base populations. This step is generally followed by sub-selection of a group of potential parental clones, which undergo further assessment. Determination of parental breeding value may be obtained through the use of a number of experimental methods (Vogel and Pedersen 1993), of which the most direct is phenotypic selection. Indirect methods include half-sib progeny testing (HSPT), which obtains information on the general combining ability (GCA) of a parental genotype, while full-sib progeny testing (FPST) measures specific combining ability (SCA). Each method displays distinct merits and demerits in terms of derived information on genetic variance, time-scale and costs.

Phenotypic recurrent selection within populations, based on visual inspection, has provided the primary method to date for white clover breeding (Woodfield and Brummer 2001). For instance, resistances to clover cyst nematode and clover root-knot nematode were established through recurrent selection over five and seven cycles, respectively, based population sizes of 1000 in each cycle (Mercer et al. 1999; Mercer et al. 2000b). However, rates of genetic gain have historically been slow (0.6–1.5% per annum for traits such as yield, stolon density and nitrogen fixation: Woodfield and Brummer 2001), suggesting that advanced breeding methods are required to accelerate the process. Schemes for inter-population recurrent selection, such as reciprocal recurrent selection (RRS) have been proposed as effective means to exploit heterosis (Brummer 1999).

When relevant genetic diversity has been absent from the primary gene pool of white clover, interspecific hybridization methods have been used to introduce new variation. Crosses between white clover and the diploid taxon *T. nigrescens* Viv. have been used to transfer genes for clover cyst nematode resistance and capacity for increased seed production (Hussain et al. 1997; Marshall et al. 1998). Hybridization with the tetraploid *T. ambiguum*, through use of a series of polyploidy bridges (Hussain and Williams 1997), has been successful in transferring traits such as rhizome formation to white clover (Abberton et al. 1998).

12.3 Structural Genomic Resources

12.3.1 EST and Genomic Libraries

High-throughput white clover gene discovery was initially performed by sequencing of expressed sequence tags (ESTs) from 16 cDNA libraries corresponding to a broad range of plant organs, developmental stages and environmental conditions (Sawbridge et al. 2003). A collection of 42,017 white clover ESTs was generated, defining 15,989 unigenes. Each of the sequences was annotated by comparison to GenBank and SwissProt public sequence databases and automated intermediate Gene ontology (GO) annotation was obtained (Spangenberg et al. 2005). All sequences and annotation are currently maintained within components of the customized Bioinformatics and Advanced Scientific Computing (BASC) database system

(Erwin et al. 2007). An EnsEMBL genome viewer was incorporated for comparison with the complete *A. thaliana* genome sequence, draft sequences of the model legume species *Medicago truncatula* Gaertn and *Lotus japonicus* L., and expressed sequences from related legumes.

More recently, a white clover cDNA library was constructed from mRNA of aerial seedling tissue, followed by 5'-biased sequencing of 15,401 clones. Clustering subsequently defined 5,508 unigenes, as a resource for genetic marker development (Hisano et al. 2007). A further 14,527 cDNA sequences enriched for tissue-specific expression through suppression subtractive hybridization (SSH) were obtained as part of the Pastoral Genomics (PG) proprietary genome sequence database (Jones et al. 2006). This database also contains 364,537 genomic sequences enriched for genic regions by methylfiltration based on GeneThresher™ (GT) technology (Rabinowicz et al. 1999), which were assembled into 86,643 contigs; 4,360 white clover genomic DNA sequences, 59,851 ESTs; and 1,841 SSH-derived cDNA sequences from the diploid clover species *T. occidentale* (Jones et al. 2006). The PG database has also provided a resource for targeted genetic marker development (Griffiths et al. 2006).

12.3.2 BAC libraries

Complementing the EST and genomic sequence resources, large insert DNA libraries have been generated using bacterial artificial chromosome (BAC) vectors. A white clover BAC library was constructed from multiple plants of the variety Grasslands Nui by insertion of *Hind*III-generated partial DNA fragments into the pBeloBACII vector, and consists of 50,302 clones with an average insert size of 101 kb, corresponding to 6.3 genome equivalents (Spangenberg et al. 2005). Screening of this library has been performed using both macroarray hybridization and PCR analysis of microtitre plate pools. A second BAC library was constructed from a single genotype of the white clover mapping family parent R3R4 derived from the Institute for Grassland and Environmental Research (IGER), Aberystwyth, UK, using the vector pIndigoBAC-5. The library contains 37,248 clones with an average insert size of c. 85 kb, representing c. 3-fold genome coverage (Febrer et al. 2007).

The identification of sub-genome-specific genomic sequence features (see further ahead) permitted identification of paired homoeologous BACs for specific gene loci (Hand et al. 2009, 2010), and BAC sequence determination and assembly will provide the means for investigation of subgenome-specific differential gene expression (Adams et al. 2004).

12.4 Systems Biology

12.4.1 Transcriptomics

First-generation high-density cDNA microarrays representing approximately 15,000 unique genes for white clover (Sawbridge et al. 2003) were developed through robotics-based nanolitre volume spotting technology. These microarrays were applied in hybridizations with labelled total RNA isolated from a variety of genotypes, plant organs, developmental stages and growth conditions and pathogen infection states (Spangenberg et al. 2001; Rhodes 2003; Spangenberg et al. 2005; Webster et al. 2005). However, spotted arrays are vulnerable to fabrication problems due to irregular feature structure and variation between replicated slides, as well as limited sensitivity and inability to discriminate between closely-related members of multigene families. The latter issues are likely to be particularly acute for white clover, due to the whole-genome duplication inherent in allotetraploid genome architecture. The second-generation of white clover microarrays was hence based on oligonucleotide arrays. Methodologies for custom fabrication of such arrays have been developed through the use of a specially-modified complementary metal oxide semiconductor, which digitally directs highly parallel phosphoramidite-based *in situ* DNA synthesis (Maurer et al. 2006). Commercial development of this technology through the CombiMatrix (www.combimatrix.com) CustomArray™ synthesizer system permitted fabrication of arrays with 12,000 features, each containing a homogeneous population of probes 35–40 nucleotides in length, followed by a high-density format with 90,000 features per array. A CombiMatrix array for white clover was constructed containing 11,980 unigene sequences, of which 84% were sequence annotated (R.C. Chapman pers. comm.), along with probes representing the genomes of the major viral pathogens WCMV, AMV and clover yellow vein virus (CYVV). The array was validated using a range of treatments including WCMV infection, and reliable re-use of arrays for up to three additional hybridizations was demonstrated. The nature of genome-wide transcriptional changes following WCMV infection was addressed through annotation and validation of modulated features (E.J. Ludlow, pers. comm.). Global gene expression differences have also been identified in the context of proanthocyanidin biosynthesis pathways during six successive stages of flower development (Abeynayake et al. 2007; Abeynayake et al. 2010). A group of c. 5,000 modulated features included members of the MYB, MYC/bHLH and WD-40 transcription factor gene families.

In addition to microarray, cDNA-AFLP has been used to evaluate changes in gene expression during acclimation to cold temperatures, obtaining a large number of modulated features for each contrasted treatment. Cloning and sequencing of selected bands revealed sequence similarity to an *A. thaliana* vernalization control gene (Lowe et al. 2005).

The increasing resolution provided by second-generation DNA sequencing platforms provides an "open" system for expression profiling through transcriptome analysis (Torres et al. 2007; Blow 2009), which will supplant "closed" microarray systems in future applications. Combined with knowledge of sub-genome-specific sequence variation, this approach is likely to be of high value for allopolyploid species like white clover.

12.4.2 Proteomics

Relatively few proteomic studies have been performed for white clover. Proteomics of leaf senescence demonstrated the importance of nitrogen remobilization facilitated by proteolysis and chloroplast degradation (Wilson et al. 2002). However, a much broader range of studies have evaluated changes in protein profiles in a more targeted manner, either in response to an imposed stress or at a particular developmental stage.

In terms of protein metabolism, a major focus has been on study of the vegetative storage proteins (VSPs), which are implicated in nitrogen storage and over-wintering (Goulas et al. 2001). Other representative studies have analyzed changes in protein profiles in response to drought (Lee et al. 2007), phosphate starvation stress (Hunter et al. 1999) and ozone exposure (Tang et al. 1999).

12.4.3 Metabolomics

As for proteomic studies, white clover is yet to be the subject of metabolomics studies in the stringent sense of the term. Nonetheless, as for many other legumes species, white clover displays a complex and diverse secondary metabolism (Carlsen and Fomsgaard 2008). In particular, aspects of flavonoid metabolism important for interaction with *Rhizobia* and other micro-organisms are of considerable importance, and have been the focus of many studies (e.g., Carlsen et al. 2002). The spatial-temporal profile of proanthocyanidin biosynthesis has been assessed in different organs, tissues, cells and cell compartments of flowers at different developmental stages (Abeynayake et al. 2010). Another key issue is the biological activity of metabolites during their passage through the rumen and bloodstream of grazing livestock, which may be important for animal physiology and also quality and flavour of derived meat and milk (Schreurs et al. 2007).

12.5 Molecular Genetic Marker Technology

12.5.1 SSR Marker Development

The development of molecular genetic markers for white clover has been reviewed by Forster et al. (2001, 2008). A comprehensive set of unique white clover genomic DNA-derived simple sequence repeat (TRSSR) markers was developed using enrichment library construction technology (Kölliker et al. 2001a). For this purpose, 1,123 clones from a genomic library enriched for (CA)$_n$ repeats by a modification of the method of Edwards et al. (1996) yielded 793 SSR loci. Exclusion of redundant and truncated sequence provided 397 unique loci suitable for genetic analysis, of which 117 were initially characterized for amplification efficiency and polymorphism rate, and a total of c. 250 have been further characterized to date. Cross-species amplification was assessed for a sub-set of primer pairs, revealing substantial transfer to closely related taxa, but limited transfer to other genera within the Trifolieae tribe, such as *Medicago* and *Melilotus*.

The white clover EST library (Sawbridge et al. 2003) was also used to develop EST-SSR primer pairs (Barrett et al. 2004). A total of 2,086 SSR loci were identified in 26,480 sequence accessions. A sub-set of 792 primer pairs were evaluated for amplification efficiency and polymorphism detection within or between the parental genotypes of a two-way pseudo-testcross mapping family (F$_1$ [Sustain 6525-2 x NRS 364-7]). Of these, 566 reliably produced amplicons, and 379 detected allelic variation. In addition to the EST-SSR primer pairs, a total of 107 genomic DNA-derived SSRs were obtained from enrichment libraries, of which 30 provided efficiently amplified polymorphic products.

More recently, the fast isolation by AFLP of sequences containing repeats (FIASCO) technique (Zane et al. 2002) was used to isolate SSR-containing clones through deployment of (CA)$_n$ and (ATG)$_n$ motif-containing probes (Zhang et al. 2008a). Clone sequences were assembled into 1,698 contigs, of which 32% currently represent novel sequences based on BLASTN analysis of GenBank. The majority of identified sequences matched the capture motifs, although other simple and compound SSR types were also isolated. Primer pairs were designed to match 859 unique sequences, of which 191 were evaluated for amplification efficiency (92%) and polymorphism detection (66%) in the F$_1$ (GA43 x SRVR) two-way pseudo-testcross mapping family. Full sequence information has been made available through GenBank submission.

12.5.2 SNP Marker Development

Although SSR markers have proved valuable for framework genetic map construction and DNA profiling studies in white clover, single nucleotide polymorphism (SNP) markers are likely to prove more valuable for high-density mapping and marker implementation activities. Functionally-associated SNPs derived from candidate genes (Andersen and Lübberstedt 2003) have been proposed to be of particularly high value for obligate outbreeding species such as white clover (Dobrowolski and Forster 2007; Forster et al. 2008), due to limitations in the use of linked markers and desirability of diagnostic markers for selection of superior allele content (Sorrells and Wilson 1997).

Candidate gene selection in white clover may be achieved through functional annotation of homologous genomic resources, such as EST and GT libraries, and through translational genomics from related species, especially model legumes. The model legume species *M. truncatula* shares a common ancestor relatively recently in evolutionary time with white clover. Translational genomics based on whole genome sequencing of *M. truncatula* (Young et al. 2005; Zhu et al. 2005) is consequently anticipated to be highly efficient for members of the *Trifolium* genus. The other model legume species, *Lotus japonicus*, is also a Galegoid legume located in a separate tribe (Loteae). Relevant data may also be obtained through the whole genome sequence of soybean (*Glycine max* L.), which is located in the corresponding Phaseoloid clade of the Papilinoideae, and, more distantly, from *Arabidopsis thaliana*.

Methods for candidate gene-associated SNP discovery in white clover have been implemented through a two-part strategy. An *in silico* discovery component (Buetow et al. 1999; Picoult-Newberg et al. 1999) was based on computational identification of predicted SNPs in EST sequence contigs, while the *in vitro* discovery component was based on cloning and sequencing (Zhang and Hewitt 2003) of gene-specific amplicons derived from the heterozygous parents of two-way pseudo-testcross mapping families (Cogan et al. 2006b).

For the *in silico* strategy, analysis was based on EST contigs derived from cDNAs generated from multiple heterogeneous individuals of the white clover variety cultivar Grasslands Huia (Sawbridge et al. 2003). This approach is expected to sample moderate allelic complexity in candidate genes, due to the relatively non-restricted base of this population (Kölliker et al. 2001b). However, the white clover EST resource was generated from a limited number of genotypes and hence may fail to capture significant proportions of SNP variation, including diversity relevant to specific germplasm, especially in specific candidate genes. A sample of 236 EST contigs was selected for validation of predicted polymorphisms using the parents and progeny of the two-way pseudo-testcross mapping families F_1 (Haifa$_2$ x LCL$_2$) and F_1(S184$_6$ x LCL$_6$) (Cogan et al. 2007). A total of 106 of the clusters (45%) were initially identified, and more detailed analysis confirmed 58 clusters as containing validated segregating SNPs. The SNP-containing genes belong to a range of predicted functional categories, including ribosomal proteins, heat-shock proteins, calmodulins and lipid and organic acid biosynthesis enzymes. Bioinformatic alignment of validated SNP-containing EST contigs with model legume genome drafts identified a small proportion (10%) matching *L. japonicus* sequence, but a larger proportion (55%) matched *M. truncatula* genomic regions, attributable to each of the eight chromosomes of this species, hence providing resources for computational comparative genetics.

The *in vitro* SNP discovery process for outbreeding forage species was initially developed for the diploid grass perennial ryegrass (*Lolium perenne* L.). Although this method is costly and time-consuming compared to direct sequencing of PCR products, it provides several important advantages. Ambiguities in sequence traces due to heterozygous indels may be readily distinguished, and haplotype structures within amplicons are directly determined (Cogan et al. 2006b). In addition, paralogous sequences amplified by conserved primers may be discriminated, and used to distinguish between members of multigene families (Edwards et al. 2007). For white clover, *in vitro* SNP discovery is expected to be influenced by the allopolyploid genetic constitution of this species. Previous studies of allopolyploids have been largely restricted to inbreeding crops like bread wheat (Caldwell et al. 2004), for which SNP discovery in single homozygous genotypes in concert with aneuploid lines can provide confident locus-specific assignment. White clover, in contrast, exhibits high levels of intragenotype heterozygosity (Kölliker et al. 2001b; Jones et al. 2003) and intrapopulation genetic diversity (George et al. 2006), and aneuploid lines for one-step chromosome assignment are not available. The combination of variability between paralogous, homoeologous and homologous sequences is hence likely to complicate *in vitro* SNP discovery and identification of both homoeologous sequence variants (HSVs) and paralogous sequence variants (PSVs).

In a preliminary study (Lawless et al. 2008), a total of 43 white clover cDNAs were selected from public databases and from the EST unigene resource, including genes for flavonoid biosynthesis (relevant to bloat safety) and organic acid biosynthesis (relevant to aluminium tolerance and phosphorus acquisition). The F_1 (Haifa$_2$ x LCL$_2$) and F1(S184$_6$ x LCL$_6$) families provided the parental DNA templates for *in vitro* SNP discovery. DNA sequence suitable for SNP discovery was obtained from multiple amplicons of 35 genes, corresponding to total of 29.4 kb of consensus resequenced genomic DNA, at an average of 840 bp per template gene. High levels of haplotypic complexity were observed for the majority of template genes, at levels in excess of those predicted on the basis of homologous sequence amplification. Elevated rates of failed validation for predicted SNPs were also observed (82% compared to 25% in perennial ryegrass, across comparable sample sizes). Nonetheless, individual SNP loci suitable for genetic mapping were validated for selected genes. Although higher levels of intragenomic paralogy could account for the lower efficiency of *in vitro* SNP discovery in white clover compared to perennial ryegrass, homoeologous gene amplification (Cronn and Wendel 1998) is a more likely explanation.

Improved methods were required to improve the efficiency of *in vitro* SNP discovery, and to allow discrimination of genome- and gene-specific nucleotide variation sequences. The identification of the contemporary taxa *T. occidentale* and *T. pallescens* as putative diploid progenitor species (Ellison et al. 2006) provided a method based on production and subtraction of related sequence haplotypes. To exemplify this approach, 9 ESTs corresponding to genes associated with tolerance to abiotic stresses (drought and salinity) were selected, including members of the dehydration –responsive element binding (DREB) and late embrogenesis abundant protein (LEA) gene families. A total of 7, 290 bp of genomic DNA was resequenced across the target genes from the parental genotypes of the F_1 (Haifa$_2$ x LCL$_2$) and F1(S184$_6$ x LCL$_6$) genetic mapping families and single genotypes of both *T. occidentale* and *T. pallescens* (Hand et al. 2008). White clover-derived sequence haplotypes were separated into two putative sub-genome-specific clusters, one of which closely resembled the sequences from *T. occidentale*, the other being more distantly related to both *T. occidentale* and *T. pallescens*. The two subgenomes were hence designated O and P', respectively. Detailed sequence comparison and phenetic clustering analysis confirmed the divergent status of *T. pallescens* and suggested that a third, as yet uncharacterized taxon contributed the P' subgenome. Resolution of homoeologous sequence clusters permitted assessment of SNP frequency within the subgenomes (1 per 87 bp and 1 per 97 bp, respectively), and an HSV frequency of 1 per 27 bp. A large proportion of SNPs colocated with HSVs, and in majority of instances (88%), the coincident SNP-HSV shared the same nucleotide variant, providing a partial explanation for the very high rate of failed SNP validation observed in the absence of this information. The validation rate was restored to levels comparable with those obtained for perennial ryegrass (61%). Identification of two distinct copies related to the *Tr*LEAa EST also permitted resolution of PSVs at a frequency of 1 per 14 bp. Complementary SNPs for both sub-genomes were assigned to the relevant homoeologous linkage groups (Fig. 12-1).

Progenitor comparison provided a highly improved system for *in vitro* SNP discovery in white clover, at least for specific mapping family

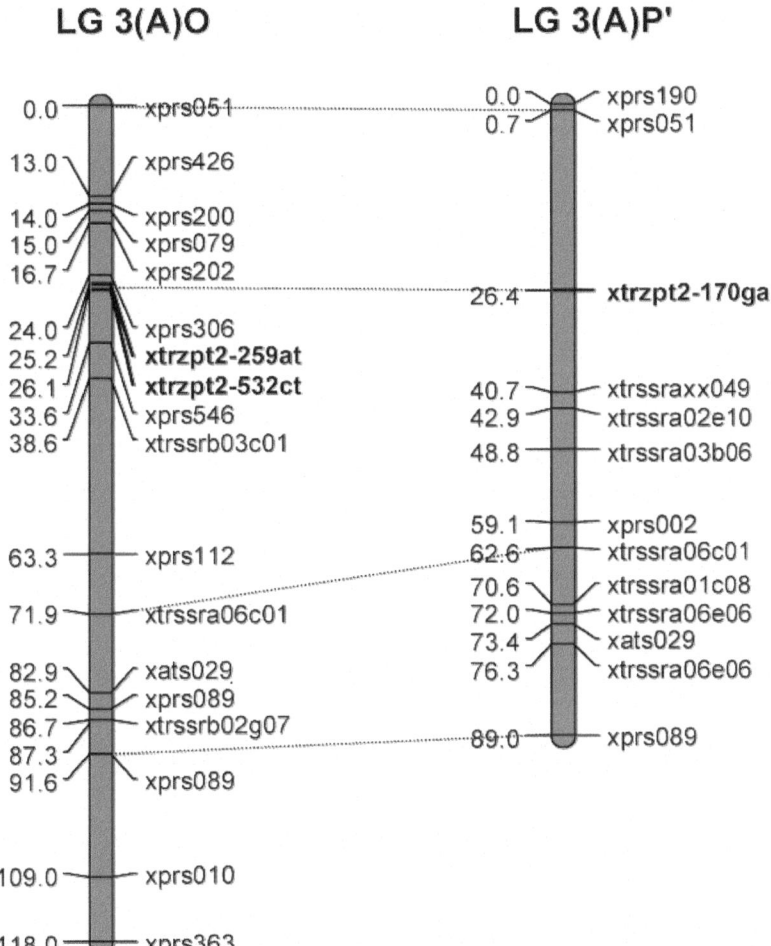

Figure 12-1 Location of *Tr*ZPT2 SNP loci on Haifa₂ parental genetic map homoeologous LGs. HG designation is based on both the A-H nomenclature of Barrett et al. (2004) and the proposed numerical system of George et al. (2006). Homoeology was inferred on the basis of common EST-SSR (xprs) loci, as described by Barrett et al. (2004). Nomenclature for genomic DNA-derived SSR (xtrssr and xats) loci is as described by Jones et al. (2003) and Barrett et al. (2004), and SNP loci are named according to the system xtr-gene name abbreviation-nucleotide coordinate-SNP identity, as for perennial ryegrass (Cogan et al. 2006). The minor difference in map positions (0.9 cM) for multiple SNPs on LG 3O is attributable to small numbers of missing genotypic datapoints.

sib-ships. However, both SNP-HSV coincidence, and the presence of identical nucleotide variants at equivalent locations between the sub-genomes present ongoing challenges for discovery in large-scale germplasm collections (Hand et al. 2008). The rate of *in vitro* discovery is inherently limited by the cloning and sequencing step, but this limitation has been addressed through massively-parallel picrolitre volume microreactor-based pyrosequencing (Margulies et al. 2005) on the Roche GSFLX platform, by pooling of amplicons from multiple genes (c. 200) after unimolecular amplification reactions, followed by sequestration of the template in genotype-specific sectors. The products of the sequencing reaction were also compared to data derived from direct sequencing of the inbred *T. occidentale* reference sample (Cogan et al. 2008). Prior knowledge of base variant (SNP and HSV) frequency was in used concert with computational sequence assembly to identify a total of 9,610 variable loci. The identified variant bases were submitted to Illumina Inc. (www.illumina.com) to assess the feasibility of design for a GoldenGate™ multiplexed SNP genotyping panel, and a total of 3317 variant bases were identified (N.O.I. Cogan, unpubl.).

12.5.3 Genetic Map Construction

Initial genetic map development in white clover was performed using a combination of genomic DNA-derived SSR and AFLP markers. The reference mapping population was the F_2 (I.4R x I.5J) family that was developed at IGER, parental genotypes being derived from fourth and fifth generation inbred lines descended from plants containing the rare self-fertile (S_f) allele. A single F_1 plant was self-pollinated to generate an F_2 population of 150 individuals (Michaelson-Yeates et al. 1997). The level of genetic polymorphism between the inbred parents, as assessed with TRSSR markers, was 48% of those markers showing efficient amplification. The F_2 (I.4R x I.5J) map contained 135 loci (78 TRSSR and 57 AFLP) on 18 linkage groups (LGs) (two more than the karyotypic number), with a total map length of 825 cM. The extent of map construction was limited by high levels of segregation distortion, affecting 39% of the TRSSR loci, the majority being distorted towards the heterozygous genotypic class (Jones et al. 2003).

A higher-resolution genetic map largely based on EST-SSR markers was constructed using the F_1 (Sustain 6525-2 x NRS 364-7) mapping family (Barrett et al. 2004). A total of 335 EST-SSR and 30 genomic DNA-derived SSR primer pairs detected polymorphism and permitted assignment of 493 loci to a genetic map containing 16 LGs, with a total map length of 1,144 cM. The EST-SSR markers detected homoeologous locations between the ancestral genomes at high frequency, and provided the basis for development of a standard homoeologous group (HG) nomenclature (A-H). A third genetic map, incorporating information derived from genotyping of both genomic DNA-derived SSR and EST-SSR markers, was developed using the F_1 (GA43 x SRVR) population (Zhang et al. 2007), containing 415 SSR loci (including primer pairs derived from white clover, red clover (*T. pratense* L.), soybean and *M. truncatula*). LGs for all eight HGs were assigned based on EST-SSR markers in common with those assigned to the F_1 (Sustain 6525-2 x NRS 364-7) map.

Parental maps for the F_1(Haifa$_2$ x LCL$_2$) and F1(S184$_6$ x LCL$_6$) genetic mapping families have been constructed using TRSSR, white clover EST-SSR, red clover EST-SSR (Sato et al. 2005) and SNP markers derived from both the *in silico* and *in vitro* discovery methods (Drayton et al. 2007). LGs were attributed to the O and P' subgenomes through mapping of specific SNP loci, and this process also permitted retrospective identification of sub-genome-specific SSR loci (Wang et al. 2010).

Additional genetic mapping populations have been constructed for the purposes of trait-dissection (Jahufer et al. 2008; Casey et al. 2010; Tashiro et al. 2010) and are described later.

12.5.4 Comparative Genetic Mapping

Computational analysis has been used to align the white clover genetic map with the genome drafts of the model legume species (George et al. 2008). Data with higher informative value was obtained for *M. truncatula* as compared to *L. japonicus*, due to a closer taxonomic affinity. Sequences from 243 SSR-containing ESTs were compared with a draft estimated to comprise c. 77% of the *M. truncatula* gene space, of which 159 (65%) obtained highest BLASTN matches against known chromosomal locations. Six of the eight HGs revealed predominant correspondences with single *Mt* chromosomes, while the two remaining groups (F and H, based on the nomenclature of Barrett et al. [2004]) may have participated in an evolutionary translocation between *Mt*2 and *Mt*6. This interpretation is consistent with a parallel study based on alignment of 163 map-assigned SSR-containing ESTs and 263 map-assigned SSR-containing hypomethylated genomic fragments from white clover with 1,547 phase 3 *M. truncatula* BAC clones (Griffiths et al. 2007). The inclusion of *M. truncatula*-derived SSR loci in the F_1 (GA43 x SVRR) genetic map also provided empirical support for the status of the unequivocal HGs. On this basis, a new HG nomenclature system for allotetraploid white clover was proposed, which could be directly related to the alphabetical system of Barrett et al. (2004) (Table 12-1). HGs of the F_1 (Haifa$_2$ x LCL$_2$) (Wang et al. 2010) and F_1 (S184$_6$ x LCL$_6$) maps (Drayton et al. 2007) have been named accordingly, permitting assignment of markers and quantitative loci to LGs with specific designation by HG and sub-genome (e.g., 3O and 3P': Hand et al. 2008). Through incorporation of both TRSSR and EST-SSR markers in the F_1(Haifa$_2$ x LCL$_2$) and F1(S184$_6$ x LCL$_6$) genetic maps, the numerical HG nomenclature was also related to the system for F_2 (I.4R x I.5J) LGs (Table 12-1). These relationships have also been partially confirmed by empirical mapping in the F_1 (GA43 x SVRR) population (Zhang et al. 2007).

Table 12-1 Summary information for relationships between homoeologous group nomenclatures based on different genetic mapping studies.

HG nomenclature		
Based on George et al. (2008)	Based on Barrett et al. (2004)	Based on Jones et al. (2003)
1	E	1/11
2	F	7
3	A	3/4/17
4	D	9
5	G	12
6	H	2
7	C	5
8	B	7

Microsynteny with the *M. truncatula* genome was investigated through alignment of BAC end sequences (BES). A total of 1474 BES from the R3R4-derived BAC library (Febrer et al. 2007) detected significant (E > 10⁻¹⁰) BLASTN matches to the *M. truncatula* draft, of which 1,450 defined 725 paired-end sequences. Of these, 204 contained ends which both matched the model genome, but only 14 (7%) identified equivalent *M. truncatula* sequence pairs within a span of 20–200 kb on the same chromosome. This result suggests that although macrosynteny between white clover and *M. truncatula* is extensive, microcolinearity may be limited by processes of repetitive DNA expansion and segmental duplication/deletion, compatible with relationships between other closely-related plant genomes (Bennetzen and Ramakrishna 2002). In contrast, comparisons between white clover BAC sequences and draft genome sequences from both model species (Hand et al. 2009; Hand et al. 2010) revealed substantial microsynteny in both instances, exceptions being largely attributable to incomplete assembly of the *M. truncatula* sequence. Similar results were obtained by Febrer et al. (2010), suggesting that translational genomics from *M. truncatula* may be useful not only for structured genetic map enhancement and orthologous gene sequence prediction, but also for comparative physical mapping and sequence assembly.

Genome structure comparisons over extended physical distances have also been performed for the subgenomes of white clover. Discrimination of O and P' subgenome-specific SNP haplotypes permitted identification of five pairs of homoeologous BACs. Although gene content and order was largely conserved between subgenomes, some differences of gene presence and orientation were observed, as well as retrotransposon cluster expansion (Hand et al. 2009, 2010).

12.6 Molecular Breeding Strategies

12.6.1 Implementation of Molecular Breeding Strategies

Genomic resources may be exploited through utilization of two major complementary strategies. The first method is based on the use of gene- or non-gene-associated sequence polymorphisms as indirect selection indices during breeding improvement. Alternatively, homologous or heterologous candidate genes may be deployed as transgenes for modification of appropriate traits. Targets for transgenic modification in white clover, linked to the key agronomic characters described earlier, include: virus resistance; pest resistance; aluminium toxicity in the rhizosphere; enhanced phosphorus acquisition; delayed plant senescence; improved seed yield; and bloat safety, based on modification of flavonoid biosynthesis. This chapter, however, focuses solely on the non-transgenic component of molecular breeding.

12.6.2 Trait-dissection

The F_2 (I.4R x I.5J) genetic map was exploited for QTL analysis of a number of vegetative morphogenesis, reproductive morphogenesis and reproductive development traits (Cogan et al. 2006a). Target traits were measured using clonally replicated sets of plants at different geographical sites in Wales and Scotland and across a number of years of evaluation (at IGER in 1999 and 2001, and at East Craigs, Scotland in 2001). Coefficients of correlation between traits were calculated, revealing results consistent with previous studies (such as the positive correlation between plant height and green weight, and negative correlations between flowering date and vegetative characters), while other relationships (such as between internode length and leaf dimensions) were less consistent with former analysis.

Individual environment analyses detected a large number of QTLs for each trait. For vegetative morphogenesis, 62 QTLs were detected for nine independently-evaluated traits (plant height, plant spread, leaf length, leaf width, petiole length, leaf area, internode length, growth score and green weight). Large clusters of coincident QTLs were identified on LGs 2, 3, 7, 11 and 12, including those for highly correlated traits, with some commonality across datasets from different environments.

For reproductive traits, multiple QTLs were detected for flowering date, height of tallest flower, peduncle length and girth, number of flowers, number of florets, number of seeds per flower, fertility score (number of seeds per floret), SY per plant and thousand seed weight (TSW) per plant. High trait correlation coefficients were observed for fertility score, seed per flower and SY, but these relationships were not reflected by high correlations with TSW. QTL x E (environment) variation was observed for the flowering date trait. Clusters of coincident QTLs were observed on LGs 2 and 3: for instance, SY and flower number QTLs colocated in the upper region of LG3. The two SY QTLs were of relatively large effect (accounting for 56 and 61% of the phenotypic variance [V_p], respectively), providing the basis for effective selection for this trait (Cogan et al. 2006a).

Multi-environment combined analysis was also performed for traits assessed in the F_2 (I.4R x I.5J) population, leading to identification of 25 genomic locations, with major clusters for vegetative morphogenesis traits on LGs 7 and 12. Comparative analysis with aerial morphogenesis trait data from *M. truncatula* (Julier et al. 2007) has identified QTL clusters on *Mt2* and *Mt5*, in regions of putative conserved synteny with those of white clover.

The F_1 (Sustain 6525-2 x NRS 364-7) white clover mapping population has also been used for QTL analysis, specifically through evaluation for SY and the component traits of inflorescence density (ID), yield per inflorescence (YI) and TSW (Barrett et al. 2005). Data were obtained from plants grown in the field at Lincoln, NZ during three full growing seasons completed in 2002, 2003 and 2004. A total of 11 QTLs was identified on nine of the 16 LGs. Single SY QTLs were detected for each of the three years, with coincidence between the regions detected from the 2003 and 2004 datasets. SY QTLs accounted for 19.5–23.2% of total V_p. Coincidence was also observed between ID QTLs from 2002 and 2003 on both LGs C2 and E1, and TSW QTLs from 2002 and 2003 on both LGs D2 and G1. Most QTLs for a given trait were observed on only one of a given homoeologous chromosome pair. The largest cluster of QTLs was observed in the lower region of LG D2, including effects for all of the measured traits. The effects of YI and TSW genotype class in this genomic region were inversely correlated, possibly due either to pleiotropy or repulsion phase linkage. The temporal stability and relatively independent genetic control of traits observed in this study suggest that substantial improvement through molecular marker-based breeding is achievable.

The reproductive trait QTL clusters on F_2 (I.4R x I.5J) LGs 2 and 3 do not correspond to the major regions for equivalent characters (LGs C2, D2, E1 and G1) identified on the F_1 (Sustain 6525-2 x NRS 364-7) parental maps (Table 12.1). However, *M. truncatula* chromosomes *Mt1*, *Mt5*, *Mt7* and *Mt8*, which contain flowering time QTLs (Julier et al. 2007), have been deduced to correspond to homoeologous groups E, G, C and B, respectively (George et al. 2008). As F_2 (I.4R x I.5J) LG 12, which also contains flowering date QTLs, has been tentatively aligned with group G and hence *Mt5*, orthologous QTLs may be located on these syntenic chromosomes. Flowering time QTLs were also identified on *L. japonicus* chromosome 1 (*Lj1*), which corresponds to segments of both *Mt3* and *Mt7*. The seed trait QTLs located on *Lj4*, 5 and

6 may also correspond to the equivalent regions identified on F_1 (Sustain 6525-2 x NRS 364-7) LGs D2, E1 and A1, respectively. Although these initial inferences are highly preliminary, more detailed map melding and comparative analysis will permit subsequent refinement (Gendall and Forster 2007).

QTLs associated with morphological and agronomic traits such as leaf length and width, petiole length, stolon inter-node length, plant spread, plant height and stolon number were also obtained for the F_1 (GA43 x SRVR) population (Zhang et al. 2008b). Data was obtained from multiple locations over different years, and as for F_2 (I.4R x I.5J), significant correlations were identified between multiple traits. Genotypic analysis with 343 SSR primer pairs permitted identification of 37 QTLs on 8 LGs, with consistent location of QTLs for highly-correlated traits across locations and years, indicating the potential for breeding improvements through marker-selection.

A mapping population for root trait morphology and architecture has been generated by pair-crossing two unrelated genotypes selected from breeding populations with distinct trait divergence (Jahufer et al. 2008). Phenotypic analysis of 386 F_1 progeny identified significant genotypic variance and phenotypic correlation for traits such as root diameter, number of root forks, root length, root surface area, number of root tips, root volume, root dry weight and specific root length. PG-derived molecular markers (Jones et al. 2006) are currently being used to genotype this population for QTL detection.

The F_1 (Haifa$_2$ x LCL$_2$) population has been analyzed to detect QTLs for salinity tolerance, assessed through vegetative growth under control and stress conditions. The LCL population is a salt-tolerant selection (Rogers et al. 1997) from the Israeli cultivar Haifa. Two parental genetic maps consisting of 203 and 159 marker loci and spanning 1973.0 and 1837.6 cM, respectively, were constructed using SSR and SNP markers. A total of 51 QTLs were identified, and coincidence was employed to identify unique QTL-containing regions. A total of eight unique genomic regions on 8 LGs of the Haifa$_2$ parental map and six unique regions on 5 LGs of the LCL$_2$ parental map were associated with plant growth under salt stress and relative growth under stress, as compared to control conditions. Identification of homoeologous groups using specific SNPs allowed regions to be categorized as common between sub-genomes, or exclusive to individual sub-genomes (Wang et al. 2010).

A genetic mapping population was constructed by crossing the GA43 parental genotype, which is common with F_1 (GA43 x SRVR), and 05-O-34, a genotype that exhibits several traits of ornamental value. The resulting F_1 population showed variation for eight morphological traits, including a series of leaf marks and number of leaflets, and was evaluated in both winter and summer at two different locations (Tashiro et al. 2010). A validation population was generated by self-pollination of 05-O-34. Genes controlling qualitative characters were assigned to LGs classified according to the nomenclature of Barrett et al. (2004). Two genes in close linkage on LG B1 controlled the red mid-rib and red fleck traits, respectively, while the trifoliolate character was controlled by a gene on LG H1.

The F_1(S1S4 x R3R4) mapping population was genotyped using a combination of AFLP and SSR markers to construct parental genetic maps (Casey et al. 2010). Cross-pollination procedures were used to obtain data used to locate the SI locus *S* to LG E1 (Casey et al. 2010). Comparative analysis with *M. truncatula* in this region provided possible candidate gene functions.

12.6.3 Implementation of Genetic Markers in Molecular Breeding

A preliminary study of marker-assisted selection (MAS) in white clover has been described by Barrett et al. (2009). The SY QTL locus at the distal end of LG D2 was targeted for analysis in 12 multi-parent complex breeding pools in a single year of evaluation. A total of 90 individuals were sampled from each pool through unreplicated field trial-based phenotyping, and the allelic contents of three SSR loci closely linked to the QTL (Barrett et al. 2005) were surveyed. Significant marker locus-trait associations were observed for eight pools, although specific marker identity, allelic identity and direction of effect were not conserved across pools. Such results are consistent with decay of linkage disequilibrium (LD) over relatively short genomic distances in outbreeding forage species (Dobrowolski and Forster 2007; Ponting et al. 2007; Forster et al. 2008).

Future progress in MAS implementation for species such as white clover is likely to depend on provision of large-scale SNP collections, capable of detecting enduring associations with trait-associated genes, in concert with methods for use of genome-wide polymorphisms such as genomic selection (GS) (Meuwissen et al. 2001; Heffner et al. 2009).

12.7 International Collaboration and Coordination

White clover is an important crop. However, global resources dedicated to breeding, and particularly to genetics and genomics, are currently limited. At the same time, developments in the genomics and system biology of *M. truncatula* (see earlier) offer significant opportunities for translation of genomic information and tools from model species to related crops. In recognition of these two issues, the International *Trifolium* Network (ITN) was established following at a two-day workshop held in Aberystwyth, Wales, UK following the Molecular Breeding of Forage and Turf conference in July 2005 (http://www.trifoliumnetwork.org). The initial membership consisted of 23 individuals from organizations around the world, and a nominated steering group. ITN is hence a consortium aimed at co-ordination, integration and progression of genetic and genomic approaches for *Trifolium* species, and translation of tools, resources and approaches from model legumes, particularly *M. truncatula*. ITN is an open community, and tools and approaches are available to all parties.

Four working groups were established, dedicated to: trait dissection and genetic map alignment (including common nomenclature for HGs and LGs, map information integration and map alignment, development and use of common populations for phenotyping and mapping studies, development and use of common genetic markers); informatics (development of common web-based resources and comparative genomics tools); ESTs and microarrays (development of integrated, common resources with respect to production and use of EST libraries and transcriptome arrays in *Trifolium*); and genome sequencing (BAC (or fosmid) end sequencing, and initiation and co-ordination of funding bids for physical and genetic mapping in *Trifolium*). Significant progress has already been made through intracommunity sharing of information on c. 200 genomic DNA-derived SSRs, an intensive audit of trait-specific genetic map information, and promotion of the rationalized HG and sub-genome-specific LG nomenclature (see earlier).

12.8 Nonforage Uses of White Clover

A range of other applications for forage legumes are well-established, including use as living mulches in agroforestry systems (Alley et al. 1998) and through intercropping with cereals (Thorsted 2002; Bergkvist 2003). A role in phytoremediation of contaminated soils (Palmroth et al. 2002) may become more prominent in the future. The impact of forage legumes such as white clover may not be beneficial, however, in all circumstances, as aggressive colonization and growth stimulation due to N fixation may result in an undesirable loss of biodiversity in schemes intended to "restore" grassland habitats (Warren 2000). Apart from these agronomic considerations, white clover has also been developed as an ornamental species, particularly due to the variety and aesthetic qualities of leaf markings (Fig. 12-2).

12.9 Conclusions

Although white clover does not belong to the group of most economically-significant crop species internationally, investment in large-scale species-specific genomics activities has proven highly effective and is likely to lead to effective molecular breeding improvement. Translational genomics based in comparison with *M. truncatula* and *L. japonicus* provides valuable support

Figure 12.2 Ornamental varieties of white clover, displaying various forms of leaf marking and colour (images kindly supplied by Dr. Wayne Parrott, University of Georgia, USA).

Color image of this figure appears in the color plate section at the end of the book.

to these homologous resources. The capacity to identify sub-genome-specific sequence variants will be crucial for genetic mapping, trait-dissection and is also likely to make important contributions to homoeolocus selection for transgene deployment and gene expression studies. Advances in white clover genomics will also be of benefit to related forage species within the Trifolieae, including red clover, subterranean clover and alfalfa.

Acknowledgments

The authors thank Dr. Wayne Parrott (University of Georgia, Athens, Georgia, USA), Prof. E. Charles Brummer (Samuel Roberts Noble Foundation, Ardmore, Oklahoma, USA), Prof. Michael Hayward (Rhydgoch Genetics, Aberystwyth, United Kingdom) and Dr. Aidyn Mouradov and Prof. German Spangenberg (Victorian Department of Environment and Primary Industries) for helpful critical comments.

References

Abberton MT, Michaelson-Yeates TPT, Marshall AH, Holdbrook-Smith K, Rhodes I (1998) Morphological characteristics of hybrids between white clover, *Trifolium repens* L., and Caucasian clover, *Trifolium ambiguum* M. Bieb. Plant Breed 117: 494–496.

Abberton MT, Marshall AH (2005) Progress in breeding perennial clovers for temperate agriculture. J Agri Sci 143: 117–135.

Abeynayake S, Panter S, Chapman R, Webster T, Mouradov A, Spangenberg G (2007) Molecular dissection of flavonoid pathways in developing flowers of white clover (*Trifolium repens* L.) using CombiMatrix oligonucleotide array expression analysis. Proceedings of the Fifth International Symposium on the Molecular Breeding of Forage and Turf, Sapporo, Japan, July 1st–6th 2007, p 61.

Abeynayake S, Panter S, Rochfort S, Drayton M, Hand M, Cogan N, Forster JW, Mouradov A, Spangenberg GC (2010) Spatio-temporal profile of proanthocyanidin biosynthesis in white clover (*Trifolium repens* L.). Proceedings of the Sixth International Symposium on the Molecular Breeding of Forage and Turf, Buenos Aires, Argentina, March 15th–19th 2010, p 207.

Adams KL, Cronn R, Percifield R, Wendel JF (2003) Genes duplicated by polyploidy show unequal contributions to the transcriptome and organ-specific reciprocal silencing. Proc Natl Acad Sci USA 100: 4649–4654.

Alley JL, Garrett HE, McGraw RL, Dwyer JP, Blanche CA (1998) Forage legumes as living mulches for trees in agroforestry practices—preliminary results. Agrofor Syst 44: 281–291.

Andersen JR, Lübberstedt T (2003) Functional markers in plants. Trends Plant Sci 8: 554–560.

Annicchiarico P (1993) Variation for dry matter yield, seed yield and other agronomic traits in Ladino white clover landraces and natural populations. Euphytica 71: 131–141.

Annicchiarico P, Piano E, Rhodes I (1999) Heritability of, and genetic correlations among, forage and seed yield traits in Ladino white clover. Plant Breed 118: 341–346.

Ansari HA, Ellison NW, Reader SM, Badaeva ED, Friebe B, Miller TE, Williams WM (1999) Molecular cytogenetic organisation of 5S and 18S–28S rDNA loci in white clover (*Trifolium repens* L.) and related species. Ann Bot 83: 199–206.

Attwood SS (1940) Genetics of cross-incompatibility among self-incompatible plants *of Trifolium repens*. J Am Soc Agron 32: 955–968.

Attwood SS (1941) Controlled self- and cross-pollination of *Trifolium repens*. J Am Soc Agron 33: 538–545

Attwood SS (1942a) Oppositional alleles causing self-incompatibility in *Trifolium repens*. Genetics 27: 333–338.

Attwood SS (1942b) Genetics of pseudo-self-incompatibility and its relation to cross-incompatibility in *Trifolium repens* L. J Agri Res 64: 699–709.

Ayres E, Dromph KM, Cook R, Ostle N, Bardgett RD (2007) The influence of below-ground herbivory and defoliation of a legume on nitrogen transfer to neighbouring plants. Funct Ecol 21: 256–263.

Barbour M, Caradus JR, Woodfield DR, Silvester WB (1996) Water stress and water use efficiency of ten white clover cultivars. In: Woodfield DR (ed) White Clover: New Zealand's Competitive Edge. Grassland Research and Practice Series 6: 159–162. New Zealand Grassland Association, Palmerston North, New Zealand.

Barrett B, Griffiths A, Schreiber M, Ellison N, Mercer C, Bouton J, Ong B, Forster J, Sawbridge T, Spangenberg G, Bryan G, Woodfield D (2004) A microsatellite map of white clover (*Trifolium repens* L.). Theor Appl Genet 109: 596–608.

Barrett BA, Baird IJ, Woodfield DR (2005) A QTL analysis of white clover seed production. Crop Sci 45: 1844–1850.

Barrett B, Baird I, Woodfield D (2009) White clover seed yield: a case study in marker-assisted selection. In: Yamada T, Spangenberg G (eds) Molecular Breeding of Forage and Turf: The Proceedings of the 5th International Symposium on the Molecular Breeding of Forage and Turf. Springer, New York, pp 241–250.

Bennetzen JL, Ramakrishna W (2002) Numerous small rearrangements of gene content, order and orientation differentiate plant genomes. Plant Mol Biol 48: 821–827.

Bergkvist G (2003) Effect of white clover and nitrogen availability on the grain yield of winter wheat in a three-season intercropping system. Acta Agri Scand B—Soil Plant Sci 53: 97–109.

Brink GE, Pederson GA (1998) White clover response to a water application gradient. Crop Sci 38: 771–775.

Brock JL, Hay MJM (1996) A review of the role of grazing management on the growth and performance of white clover cultivars in lowland New Zealand pastures. In: Woodfield DR (ed) White Clover: New Zealand's Competitive Edge. Grassland Research and Practice Series 6: 65–70. New Zealand Grassland Association, Palmerston North, New Zealand.

Blow N (2009) Transcriptomics: the digital generation. Nature 458: 239–242.

Brummer EC (1999) Capturing heterosis in forage crop cultivar development. Crop Sci 39: 943–954.

Buetow KH, Edmonson MN, Cassidy AB (1999) Reliable identification of large numbers of candidate SNPs from public EST data. Nat Genet 21: 323–325.

Caldwell KS, Dvorak J, Lagudah ES, Akhunov E, Luo M-C, Wolters P, Powell W (2004) Sequence polymorphism in polyploid wheat and their D-genome diploid ancestor. Genetics 167: 941–947.

Caradus JR (1994) Selection for improved adaptation of white clover to low phosphorus and acid soils. Euphytica 77: 243–250.

Caradus JR, Chapman DF (1996) Selection for and heritability of stolon characteristics in two cultivars of white clover. Crop Sci 36: 900–904.

Caradus JR, Christie BR (1998) Winter-hardiness and artificial frost tolerance of white clover ecotypes and selected breeding lines. Can J Plant Sci 78: 251–255.

Caradus JR, Crush JR, Ouyang L, Fraser W (2001) Evaluation of aluminium tolerant white clover (*Trifolium repens*) selections on East Otago upland soils. NZ J Agri Res 44: 141–150.

Caradus JR, Kennedy LD, Dunn A (1998) Genetic variation for the ratio of inorganic to total phosphorus in white clover leaves. J Plant Nutr 21: 2265–2272.

Caradus JR, Woodfield DR, Stewart AV (1996) Overview and vision for white clover. In: Woodfield DR (ed) White Clover: New Zealand's Competitive Edge. Grassland Research and Practice Series 6: 159–162. New Zealand Grassland Association, Palmerston North, New Zealand.

Carlsen SCK, Fomsgaard IS (2008) Biologically active secondary metabolites in white clover (*Trifolium repens* L.)—a review focusing on contents in the plant, plant-pest interactions and transformation. Chemoecology 18: 129–170.

Carlsen SCK, Understrup A, Fomsgaard IS, Mortensen AG, Ravnskov S (2008) Flavonoids in roots of white clover: interaction of arbuscular mycorrhizal fungi and a pathogenic fungus. Plant Soil 302: 33–43.

Casey NM, Milbourne D, Barth S, Febrer M, Jenkins G, Abberton MT, Jones C, Thorogood D (2010) The genetic location of the self-incompatibility locus in white clover (*Trifolium repens* L.), Theor Appl Genet 121: 567–576.

Cogan NOI, Abberton MT, Smith KF, Kearney G, Marshall AH, Williams A, Michaelson-Yeates TPT, Bowen C, Jones ES, Vecchies AC, Forster JW (2006a) Individual and multi-environment combined analyses identify QTLs for morphogenetic and reproductive development traits in white clover (*Trifolium repens* L.). Theor Appl Genet 112: 1401–1415.

Cogan NOI, Ponting RC, Vecchies AC, Drayton MC, George J, Dobrowolski MP, Sawbridge TI, Spangenberg GC, Smith KF, Forster JW (2006b) Gene-associated single nucleotide polymorphism (SNP) discovery in perennial ryegrass (*Lolium perenne* L.). Mol Genet Genom 276: 101–122.

Cogan NOI, Drayton MC, Ponting RC, Vecchies AC, Bannan NR, Sawbridge TI, Smith KF, Spangenberg GC, Forster JW (2007) Validation of in silico-predicted genic single nucleotide polymorphism in white clover (*Trifolium repens* L.). Mol Genet Genom 277: 413–425.

Cogan NOI, Sawbridge TI, Dobrowolski MP, Spangenberg GC, Smith KF, Forster JW (2008) Novel strategies for characterization of functionally-associated polymorphisms in temperate pasture species. Plant and Animal Genome XVI Conference, San Diego, California, CA. W268.

Collins RP, Abberton MT, Michaelson-Yeates TPT, Rhodes I (1997) Response to divergent selection for stolon characters in white clover (*Trifolium repens*). J Agri Sci 129: 279–285.

Collins RP, Helgadottir A, Fothergill M, Rhodes I (2002) Variation amongst survivor populations of white clover collected from sites across Europe: Growth attributes and physiological responses to low temperature. Ann Bot 89: 283–292.

Collins RP, Rhodes I (1994) Influence of root competition on compatibility between white clover and perennial ryegrass populations during seedling establishment. Grass Forage Sci 49: 506–509.

Coulman BE, Lambert M (1995) Selection for resistance to root rot caused by *Fusarium* spp. in red clover (*Trifolium pratense* L.). Can J Plant Sci 75: 141–146.

Cronn RC, Wendel JF (1998) Simple methods for isolating homoeologous loci from allopolyploid genomes. Genome 41: 756–762.

Crush JR (1995) Effect of VA mycorrhizas on phosphorus uptake and growth of white clover (*Trifolium repens* L.) growing in association with ryegrass (*Lolium perenne* L.). NZ J Agri Sci 38: 303–307

Damsteegt VD, Stone AL, Russo AJ, Luster DG, Gildow FE, Smith OP (1999) Identification, characterization, and relatedness of luteovirus isolates from forage legumes. Phytopatholology 89: 374–379.

Dixon RA, Sumner LW (2003) Legume natural products: Understanding and manipulating complex pathways for human and animal health. Plant Physiol 131: 878–885.

Dobrowolski MP, Forster JW (2007) Chapter 9: Linkage disequilibrium-based association mapping in forage species. In: Oraguzie NC et al. (eds) Association Mapping in Plants. Springer, New York, pp 197–209.

Doyle JJ, Luckow MA (2003) The rest of the iceberg. Legume diversity and evolution in a phylogenetic context. Plant Physiol 131: 900–910.

Drayton M, George J, Cogan N, Hand M, Ponting R, Trigg P, Wilkinson T, Sawbridge T, Spangenberg G, Smith K, Forster J (2007) Trait-specific genetic map construction and reference map integration in white clover (*Trifolium repens* L.). Proceedings of the Fifth International Symposium on the Molecular Breeding of Forage and Turf, Sapporo, Japan, 1–6 July 2007, p 73.

Edwards KJ, Barker JHA, Daly A, Jones C, Karp A (1996) Microsatellite libraries enriched for several microsatellite sequences in plants. Biotechniques 20: 758–759.

Edwards D, Forster JW, Cogan NOI, Batley J, Chagné D (2007) Chapter 4: Single nucleotide polymorphism discovery in plants. In: Oraguzie NC et al. (eds) Association Mapping in Plants. Springer, New York, pp 53–76.

Ellison NW, Liston A, Steiner JJ, Williams WM, Taylor NL (2006) Molecular phylogenetics of the clover genus (*Trifolium*–Leguminosae). Mol Phylogenet Evol 39: 688–705.

Erwin TA, Jewell EG, Love CG, Lim GAC, Li X, Chapman R, Batley J, Stajich JE, Mongin E, Stupka ER, Spangenberg G, Edwards D (2007) BASC: an integrated bioinformatics system for *Brassica* research. Nucl Acids Res 35: 870–873.

Evans DR, Williams TA (1987) The effects of cutting and grazing managements on dry-matter yield of white clover varieties (*Trifolium repens*) when grown with S23 perennial ryegrass. Grass Forage Sci 42: 153–159.

Evans DR, Williams TA, Jones S, Evans SA (1995) The effect of blending white clover varieties and their contribution to a mixed grass/clover sward under continuous sheep stocking. Grass Forage Sci 50: 10–15.

Evans DR, Williams TA, Evans SA (1996) Breeding and evaluation of new white clover varieties for persistency and higher yields under grazing. Grass Forage Sci 51: 403–411.

Febrer M, Cheung F, Town CD, Cannon SB, Young ND, Abberton MT, Jenkins G, Milbourne D (2007) Construction, characterization, and preliminary BAC-end sequencing analysis of a bacterial artificial chromosome library of white clover (*Trifolium repens* L.). Genome 50: 412–421.

Febrer M, Abberton MT, Jenkins G, Milbourne D (2010) Exploring the potential for translational genomics approaches in forage legumes: regions of highly conserved microsynteny between white clover and *Medicago truncatula* revealed by BAC Sequencing. In: Huyghe C (ed) Sustainable use of genetic diversity in forage and turf breeding. Springer, New York, pp 414–419.

Finne MA, Rognli OA, Schjelderup I (2000) Genetic variation in a Norwegian germplasm collection of white clover (*Trifolium repens* L.). Euphytica 112: 57–68.

Forster JW, Jones ES, Kölliker R, Drayton MC, Dumsday J, Dupal MP, Guthridge KM, Mahoney NL, van Zijll de Jong E, Smith KF (2001) Development and implementation of molecular markers for forage crop improvement. In: Spangenberg G (ed) Molecular Breeding of Forage Crops. Kluwer Academic Press, Dordrecht, Netherlands, pp 101–133.

Forster JW, Cogan NOI, Dobrowolski MP, Francki MG, Spangenberg GC, Smith KF (2008) Functionally-associated molecular genetic markers for temperate pasture plant improvement. In: Henry RJ (ed) Plant Genotyping II: SNP Technology. CABI Press, Wallingford, Oxford, pp 154–187.

Frame J, Charlton JFL, Laidlaw AS (1998) Temperate Forage Legumes. CAB International, Wallingford, Oxon, United Kingdom.

Frame J, Laidlaw AS (1998) Managing white clover in mixed swards: principles and practice. Pastos 28: 5–13.

Frame J, Newbould P (1986) Agronomy of white clover. Adv Agron 40: 1–88.

Gendall AR, Forster JW (2007) Genetics of reproductive development in forage legumes. Int J Plant Dev Biol 1: 245–252.

George J, Dobrowolski MP, van Zijll de Jong E, Cogan NOI, Smith KF, Forster JW (2006) Assessment of genetic diversity in cultivars of white clover (*Trifolium repens* L.) detected by simple sequence repeat polymorphism. Genome 49: 919–930.

George J, Sawbridge TI, Cogan NOI, Gendall AR, Smith KF, Spangenberg GC, Forster JW (2008) Comparison of genome structure between white clover and *Medicago truncatula* Gaertn. supports homoeologous group nomenclature based on conserved synteny. Genome 51: 905–911.

Gerard PJ, Hackell DL, Bell NL (2007) impact of clover root weevil *Sitona lepidus* (Coleoptera: Curculionidae) larvae on herbage yield and species composition in a ryegrass-white clover sward. NZ J Agri Res 50: 381–392.

Goulas E, Le Dily F, Teissedre L, Corbel G, Robin C, Ourry A (2001) Vegetative storage proteins in white clover (*Trifolium repens* L.): Quantitative and qualitative features. Ann Bot 88: 789–795.

Griffiths AG, Bickerstaff P, Anderson CB, Franzmayr BK (2006) Threshing the white clover genome for gene-associated molecular markers. In: Mercer CF (ed) Proceedings of the 13th Australasian Plant Breeding Conference, Christchurch, New Zealand, 18–21 April 2006, pp 817–821.

Griffiths A, Barrett, B, Simon D, Anderson C, Somerville D, Lawn J, Warren J, Khan A, Jones C (2007) A consensus map of white clover with *in silico* alignment to *Medicago* indicates a translocation. Proceedings of the Fifth International Symposium on the Molecular Breeding of Forage and Turf, Sapporo, Japan, 1–6 July 2007, p 115.

Gustine DL, Huff DR (1999) Genetic variation within and among white clover populations from managed permanent pastures of the Northeastern USA. Crop Sci 39: 524–531.

Gustine DL, Voigt EC, Brummer EC, Papadopoulos YA (2002) Genetic variation of RAPD markers for North American white clover collections and cultivars. Crop Sci 42: 343–347.

Guthridge KM, Dupal MD, Kölliker R, Jones ES, Smith KF, Forster JW (2001) AFLP analysis of genetic diversity within and between populations of perennial ryegrass (*Lolium perenne* L.). Euphytica 122: 191–201.

Hand ML, Ponting RC, Drayton MC, Lawless KA, Cogan NOI, Brummer EC, Sawbridge TI, Spangenberg GC, Smith KF, Forster JW (2008) Identification of homologous, homoeologous and paralogous sequence variants in an outbreeding allopolyploid species based on comparison with progenitor taxa. Mol Genet Genom 280: 293–304.

Hand ML, Cogan NOI, Smith KF, Spangenberg GC, Forster JW (2009) Comparative structural analysis of gene clusters between sub-genomes of white clover and with model legume species. Plant and Animal Genome XVII Conference, San Diego, California, p 353.

Hand ML, Cogan NOI, Sawbridge TI, Spangenberg GC, Forster JW (2010) Comparison of homoeolocus organisation in paired BAC clones from allotetraploid white clover (*Trifolium repens* L.) and microcolinearity with model legume species. BMC Plant Biol 10: 94.

Heagle AS, Rebbeck J, Shafer SR, Lesser VM, Blum U, Heck WW (1989) Effects of long-term O3 exposure and soil moisture deficit on growth of a white clover—tall fescue pasture. Phytopathology 79: 128–136.

Heagle AS, McLaughlin MR, Miller JE, Joyner RL, Spruill SE (1991) Adaptation of a white clover population to ozone stress. New Phytol 119: 61–68.

Heffner EL, Sorrells ME, Jannink J-L (2009) Genomic selection for crop improvement. Crop Sci 49: 1–12.

Helgadottir A (1997) Legume breeding in Iceland. Buvisindi 11: 29–39.

Hisano H, Isobe S, Sasamoto S, Wada T, Sato S, Okumura K, Tabata S (2007) Development of microsatellite markers of *Trifolium* species for a white clover linkage map. Abstracts of the Fifth International Symposium on the Molecular Breeding of Forage and Turf, Sapporo, Japan, 1–6 July 2007, p 113.

Hogh-Jensen H (2006) The nitrogen transfer between plants: an important but difficult flux to quantify. Plant and Soil 282: 1–5.

Holtham DAL, Matthews GP, Scholefield DS (2007) Measurement and simulation of void structure and hydraulic changes caused by root induced soil structuring under white clover compared to ryegrass. Geoderma 142: 142–151.

Hunter DA, Watson LM, McManus MT (1999) Cell wall proteins in white clover: Influence of plant phosphate status Plant Physiol Biochem 38: 259–270.

Hussain SW, Williams WM (1997) Development of a fertile genetic bridge between *Trifolium ambiguum* M. Bieb. and *T. repens* L. Theor Appl Genet 95: 678–690.

Hussain SW, Williams WM, Mercer CF, White DWR (1997) Transfer of clover cyst nematode resistance from *Trifolium nigrescens* Viv. to *T. repens* L. by interspecific hybridisation. Theor Appl Genet 95: 1274–1281.

Jahufer MZZ, Gawler FI (2000) Genotypic variation for seed yield components in white clover (*Trifolium repens* L.). Aust J Agri Res 51: 657–663.

Jahufer MZZ, Cooper M, Ayres JF, Bray RA (2002) Identification of research to improve the efficiency of breeding strategies for white clover in Australia: a review. Aust J Agri Res 53: 239–257.

Jahufer MZZ, Cooper M, Lane LA (1995) Variation among low rainfall white clover (*Trifolium repens* L.) accessions for morphological attributes and herbage yield. Aust J Agri Res 35: 1109–1116.

Jahufer MZZ, Nichols SN, Crush JR, Ouyang L, Dunn A, Ford JL, Care DA, Griffiths AG, Jones CS, Jones CG, Woodfield DR (2008) Genotypic variation for root trait morphology in a white clover mapping population grown in sand. Crop Sci 48: 487–494.

Jin L, Wei Y, Barrett B, Woodfield D, Brummer EC (2010) Population genetic structure based on SSR markers in white clover (*Trifolium repens* L.). Plant and Animal Genome XVIII Conference, San Diego, California, p 428.

Jones CS, Williams WM, Hancock KR, Ellison NW, Scott AG, Collette VE, Jahufer MZZ, Richardson KA, Hay MJM, Rasmussen S, Jones CG, Griffiths AG (2006) Pastoral Genomics —a foray into the clover genome. In: Mercer CF (ed) Proceedings of the 13th Australasian Plant Breeding Conference, Christchurch, New Zealand, 18–21 April 2006, pp 928–931.

Jones ES, Hughes LJ, Drayton MC, Abberton MT, Michaelson-Yeates TPT, Bowen C, Forster JW (2003) An SSR and AFLP molecular marker-based genetic map of white clover (*Trifolium repens* L.). Plant Sci 165: 447–479.

Joyce TA, Abberton MT, Michaelson-Yeates TPT, Forster JW (1999) Relationships between genetic distance measured by RAPD-PCR and heterosis in inbred lines of white clover (*Trifolium repens* L.). Euphytica 107: 159–165.

Julier B, Huguet T, Chardon F, Ayadi R, Pierre JB, Prosperi JM, Barre P, Huyghe C (2007) Identification of quantitative trait loci influencing aerial morphogenesis in the model legume *Medicago truncatula*. Theor Appl Genet 114: 1391–1406.

Kölliker R, Jones ES, Drayton MC, Dupal MP, Forster JW (2001a) Development and characterisation of simple sequence repeat (SSR) markers for white clover (*Trifolium repens* L.). Theor Appl Genet 102: 416–424.

Kölliker R, Jones ES, Jahufer MZZ, Forster JW (2001b) Bulked AFLP analysis for the assessment of genetic diversity in white clover (*Trifolium repens* L.). Euphytica 121: 305–315.

Laidlaw AS, Chrissie P, Lee HW (1996) Effects of white clover cultivar on apparent transfer from clover to grass and estimation of relative turnover rates in roots. Plant Soil 179: 243–253.

Laidlaw AS, Teuber N (2001) Temperate forage grass-legume mixtures: advances and perspectives. Proc XIX International Grassland Congress, 11–21 February 2001, Sao Paulo, Brazil, pp 85–92.

Lane LA, Ayres JF, Lovett JV (1997) A review of the introduction and use of white clover (*Trifolium repens* L.) in Australia—significance for breeding objectives. Aust J Exp Agri 37: 831–839.

Lawless KA, Drayton MC, Hand MC, Ponting RC, Cogan NOI, Sawbridge TI, Smith KF, Spangenberg GC, Forster JW (2008) Interpretation of SNP haplotype complexity in white clover (*Trifolium repens* L.), an outbreeding allotetraploid species. In: Yamada T, Spangenberg G (eds) Molecular Breeding of Forage and Turf: The Proceedings of the 5th International Symposium on the Molecular Breeding of Forage and Turf. Springer, New York, pp 211–221.

Ledgard SF, Penno JW, Sprosen MS (1999) Nitrogen inputs and losses from clover/grass pastures grazed by dairy cows, as affected by nitrogen fertiliser. J Agri Sci 132: 215–225.

Lee MRF, Harris LJ, Dewhurst RJ, Merry RJ, Scollan ND (2003) The effect of clover silages on long chain fatty acid rumen transformations and digestion in beef steers. Animal Sci 76: 491–501.

Lee B-R, Jung W-J, Lee B-H, Avice J-C, Ourry A, Kim, T-H (2007) Kinetics of drought -induced pathogenesis –related proteins and its physiological significance in white clover leaves. Physiol Plant 132: 329–336.

Lowe M, Collins RP, Abberton MT (2005) Changes in gene expression during acclimation to cold temperatures in white clover (*Trifolium repens* L.). Proceedings of the Fourth International Symposium on the Molecular Breeding of Forage and Turf, Aberystwyth, Wales, UK, 3rd–7th July 2005, p 151.

Margulies M, Egholm M, Altman WE, Attiya S, Bader JS, Bemben LA, Berka J, Braverman MS, Chen Y-J, Chen Z, Dewell SB, Du L, Fierro JM, Gomes XV, Godwin BC, He W, Helgesen S, Ho CH, Irzyk GP, Jando SC, Alenquer MLI, Jarvie TP, Jirage KB, Kim J-B, Knight JR, Lanza

JR, Leamon JH, Lefkowitz SM, Lei M, Li J, Lohman KL, Lu H, Makhijani VB, McDade KE, McKenna MP, Myers EW, Nickerson E, Nobile JR, Plant R, Puc BP, Ronan MT, Roth GT, Sarkis GJ, Simons JF, Simpson JW, Srinivasan M, Tartaro KR, Tomasz A, Vogt KA, Volkmer GA, Wang SH, Wang Y, Weiner MP, Yu P, Begley RF, Rothberg JM (2005) Genome sequencing in microfabricated high-density picolitre reactors. Nature 437: 376–380.

Marshall AH (1995) Peduncle characteristics, inflorescence survival and reproductive growth of white clover (*Trifolium repens* L.). Grass Forage Sci 50: 324–330.

Marshall AH, Holdbrook-Smith K, Michaelson-Yeates TPT, Abberton MT, Rhodes I (1998) Growth and reproductive characteristics in backcross hybrids derived from *Trifolium repens* L. x *T. nigrescens* Viv. interspecific crosses. Euphytica 104: 61–66.

Mather RDJ, Melhuish DT, Herlihy M (1996) Trends in the global marketing of white clover cultivars. In: Woodfield DR (ed) White Clover: New Zealand's Competitive Edge. Grassland Research and Practice Series 6: 7–14. New Zealand Grassland Association, Palmerston North, New Zealand.

Maurer K, Cooper J, Caraballo M, Crye J, Suciu D, Ghindilis A, Leonetti JA, Wang W, Rossi FM, Stover AG, Larson C, Gao H, Dill K, McShea A (2006) Electrochemically-generated acid and its containment to 100 micron reaction areas for the production of DNA microarrays. PLoS ONE 20: 1–7.

Mayne CS, Wright IA, Fisher GEJ (2000) Grassland management under grazing and animal response. In: Hopkins A (ed) Grass—its Production and Utilization, 3rd Edn. British Grassland Society, Blackwell, Oxon, pp 247–291.

McNeill AM, Wood M (1996) ^{15}N estimates of nitrogen fixation by white clover (*Trifolium repens* L.). Plant and Soil 128: 265–273.

McLaughlin MR, Pederson GA, Evans RR, Ivy L (1992) Virus diseases and stand decline in white clover pasture. Plant Dis 76: 158–162.

Mercer CF, van den Bosch J, Miller KJ (1999) Effectiveness of recurrent selection of white clover (*Trifolium repens*) for resistance to New Zealand populations of clover cyst nematode (*Heterodera trifolii*). Nematology 1:449–455.

Mercer CF, van den Bosch K, Miller KJ, Woodfield DR (2000a) Genetic solutions for two major New Zealand pasture pests. Proceedings of 16th *Trifolium* Conference. Pipestem, WV, 20–22 June 2000, p 13.

Mercer CF, Miller KJ, van den Bosch J (2000b) Progress in recurrent selection and in crossing cultivars with white clover resistant to the clover root-knot nematode *Meloidogyne trifoliophila*. NZ J Agri Res 43: 41–48.

Meuwissen THE, Hayes BJ, Goddard ME (2001) Prediction of genetic value using genome-wide dense genetic map. Genetics 157: 1819–1829.

Michaelson-Yeates TPT, Marshall A, Abberton MT, Rhodes I (1997) Self-incompatibility and heterosis in white clover (*Trifolium repens* L.). Euphytica 94: 341–348.

Mytton LR, Cresswell A, Colbourn P (1993) Improvement in soil structure associated with white clover. Grass Forage Sci 48: 84–90.

Palmroth MRT, Pichtel J, Puhakka JA (2002) Phytoremediation of subarctic soil contaminated with diesel fuel. Bioresour Technol 84: 221–228.

Pederson G A (1995a) White clover and other perennial clovers. In: Barnes RF et al. (eds) Forages Vol 1: An Introduction to Grassland Agriculture. Iowa State University Press, Ames, Iowa, pp 227–236.

Pederson GA (1995b) Registration of MSRLM red leaf mark white clover germplasm. Crop Sci 35: 1235–1236.

Pederson GA, McLaughlin MR (1995) Registration of MSRedFl red-flowered white clover germplasm. Crop Sci 35: 596.

Picoult-Newberg L, Ideker TE, Pohl MG, Taylor SL, Donaldson MA, Nickerson DA, Boyce-Jacino M (1999) Mining SNPs from EST databases. Genome Res 9: 167–176.

Ponting RC, Drayton MD, Cogan NOI, Dobrowolski MP, Smith KF, Spangenberg GC, Forster JW (2007) SNP discovery, validation, haplotype structure and linkage disequilibrium in full-length herbage nutritive quality genes of perennial ryegrass (*Lolium perenne* L.). Mol Genet Genom 278: 589–597.

Quesenberry KH, Smith RR, Taylor NL, Baltensperger DD, Parrott WA (1991) Genetic nomenclature in clovers and special-purpose legumes: I. Red and white clovers. Crop Sci 31: 861–867.

Quesenberry KH, Casler MD (2001) Achievements and perspectives in the breeding of temperate grasses and legumes. Proc XIX Intl Grassland Congress. Sao Paulo, Brazil, 11–21 February 2001, pp 517–524.

Rabinowicz PD, Schutz K, Dedhia N, Yordan C, Parnell LD, Stein L, McCombie WR, Martienssen RA (1999) Differential methylation of genes and retrotransposons facilitates shotgun sequencing of the maize genome. Nat Genet 23: 305–309.

Rhodes I, Collins RP, Evans DR (1994) Breeding white clover for tolerance to low temperature and grazing stress. Euphytica 77: 239–242.

Rhodes C-R (2003) Microarray-based analysis of gene expression in alfalfa mosaic virus (AMV)-infected white clover (*Trifolium repens* L.) and transgenic AMV-immune white clover. Honours Thesis, La Trobe University, Bundoora, Victoria, Australia.

Robin C, Hay MJM, Newton PCD, Greer DH (1994) Effect of light quality (red: far-red ratio) at the apical bud of the main stolon on morphogenesis of *Trifolium repens* L. Ann Bot 74: 119–123.

Rogers JB, Laidlaw AS, Christie P (2000) The role of arbuscular mycorrhizal fungi in the transfer of nutrients between white clover and perennial ryegrass. Chemosphere 42: 51–57.

Rogers ME, Noble CL, Nicolas ME, Halloran GM (1993) Variation in yield potential and salt tolerance of selected cultivars and natural populations of *Trifolium repens* L. Aust J Agri Res 44: 785–798.

Rogers ME, Noble CL, Halloran GM, Nicolas ME (1997) Selecting for salt tolerance in white clover (*Trifolium repens*): Chloride ion exclusion and its heritability. New Phytol 135: 645–654.

Saarijarvi K, Virkajarvi P, Heinonen-Tanski H (2007) Nitrogen leaching and herbage production on intensively managed grass and grass-clover pastures on sandy soil in Finland. Eur J Soil Sci 58: 1382–1392.

Sato S, Isobe S, Asamizu E, Ohmido N, Kataoka R, Nakamura Y, Kaneko T, Sakurai N, Okumura K, Klimenko I, Sasamoto S, Wada T, Watanabe A, Kohara M, Fujishiro T, Tabata S (2005) Comprehensive structural analysis of the genome of red clover (*Trifolium pratense* L.). DNA Res 12: 301–364.

Sawbridge T, Ong E-K, Binnion C, Emmerling M, Meath K, Nunan K, O'Neill M, O'Toole F, Simmonds J, Wearne K, Winkworth A, Spangenberg G (2003) Generation and analysis of expressed sequence tags in white clover (*Trifolium repens* L.). Plant Sci 165: 1077–1087.

Schreurs NM, Marotti DM, Tavendale MH, Lane GA, Barry TN, López-Villalobos N, McNabb WC (2007) Concentration of indoles and other rumen metabolites in sheep after a meal of fresh white clover, perennial ryegrass or *Lotus corniculatus* and the appearance of indoles in the blood. J Sci Food Agri 87: 1042–1051.

Soegaard K (1994) Agronomy of white clover. In: Mannetje LT, Frame J (eds) Proc 15th Gen Meet European Grassland Federation, Wageningen, Netherlands, June 1994. Netherlands Grassland Society, pp 515–524.

Sorrells ME, Wilson WA (1997) Direct classification and selection of superior alleles for crop improvement. Crop Sci 37: 691–697.

Spangenberg GC, Forster JW, Edwards D, John U, Mouradov A, Emmerling M, Batley J, Felitti S, Cogan NOI, Smith KF, Dobrowolski MP (2005) Future directions in the molecular breeding of forage and turf. In: Humphreys MO (ed) Molecular Breeding for the Genetic Improvement of Forage Crops and Turf. Academic Publishers, Wageningen, pp 83–97.

Swift G, Morrison MW, Cleland AT, Smith-Taylor CAB, Dickson JM (1992) Comparison of white clover varieties under cutting and grazing. Grass Forage Sci 47: 8–13.

Tang Y, Chevone B, Hess JL (1999) Ozone responsive proteins in a tolerant and sensitive clone of white clover (*Trifolium repens* L.) Environmental Pollution 106: 89–98.

Tashiro RM, Han Y, Monteros MJ, Bouton JH, Parrott WA (2010) Leaf trait coloration in white clover and molecular mapping of the red midrib and leaflet number traits. Crop Sci 50: 1260–1268.

Taylor NM (2008) A century of clover breeding developments in the United States. Crop Sci 48: 1–13.

Thorsted MD, Olesen JE, Koefoed N (2002) Effects of white clover cultivars on biomass and yield in oat/clover intercrops. J Agric Sci 138: 261–267.

Torres TT, Metta M, Ottenwälder B, Schlötterer C (2007) Gene expression profiling by massively parallel sequencing. Genome Res 18: 1–6.

van Treuren R, Bas N, Goossens PJ, Jansen J, van Soest LJ (2005) Genetic diversity in perennial ryegrass and white clover among old Dutch grasslands as compared to cultivars and nature reserves. Mol Ecol 14: 39–52.

Vogel KP, Pedersen JF (1993) Breeding systems for cross-pollinated forage grasses. Plant Breed Revs 11: 251–274.

Voigt PW, Staley TE (2004) Selection for aluminium and acid-soil resistance in white clover. Crop Sci 44: 38–48.

Wachendorf M, Collins RP, Elgersma A, Fothergill M, Frankow-Lindberg BE, Ghesquiere A, Guckert A, Guinchard MP, Helgadottir A, Luescher A, Nolan T, Nykaenen-Kurki P, Noesberger J, Parente G, Puzio S, Rhodes I, Robin C, Ryan A, Staheli B, Stoffel S, Taube F, Connolly J, (2001) Overwintering and growing season dynamics of *Trifolium repens* L. in mixture with *Lolium perenne* L.: a model approach to plant-environment interactions. Ann Bot 88: 683–702.

Wang J, Drayton MC, George J, Cogan NOI, Baillie RC, Hand ML, Kearney G, Trigg P, Erb S, Wilkinson T, Bannan N, Forster JW, Smith KF (2010) QTL analysis of salt stress tolerance in white clover (*Trifolium repens* L.). Theor Appl Genet 120: 607–609.

Warren JM (2000) The role of white clover in the loss of diversity in grassland habitat restoration. Restoration Ecol 8: 318–323.

Webster T, Nguyen N, Rhodes C, Felitti S, Chapman R, Edwards D, Spangenberg G (2005) A proposal for an international transcriptome initiative for forage and turf: microarray tools for expression profiling in ryegrass, clover and grass endophytes. Proceedings of the Fourth International Symposium on the Molecular Breeding of Forage and Turf, Aberystwyth, Wales, July 2005, p 174.

Williams TA, Abberton MT, Evans DR, Thornley W, Rhodes I (2000) Contribution of white clover varieties in high-productivity systems under grazing and cutting. J Agron Crop Sci 185: 121–128.

Williams TA, Abberton MT, Olyott O, Izen KAM, Cook R (2007) Evaluation of the effects of resistance to stem nematode (*Ditylenchus dipsaci*) in white clover (*Trifolium repens* L.) under sheep grazing and cutting. Plant Breed 126: 343–346.

Williams TA, Evans DR, Rhodes I, Abberton MT (2003) Long-term performance of white clover varieties grown with perennial ryegrass under rotational grazing by sheep with different nitrogen applications. J Agric Sci 140: 151–159.

Williams WM (1987) Genetics and Breeding. In: Baker MJ, Williams WM (eds) White Clover. CAB International, Wallingford, Oxon, UK, pp 343–319.

Williams WM, Easton HS, Jones CS (2007) Future options and targets for pasture plant breeding in New Zealand. NZ J Agric Res 50: 223–248.

Wilson KA, McManus M, Gordon ME, Jordan W (2002) the proteomics of senescence in leaves of white clover. Proteomics 2: 1114–1122.

Woodfield DR (ed) (1996) White Clover: New Zealand's Competitive Edge. White Clover: New Zealand's Competitive Edge. Grassland Research and Practice Series 6: pp 178. New Zealand Grassland Association, Palmerston North, New Zealand.

Woodfield DR, Caradus JR (1994) Genetic improvement in white clover representing six decades of plant breeding. Crop Sci 34: 1205–1213.

Woodfield DR, Brummer EC (2001) Integrating molecular techniques to maximise the genetic potential of forage legumes. In: Spangenberg G (ed) Molecular Breeding of Forage Crops. Kluwer Academic Press, Dordrecht, pp 51–67.

Yamada T, Higuchi A, Fukuoka A (1989) Recurrent selection of white clover (*Trifolium repens* L.) using self-compatible plants. I. Selection of self-compatible plants and inheritance of a self-compatibility factor. Euphytica 44: 167–172.

Young ND, Cannon SB, Sato S, Kim D, Cook DR, Town CD, Roe BA, Tabata S (2005) Sequencing the genespaces of *Medicago truncatula* and *Lotus japonicus*. Plant Physiol 137: 1174–1181.

Zane L, Bargelloni L, Patarnello T (2002) Strategies for microsatellite isolation: a review. Mol Ecol 11: 1–16.

Zhang D-X, Hewitt GM (2003) Nuclear DNA analyses in genetic studies of populations: practice, problems and prospects. Mol Ecol 12: 563–584.

Zhang X, Zhang Y-j, Yan R, Han J-g, Hong F, Wang J-h, Cao K (2010) Genetic variation of white clover (*Trifolium repens* L.) collections from China detected by morphological traits, RAPD and SSR. African J. Biotech 9: 3032–3041.

Zhang Y, Sledge M, Bouton J (2007) Genome mapping of white clover (*Trifolium repens* L.) and comparative analysis within the Trifolieae using cross-species SSR markers. Theor Appl Genet 114: 1367–1378.

Zhang Y, He J, Zhao PX, Bouton JH, Monteros MJ (2008a) Genome-wide identification of microsatellites in white clover (*Trifolium repens* L.) using FIASCO and phpSSRMiner. Plant Methods 4: 19.

Zhang Y, Bouton JH, Monteros MJ (2008b) Identification of QTL associated with morphological and agronomic traits in white clover (*Trifolium repens* L.) American Society of Agronomy Joint Annual Meeting 729–2.

Zhu H, Choi H-K, Cook DR, Shoemaker RC (2005) Bridging model and crop legumes through comparative genomics. Plant Physiol 137: 1189–1196.

Zohary M, Heller D (1984) The genus *Trifolium*. Israel Academy of Sciences and Humanities, pp 606.

Index

Color Plate Section

Chapter 1

Figure 1-2 A space-transplanted selection nursery of timothy (Kitami Agricultural Experiment Station, Kunneppu, Hokkaido, Japan).

Chapter 3

Figure 3-1 Illustration of *L. multiflorum* (right) *and L. perenne* (left). From Otto Wilhelm Thomé (1885): Flora von Deutschland Österreich und der Schweiz(Kurt Stueber. Kurt Stübers Online Library, http:// www.biolib.de).

Chapter 4

Figure 4-1 Shoot and root growth of tall fescue morphotypes; Continental (left), rhizomatous (middle), and Mediterranean (right).

Figure 4-2 A tall fescue plant in the breeding nursery of Noble Foundation, Ardmore, Oklahoma. Seed heads were collected for harvesting seed.

Figure 4-3 Severe rust infected tall fescue plant grown in field experiment at Ardmore Oklahoma, USA. Photo taken on July 16, 2012.

Chapter 8

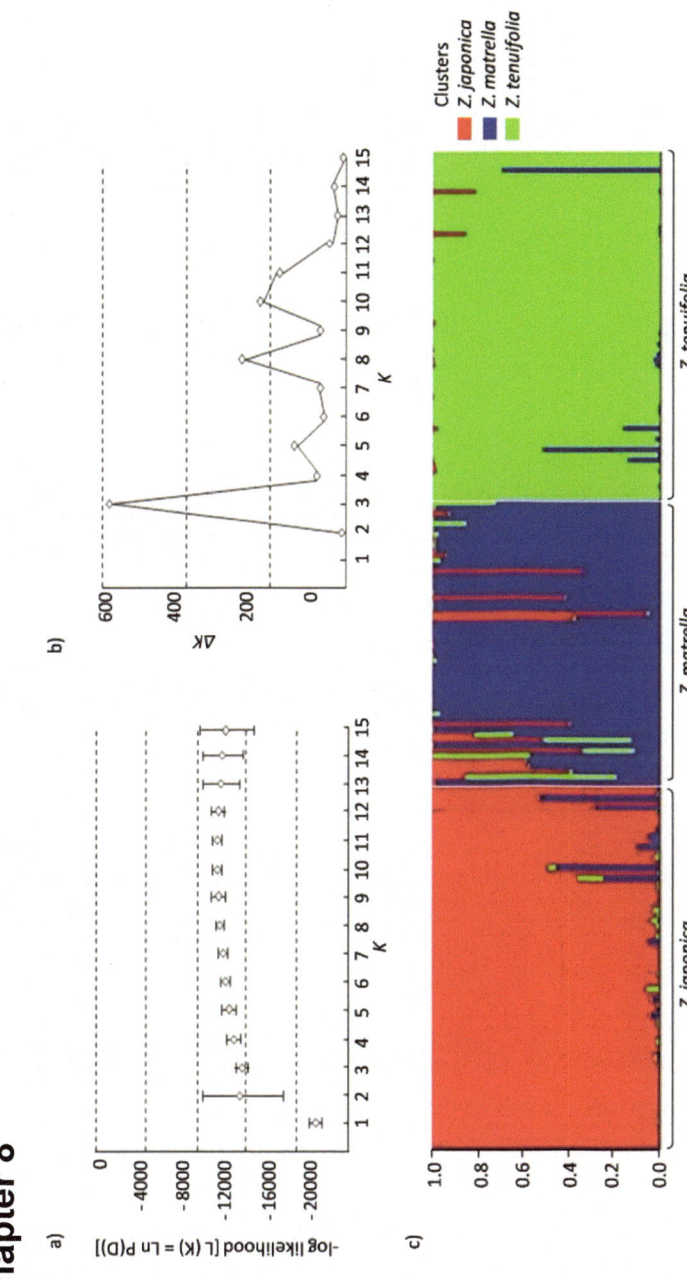

Figure 8-1 Genetic structure of 193 zoysia grass accessions identified by analysis using the STRUCTURE software based on 20 microsatellite polymorphisms. (a) Plot of the mean values (±SD) of the log likelihood of the genotyping data, for 10 runs at values of K from 1 to 15. (b) Plot of ΔK as a function of the number of clusters (K) for $K = 1$ to 15. Following calculation of the rate of change in the likelihood distribution ($L'(K) = L(K) - L(K - 1)$) and absolute values of the second-order rate of change of the likelihood distribution ($|L''(K)| = |L'(K + 1) - L'(K)|$), ΔK was calculated as $\Delta K = m |L''(K)| / s[L(K)]$. A maximum ΔK value at $K = 3$ indicated that a model with three clusters best explained the highest hierarchical level of genetic structure. (c) Hierarchical organization of 193 individuals from three *Zoysia* species identified by the STRUCTURE analysis for the $K = 3$ model.

Chapter 10

Figure 10-2 Sward plot trial of alfalfa synthetic populations and cultivars.

Figure 10-3 Alfalfa spaced plant evaluation trial.

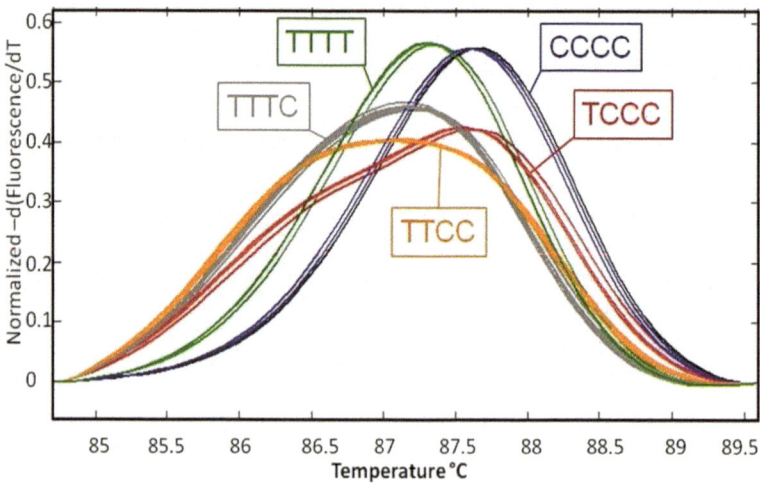

Figure 10-4 Normalized melting curves of alfalfa genotypes showing the variation in melting profile due to variation in allelic dosage at a biallelic SNP.

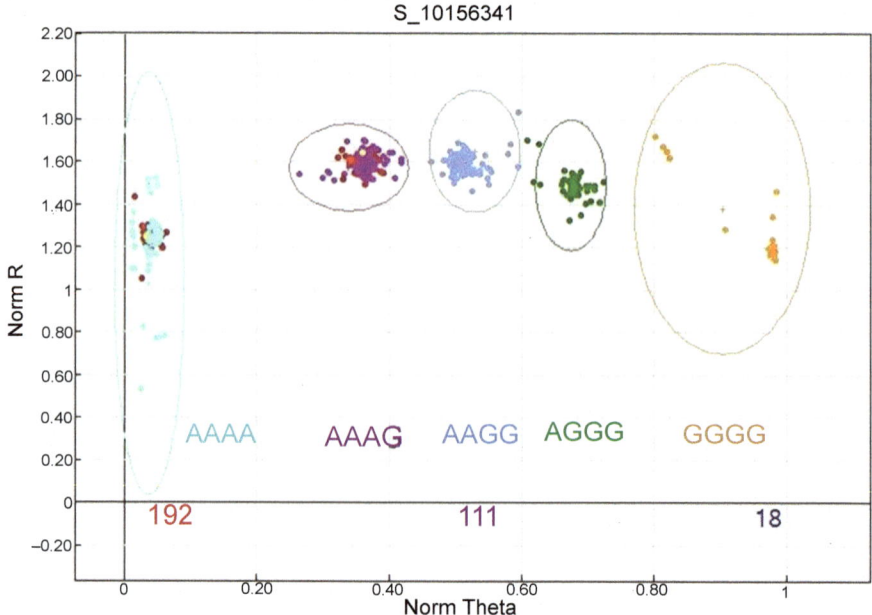

Figure 10-5 Allelic dosage variation identified in diverse alfalfa genotypes, including cultivated alfalfa, wild diploid alfalfa and other related species, using an Illumina® array.

Chapter 11

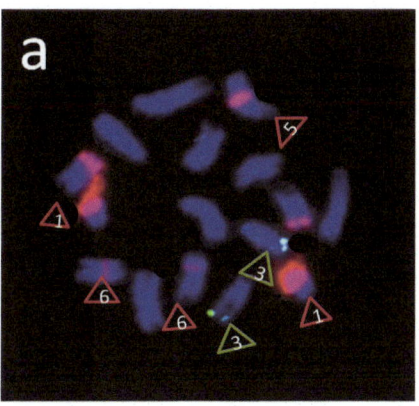

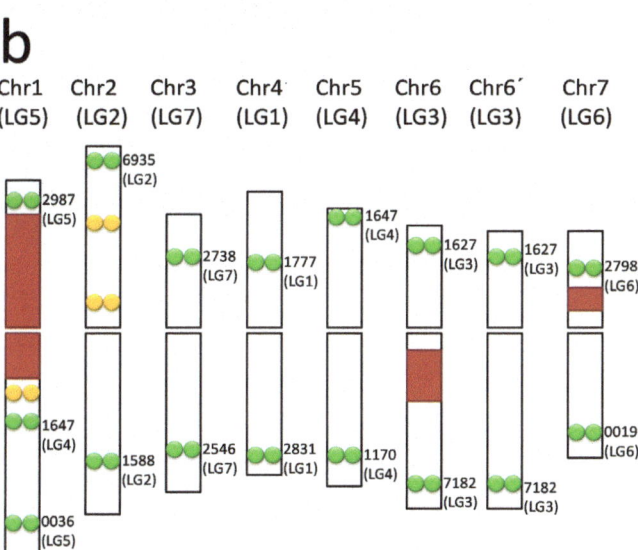

Figure 11-3 FISH analysis of red clover chromosomes. (A) FISH analysis with 28S rDNA (red signals indicated by red triangle) and a red clover BAC (RCS2546, green signals indicated by green triangle). Numbers on triangles indicate chromosome number. (B) Chromosome map of red clover. Green circles, loci of BACs corresponding to LG-specific markers; red boxes, 28S rDNA loci; yellow circles, 5S rDNA loci. Numbers represent BAC clones containing microsatellite sequences. The original figure and data were provided by Kataoka and Ohmido (Kobe University).

Chapter 12

Figure 12.2 Ornamental varieties of white clover, displaying various forms of leaf marking and colour (images kindly supplied by Dr. Wayne Parrott, University of Georgia, USA).